AF821194

Studies in Computational Intelligence, Volume 141

Editor-in-Chief
Prof. Janusz Kacprzyk
Systems Research Institute
Polish Academy of Sciences
ul. Newelska 6
01-447 Warsaw
Poland
E-mail: kacprzyk@ibspan.waw.pl

Further volumes of this series can be found on our homepage: springer.com

Vol. 120. George A. Tsihrintzis and Lakhmi C. Jain (Eds.)
Multimedia Interactive Services in Intelligent Environments, 2008
ISBN 978-3-540-78491-3

Vol. 121. Nadia Nedjah, Leandro dos Santos Coelho and Luiza de Macedo Mourelle (Eds.)
Quantum Inspired Intelligent Systems, 2008
ISBN 978-3-540-78531-6

Vol. 122. Tomasz G. Smolinski, Mariofanna G. Milanova and Aboul-Ella Hassanien (Eds.)
Applications of Computational Intelligence in Biology, 2008
ISBN 978-3-540-78533-0

Vol. 123. Shuichi Iwata, Yukio Ohsawa, Shusaku Tsumoto, Ning Zhong, Yong Shi and Lorenzo Magnani (Eds.)
Communications and Discoveries from Multidisciplinary Data, 2008
ISBN 978-3-540-78732-7

Vol. 124. Ricardo Zavala Yoe
Modelling and Control of Dynamical Systems: Numerical Implementation in a Behavioral Framework, 2008
ISBN 978-3-540-78734-1

Vol. 125. Larry Bull, Bernadó-Mansilla Ester and John Holmes (Eds.)
Learning Classifier Systems in Data Mining, 2008
ISBN 978-3-540-78978-9

Vol. 126. Oleg Okun and Giorgio Valentini (Eds.)
Supervised and Unsupervised Ensemble Methods and their Applications, 2008
ISBN 978-3-540-78980-2

Vol. 127. Régie Gras, Einoshin Suzuki, Fabrice Guillet and Filippo Spagnolo (Eds.)
Statistical Implicative Analysis, 2008
ISBN 978-3-540-78982-6

Vol. 128. Fatos Xhafa and Ajith Abraham (Eds.)
Metaheuristics for Scheduling in Industrial and Manufacturing Applications, 2008
ISBN 978-3-540-78984-0

Vol. 129. Natalio Krasnogor, Giuseppe Nicosia, Mario Pavone and David Pelta (Eds.)
Nature Inspired Cooperative Strategies for Optimization (NICSO 2007), 2008
ISBN 978-3-540-78986-4

Vol. 130. Richi Nayak, Nikhil Ichalkaranje and Lakhmi C. Jain (Eds.)
Evolution of the Web in Artificial Intelligence Environments, 2008
ISBN 978-3-540-79139-3

Vol. 131. Roger Lee and Haeng-Kon Kim (Eds.)
Computer and Information Science, 2008
ISBN 978-3-540-79186-7

Vol. 132. Danil Prokhorov (Ed.)
Computational Intelligence in Automotive Applications, 2008
ISBN 978-3-540-79256-7

Vol. 133. Manuel Graña and Richard J. Duro (Eds.)
Computational Intelligence for Remote Sensing, 2008
ISBN 978-3-540-79352-6

Vol. 134. Ngoc Thanh Nguyen and Radoslaw Katarzyniak (Eds.)
New Challenges in Applied Intelligence Technologies, 2008
ISBN 978-3-540-79354-0

Vol. 135. Hsinchun Chen and Christopher C. Yang (Eds.)
Intelligence and Security Informatics, 2008
ISBN 978-3-540-69207-2

Vol. 136. Carlos Cotta, Marc Sevaux and Kenneth Sörensen (Eds.)
Adaptive and Multilevel Metaheuristics, 2008
ISBN 978-3-540-79437-0

Vol. 137. Lakhmi C. Jain, Mika Sato-Ilic, Maria Virvou, George A. Tsihrintzis, Valentina Emilia Balas and Canicious Abeynayake (Eds.)
Computational Intelligence Paradigms, 2008
ISBN 978-3-540-79473-8

Vol. 138. Bruno Apolloni, Witold Pedrycz, Simone Bassis and Dario Malchiodi
The Puzzle of Granular Computing, 2008
ISBN 978-3-540-79863-7

Vol. 139. Jan Drugowitsch
Design and Analysis of Learning Classifier Systems, 2008
ISBN 978-3-540-79865-1

Vol. 140. Nadia Magnenat-Thalmann, Lakhmi C. Jain and N. Ichalkaranje (Eds.)
New Advances in Virtual Humans, 2008
ISBN 978-3-540-79867-5

Vol. 141. Christa Sommerer, Lakhmi C. Jain and Laurent Mignonneau (Eds.)
The Art and Science of Interface and Interaction Design (Vol. 1), 2008
ISBN 978-3-540-79869-9

Christa Sommerer, Lakhmi C. Jain, and Laurent Mignonneau (Eds.)

The Art and Science of Interface and Interaction Design (Vol. 1)

Professor Dr. Christa Sommerer
Institute for Media, Interface Cultures
University of Art and Industrial Design Linz
Sonnensteinstrasse 11-13
4040, Linz
Austria

Professor Dr. Laurent Mignonneau
Institute for Media, Interface Cultures
University of Art and Industrial Design Linz
Sonnensteinstrasse 11-13
4040, Linz
Austria

Professor Dr. Lakhmi C. Jain
Knowledge-Based Intelligent Engineering
Systems Centre
University of South Australia,
Adelaide,
Mawson Lakes Campus,
South Australia SA 5095,
Australia
E-mail: Lakhmi.jain@unisa.edu.au

ISBN 978-3-540-79869-9 e-ISBN 978-3-540-79870-5

DOI 10.1007/978-3-540-79870-5

Studies in Computational Intelligence ISSN 1860949X

Library of Congress Control Number: 2008926084

© 2008 Springer-Verlag Berlin Heidelberg

This work is subject to copyright. All rights are reserved, whether the whole or part of the material is concerned, specifically the rights of translation, reprinting, reuse of illustrations, recitation, broadcasting, reproduction on microfilm or in any other way, and storage in data banks. Duplication of this publication or parts thereof is permitted only under the provisions of the German Copyright Law of September 9, 1965, in its current version, and permission for use must always be obtained from Springer. Violations are liable to prosecution under the German Copyright Law.

The use of general descriptive names, registered names, trademarks, etc. in this publication does not imply, even in the absence of a specific statement, that such names are exempt from the relevant protective laws and regulations and therefore free for general use.

Typeset & Cover Design: Scientific Publishing Services Pvt. Ltd., Chennai, India.

Printed in acid-free paper

9 8 7 6 5 4 3 2 1

springer.com

Christa Sommerer
Lakhmi C. Jain
Laurent Mignonneau
(Eds.)

The Art and Science of Interface and Interaction Design (Vol. 1)

Springer

Foreword

Before we define the exact meaning of the term "interface", we will go back to its etymological roots and examine some of its historical concepts. The word interface is derived from the Latin words "inter" which means "between" in English and "facies" which means "face". One German translation of this word is "Grenzfläche". The English translation of the German word "Oberfläche" is "surface". There is an etymological and conceptual connection between "surface" and "interface". We may therefore say the study of "interface" is part of "surface science". In that sense an interface is the surface between two phases, it is for example the surface between two liquids like oil and water, which are immiscible. In a more exact sense surface science is the study of physical and chemical phenomena that occur at the interface of two phases, which includes solid-liquid interfaces, solid-gas interfaces, solid-vacuum interfaces, and liquid-gas interfaces. It includes such fields as surface chemistry and surface physics. Surface science is closely related to Interface and Colloid Science (J. Lyklema, *Fundamentals of Interface and Colloid Science*, 1995-2005).

The difference between surface and interface is intricate. Both terms contain the word "face". An example for their historical relationship is that the first surface studies were directed to the "face" or the surface of the world. Geometry is today the part of mathematics that deals with the properties of space. Initially geometry dealt with such problems as measuring the surface of the earth. The word "geometry" comes from the Greek words "geo", the earth, and "metria", to measure. This measurement of the earth is today called cartography. The word originated from Egypt. The Greek word "chartis" meant paper made from the papyrus plant. The Romans formed the Latin word "charta" from the Greek "chartis". In the 15th century it became "carte" in French or "Karte" in German. In the 17th century it became "Landkarte" in German, a "carte of the land". Cartography (from the Greek "graphein", to write) or mapmaking is therefore the study and practice of making a representation of the earth on a flat surface. One problem in creating maps is that the surface of the 3-dimensional earth which is a curved surface in three-dimensional space, must be represented as a flat surface in two dimensions. This entails a degree of distortion. This can be dealt with by utilizing a projection that minimizes the distortion in certain areas. The earth is not a regular sphere but a geoid. This is highly irregular but exactly known and has a calculable shape.

Cartography is a study of the surface of the earth, which combines science, aesthetics, and technology. Cartography is the study and representation of this system. The

approximately defined as the study of chemical reactions at interfaces. The adhesion of gas or liquid molecules to the surface is known as adsorption. Surface physics can be approximately defined as the study of physical changes that occur at interfaces. It overlaps the area of surface chemistry. The study and analysis of surfaces involves both physical and chemical analysis techniques. A renewed interest in Interface and Colloid Science, coupled with a new generation of analytical tools such as the Atomic Force Microscope (AFM), and the Scanning Tunneling Microscope (STM), it was one of the sources for the study of nanotechnology and bio-interface science. Examples of nanotechnology in modern use are the manufacture of polymers based on molecular structure, and the design of computer chip layouts based on surface science. Thirdly, in the age of computing an interface defines the communication boundary between two entities. These entities act as abstract black boxes of which only the surfaces are visible. We can call these entities or systems black boxes because the entities only provide an abstraction to the exterior. The description of the boundary is a part of the black box. Therefore these black boxes have only to know the face, which is turned to the inside to provide the means of communication. This is a new meaning of interface. The black boxes do not have eyes peering outwards. They are only able to look into the inside. Only the surfaces of the boxes have to adjust to be comparable. The internal operation may be different to the external communication. The interface provides the interconnection between the internal and the external operation. This is the meaning of interface. The interface provides a translation between the two entities or black boxes.

We have many types of interfaces. For example there is the interface in chemistry, which is a surface forming a boundary between two phases. The interface in physics is a surface forming the boundary of a body measured from the outside. There are Software interfaces and Hardware interfaces between the physical systems in computer technology. Network interfaces are points of interconnection between a terminal and a network or between networks and there are data interfaces and user interfaces. Human-Machine Interfaces or Man-Machine Interfaces are the aggregate of the means by which users interact with the machine, devices, computer programs or other complex tools or systems. These systems may be considered to interact like black boxes. The machine is a black box and the human is a black box. They do not speak the same language and an interface is necesssary. The interface translates the operations between, the hardware, the software and the user. Even when internal operations in these entities are different. Since we are dealing with black boxes, we use an input and an output. The user manipulates the input, the system reacts with the output to show the effects of the user's manipulation. In computer science the user interface controls or provides this interaction between the system or black boxes. It provides the visual, textual and acoustic output information the program presents to the user. It provides the control sequences which the user employs to control the program. In the history of the user interface from batch interface, to touch screens, from command line user interface to graphical user interface, best known is the contribution by Ivan E. Sutherland. He received the Turing Award in 1988 for his invention of *Sketchpad*, a predecessor of the graphical user interface which is ubiquitous in personal computers. Sutherland's doctoral thesis *Sketchpad, A Man-Machine Graphical Communication System*, 1963, was supervised by Claude Shannon who is considered to be the father of information theory. *Sketchpad* could draw horizontal and vertical lines and combine them into figures and shapes. Figures were able to be copied, moved, rotated, or rescaled. *Sketchpad* had the first window-drawing program and clipping algorithm. This allowed "zooming". *Sketchpad* ran on the Lincoln TX-2 computer and influenced Douglas C. Engelbart's On-Line System. Engelbart is another seminal figure in the history of interface design and in 1962 he published "Augmenting Human Intellect: A Conceptual Framework". In 1967 at the Stanford Research Institute he developed an "X-Y position indicator for a display system" now known as the "computer mouse". With the help of his student Bob Sproull Sutherland created what is considered to be the first virtual reality and augmented reality head-mounted display system in 1968. It was primitive in terms of user interface and realism. The head-mounted display worn by the user was so heavy that it had to be suspended from the ceiling, and the graphics comprising the virtual environment were simple wireframe models. In 1968 he and his friend David Evans founded the firm Evans and Sutherland. This company has done pioneering work in the field of Real-Time hardware, Accelerated 3D Computer Graphics, and Printer Languages. Former employees of Evans and Sutherland later founded companies such as Adobe (John Warnock) and Silicon Graphics (Jim Clark). Through this personal history we perceive a glimpse of the further development of interface technology. Jim Clark with Marc Andreessen founded Mosaic Communications Corporation. This firm was later known as Netscape, it released a web browser called Mosaic Netscape 0.9 in 1994.

The user interface has expanded from the map to the computer. Now with the help of the computer there is also a digital map of the earth available. *Google Earth* is a representation of the earth using satellite and computer networks. We cannot interact with the earth, with the reality, through the map, but we can interact with this digital map on the level of representation. We have a blend of surfaces and interfaces. The human system uses *Google Earth*, the machine system, as an interface in order to observe the surface of the earth, which is another system. These three systems interact like black boxes. We learn from this example that the art and science of interface and interaction design is bounded to the domain of representation. This process may be called "interfaciology". We still interact within the *Empire of Signs* (Roland Barthes, 1970), with representation. We do not interact with reality. That is, we interact with the map, not the territory. The next step is the interaction with reality using computers as an interface. Musical scores are for the future relevant interfaces. Guido of Arezzo is regarded as the inventor of modern musical notation (staff notation) that replaced neumatic notation; his text, the *Micrologus* (1025), was the second-most-widely distributed treatise on music in the middle ages (after the writings of Boethius). Guido of Arezzo is also the namesake of GUIDO Music Notation, a format for computerized representation of musical scores. Guido invented the five lines of the contemporary score to fix the tone pitch. The score is an instruction to tell a performer what he has to do with his instrument. So we have on one side a subject (in the world), on the other side an object (of the world) and a score between subject and object, which describes the transformations of phases of the object executed by the subject. We recognize: a score is an interface for the real world. The score is an interaction design.

Human-computer interactions turn into human-computer-world interactions.

Brain-computer interfaces will anticipate and enhance interactions between living biological systems. Marcus Textor's bio-interface science is the beginning of new interfaces that study the interaction between living systems and living organs instead of interactions between mechanical (machines) and biological systems (humans). To achieve this aim interfaces must become intelligent in a similar manner to humans. Jef Raskin, the so-called "Father of the Macintosh," sketched in his book *The Humane Interface. New Directions for Designing Interactive Systems* (2000) new intelligent ways for communicating information between man and computer. The head of the Interaction Design Department of The Royal College of Art in London (since 2004) Anthony Dunne has also developed new and valuable contributions to interaction design theory (*Hertzian Tales*, 1999; with Fiona Raby *Design Noir, The Secret Life of Electronic Objects*, 2001).

Interface technology has not only become a global industry but is a source for new world visions culminating in the Philosophy of Cyberspace. In a first approach the concept of an interface had to do with the transformations of states, phases or representations. The keyboard of the computer is an interface between a man and a machine. Man-machine interfaces are therefore the most popular concepts for an interface. However, parts of software can also be called interfaces. That is if it allows communication between two or more programs which have been written in different languages. An interface has not only to do with transformations but also with communication between systems or the parts of a compound system. With these definitions in mind, we then can ask whether our natural sense organs might also be interfaces between man and the natural environment. Do they provide communication between parts of a compound system? That is between us and the world? It was declared evident in the movie *Matrix*, realized in 1999 by the Wachowski brothers, that the world is a supercomputer. More precisely, the product of a supercomputer, where humans interact through machine interfaces. Three dimensional interactions in the virtual world led to the metaphor that our real world could be a computer programmed digital world where the inhabitants could also be digital simulations. Both of these systems, the inhabitants and the world, could interact three-dimensionally without realizing that they were part of a simulation as a computer game such as *Cyberia* (1994). The boundary between map and territory, representation and reality, between mechanical and organic, machine and human, simulation and real becomes blurred. That is how we look at the world from the perspective of interface theory.

The two volumes of *The Art and Science of Interface and Interaction Design* offer a survey of the newest ideas and practices of interaction design and interface technology. Cartography has combined science, aesthetics, and technology. It may be considered as the first interface, but without having interaction. Now, interface technology again combining science and art, allows for interaction. The study of representation evolves into the practice of interaction. The editors C. Sommerer, L.C. Jain and L. Mignonneau have to be praised for showing us, by their selection and commission of essays, how the "empire of signs" turned into the "empire of interfaces".

Peter Weibel

Preface

We live in a time of rapidly evolving digital systems and the way we creatively express ourselves has strongly been influenced by the media systems we use on a daily basis. Software and hardware technologies are continually updated, forcing digital creators to constantly adapt themselves to the ever new creative potentials inherent in these technologies.

Artists and creators in interactive art and interaction design have long been conducting research on human-machine interaction. Through artistic, intuitive, conceptual, social and critical projects in interaction and interface design, they have shown how digital processes are essential elements of the artistic creation process. However, resulting prototypes have also often reached beyond the art arena into areas such as mobile computing, intelligent ambiances, smart homes, intelligent architecture, fashionable technologies, ubiquitous computing and pervasive gaming.

Many of the early artist-developed interactive technologies have influenced how we interact with technology today and commercially available media products and services quite often have their roots in early artistic inventions and prototypes.

This book aims to present a snapshot of important topics in interaction and interface design. It presents articles of some pioneering practitioners and theoreticians in these fields. Additionally, it sketches out new emerging research areas where interactive art and interface design have influenced new design practices, products and services of today's media society.

We hope to present to the readers an interesting range of articles that cover some areas of current and future applications of interface and interaction design. As interactivity and connectivity are going to spread into areas such as healthcare, security, product design, shopping, entertainment and all forms of mobility, it will remain to be critical to the artists and designers to creatively deal with an increasingly detected, measured and connected world.

We wish to thank the authors and reviewers for their wonderful contribution and support.

We would also like to extend a special thank you to Springer-Verlag for working again with us on the merging of the Arts and Sciences. A special thank you should

1

Introduction to the Art and Science of Interaction and Interface Design

Christa Sommerer, Lakhmi C. Jain, and Laurent Mignonneau

Abstract. This chapter serves as an introduction to this book and will give a brief overview on the following chapters and their relationships towards the topic of artistic aspects of Interaction and Interface Design with special focus on how the notion of interactivity developed in the Arts and how early interactive artworks have left marks in later consumer products and interaction paradigms.

1.1 Introduction

The first volume of the book "The Art and Science of Interaction and Interface Design" is concerned with the question how interaction and interface design have their roots in human computer interaction engineering. They also may be seen to have parallel development in media art and more specifically in the interactive arts. The products of interactive technologies are increasingly spreading into our private and professional lives. It is interesting to see where early notions of interactivity came from. Another aspect is how media artists and media designers over the past 40 or more years already looked at the merits of interaction in their artistic and in their conceptual work. The sheer amount of publications related to interactive art and media art documents that this field of practice-based research has matured. [1-67]

Peter Weibel, one of the pioneers of media and interactive art (who also kindly contributed the foreword to this book), has already, in 1968 produced an early form of interactive artwork, the "Action Lecture No. 2," where he projected films of himself giving the same lecture projected on his own body on stage, while at the same time the audience could influence the operation of the film projector, the tape recorder and the second tape recorder which played music [3]. Weibel later, in 1989, predicted that modern art as a whole was undergoing a development towards the 'inter' principle, and he foresaw the social possibilities offered by this technology, such as participation in and interaction with the artwork as a model for emancipationist communicational forms." [4] Another pioneer and early promoter of interactivity in the Art and Sciences is Itsuo Sakane, who in 1984 and 1987 organized science and art exhibitions that included many early interactive art works, where he underlined the entertaining, educative and playful aspect of interactivity and user participation [5, 6]. The groundbreaking exhibition *Cybernetic Serendipity* organized by Jashia Reichardt in 1968 brought together early artworks using sensor technologies and early forms of interactions [1], which anticipated an art form that Cynthia Goodman in 1987 described as "Digital Vision" [2]. 40 years later these visions and predictions of Reichardt, Sakane, Goodman and Weibel have become reality in media art, interactive art, robotic art, net art and any do-it-yourself Web.02 creativity, that relies on creative interactions and exchanges.

C. Sommerer et al. (Eds.): The Art & Sci. of Interface & Inter. Des., SCI 141, pp. 1–14, 2008.
springerlink.com © Springer-Verlag Berlin Heidelberg 2008

The term "interactivity" has become so relevant these days, that a tendency is it to be obsolete [68, 69]. Subscribting to this notion would be naive and unproductive. In fact almost any media product nowadays relies on some form of interface and interaction design. This is irrespective whether it is a personal computer, multi-media mobile phones, wearable technologies, smart boards, multi-media service points, interactive guides, intelligent environments, and interactive art, entertainment and edutainment systems, for example. The study of Human-Computer Interaction and Interaction Design [69 -98] goes back to the mid 1960s where Engelbart [69], Papert [71] and Kay [70, 74] investigated how to design interfaces for better expressing human creativity. Kay was convinced that creative processes always include the use of sensormotoric and iconic forms of representations, especially in art and science. Kay's group in 1981 developed the famous first commercial workstation, the Xerox Star, which included the first graphical user interface that was easy and intuitive in its use. This workstation had an enormous impact on how we use and interact with computers today and in fact popularized graphical user interfaces (GUI) altogether.

Since then Interaction Design has become full-fledged research area and a recent stream of publications just underlines how important this topic is for designers and engineers alike [82-98]. In his seminal books "Total Interaction" Buurman [90] looks at how designers have become "the engineers of experience," who have to solve the "core problem of emulating, supporting and replacing the endless possibilities and repertoire of interaction through technological processes." Buurman also points out that this coordination and the creation of social meaning is achieved through the use of technology and the exploitation of new potentials in diverse contexts. We completely subscribe to this notion that "interaction is a social construction". We consider the work of media artists dealing with interaction, which is as yet another valuable form of creating social meaning obtained by interactive design and the creation of social experiences.

When we ask the question as to when artists started to use interaction in artworks, it is necessary to distinguish between interaction as a social form or for participation and execution of instructions, as it was already practiced in the participatory artworks of Dada (1916-1920) [99], Fluxus (about 1960s) [100] and Conceptual Art (1960s) [101]. The interaction was achieved through technological means. In the later category the early works by Schoeffer, Seawright, Ihnatowicz and Martin are the key works to consider [described in 1, 102] as these projects have already reacted to the environment and to viewers through the use of various sensor technologies. Paul, Gere and Rellie [103] put together a comprehensive exhibition and catalogue which deals with the connection of early instruction based artworks and art responding to input and its environment. From this compendium we can learn that artists, since the 1920s, have considered interaction and participation as a means of engaging in a social dialogue.

It was in 1990 that Interactive Art became a widely used term. This was when the largest and oldest media art festival in the world, Ars Electronica Festival in Linz, Austria started to include "Interactive Art" as a new category in its international competition, the Prix Ars Electronica [8]. Krueger [104] is often quoted as having originally coined the term and, not surprisingly, he was the main award winner in this category in 1990. The head of this category's jury was Roger F. Malina and he described "Interactive Art" as a new emerging art form. This was because it changes the status of the observer. The artists have started to use the viewer as an active component to their artwork [105]. The award winning artworks by Krueger, White and Shaw according to Malina "break down the traditional role of the observer as being external to the artwork." One could strictly say that this breaking up of boundaries between the observer and the artwork had been achieved already by Dada [99], Fluxus [100], and in Conceptual Art [101]. The way in which this involvement of the viewer into the development of the artwork was through computer algorithms and sensing technologies was new.

What is relevant to our discussion on interactivity and interaction design in the arts is the fact that in the early 1990s and perhaps long before (Myron Krueger [104, 106]), artists started to be unsatisfied with the interfaces to computers provided by hardware manufacturers. Malina sees a logical causality in this as "the artist who is used to the degree of control permitted by a paintbrush or pencil, is often disappointed by the usual computer graphics system" [105]. Similar arguments from music could state that "musicians who have learned to play musical instruments using ten fingers, two feet, body motions as well as their lips and breath are frustrated by being forced to use a keyboard" [105]. In fact Malina's statement is still valid today and the need to design new artistic user experiences and new specific interfaces in hardware or software has kept hundreds of media artists in Interactive Art busy since the 1990s. An excellent overview of these different types of artistic interfaces and interactive systems is provided in the yearly Prix Ars Electronica catalogues [8, 9, 11, 15, 17, 21, 25, 32, 35, 37, 44, 47, 53, 55, 56, 57, 58, 63] in the category of Interactive Art since 1990, as well as the Next Idea categories (from 2004 on) and the Hybrid Art category (from 2007 onwards).

Let us now return to the initial argument that interface design and interaction design for artistic purposes are valuable ways of creating social meaning through the creation of social experiences. We now show examples of how this can be done by introducing various authors and experts. The selection of topics is based on the need to extract sub areas of interaction and interface design and its interconnection with and to interactive art. We focus on areas where the notion of interactivity is currently being discussed and designed. It considers where it has spread beyond the pure art context into wider fields and application. These include architecture, fashion, gaming, communication and design. We are aware that the selection presents a very small fraction of the areas and individuals who advance these fields. However knowing that "interaction" itself is a term that is not even defined and has different significance in different disciplines, we allow ourselves to propose topics that bring artistic and scientific principles and approaches together.

It is not the aim of this book to give a survey of the field of interaction and interface design. This would be impossible due to interdisciplinary nature of the field and many other books have done this already, see [69-98]). It is hoped to outline currently discussed topics in interaction design and interactive art. The term interactivity and interaction have become important motivations in this field of art and research practice and production.

1.2 Chapters Included in This Book

The book contains 11 chapters. Chapter 1 is the introduction. Chapter 2 was written by Dr. Katja Kwastek and it provides a comprehensive overview of the term

which combined speech recognition, intuitive image browsing and sound output to let users intuitively browse the Internet through visualized conversations. Interestingly, this idea resurfaced one year after the presentation of "Riding the Net" at Siggraph 2000, in the science fiction thriller "Minority Report" [122], where the main protagonist interacts with projected images and sounds in a simulated multi-touch interactive web browsing system, extremely similar to what Sommerer and Mignonneau presented a year earlier. Be it just coincidence or not, these cases as well as that of "T_Vision" and "Google Earth," just underline how early developments in interactive art have left significant traces in today's consumer products, interaction strategies and even popular media culture. As stated earlier, interface design and interaction design for artistic purposes do in create social experiences. These clearly expand beyond the art arena into areas such as entertainment, consumer products and other forms of socially shared media experiences. Since 2004 Sommerer and Mignonneau develop interactive systems for public spaces. In their article in this volume they present several newest projects that bring interactivity to public buildings and architectural facades to engage the public in artwork experiences.

Chapter 7 is written by the Japanese researcher and media director Inakage. He directs a group of researchers in interaction design at Keio University in Japan. His chapter fits well into the argument that interactivity has become absorbed into everyday experiences. He argues that ubiquitous content creation is an emerging genre that uses everyday and media as a platform for creative content to achieve emotional and entertaining experiences. Inakage observes that the Industrial Society has developed to the Creative Society where personal and everywhere media will be not only important for art content creation but it also lets designers and artists embed ubiquitous media content and emotion in the artifacts and environment of our daily lives [123, 124]. Inakage gives several examples of ubiquitous computing prototypes which have been developed at his laboratories.

Chapter 8 is a logic continuation of this line of thought. The authors Martins et al. show how an ubiquitous gaming interface can reshape the idea of game play by bringing the gaming experience into a daily life scenario. The authors argue that when computers blend themselves into our daily environment, games will also move. They will become part of the new, ubiquitous paradigm of computation. Ubiquitous game design presents a new and exciting challenge in interface design. This because the designers and developers must break free from the limitations of traditional screens, mice and key pads and must instead reclaim the physical space by repurposing real objects and rediscovering social bonds. The authors show several examples of such ubiquitous game experiences given by various authors [125, 126]. They then continue to present their own prototype. This is a wearable *Gauntlet* interface [127] that allows ubiquitous, gesture-driven interaction with real objects and spaces without the need for a complicated setup.

Chapter 9 also deals with wearable technologies for ubiquitous computing environments. In this instance the focus is on wearable technology embedded into fashionable garments. The authors, Seymour and Beloff coined the term Fashionable Technology [128] for this blend of next generation wearable. The concept of wearable technology goes back to Mann who in the 1980s developed a series of wearable systems with body mounted cameras, lighting equipment and head mounted displays [129]. From this initially bulky and heavy portable equipments strapped to the users body, wearable technology has greatly developed. Seymour and Beloff argue that the worlds of fashion, ubiquitous computing, design, science, and wearable technologies are rapidly converging. With the appearance of transitive materials like electronic textiles, shape-memory alloys, and technologies such as Skinplex, novel interfaces for the body can be created. These will give designers and creators new possibilities to investigate new functions, new properties and new fashion statements. The garments in this case may be understood as an immediate interface to the environment and as a constant transmitter and receiver of messages, emotions, and experiences. To demonstrate this development, the authors give several examples of such Fashionable Technology prototypes.

Chapter 10 presents another design field that applies the ubiquitous computing paradigm in an artistic way. The authors Boj and Diaz, propose an augmented reality system for visualizing information and communication systems inside a physical public space. They argue that technological devices are increasingly assimilated into the urban space, which needs designers to respond to such transformations of cities. The ability to respond to the resulting integration of real and synthetic urban spaces is required. After a brief introduction to the basics of Mixed Reality paradigm (Milgram et al. [130]) in a continuation of Sutherlands approach on Virtual Reality [131] Boj and Diaz show how artists and designers could use mixed reality as a way to visualize invisible networks of digital data flows between wireless networks. The data is visualized and presented inside a wearable computing system consisting of a head mounted display unit that overlays the visualized network data in form of data packages which are transmitted and visualized in the physical space of the cities. While this is an artistic experiment, it can create more awareness of reality as it is being continuously enriched with digital information. This process is visible in products such as Google Earth [114], wikimaps [132], Stickymap [133] and OpenStreetMap [134] for example.

In Chapter 11 the media art historian Paul analyses how digital technologies have expanded into the public space and how artists have redesigned flows of data networks to create new forms of communication and interaction. Paul presents various artists and projects that use electronic networks to redefine the notion of "public art" through audience participation and agency. She argues that the networked commons as artistic field of exploration goes back to the 60s when many-to-many distribution networks were explored by artists such as Ascott [135], Adrian X. [136], or Loeffler [137]. They used radios, satellites, faxes, slow-scan TV or computers to communicate and transmit artistic content. This was long before networked digial art was developed. Paul then proceeds to show recent art projects that analyze the notion of networked commons as a free and open site of social and cultural space where ideas are shared in the broadest sense. She presents several examples of public artworks that discuss artistic engagement, agency, and conflicting authorities where activist art practice in the public space of the networked commons redefine what "public art" is and can be. This brings us full-swing back to the arguments presented by Daniels in Chapter 3. As he pointed out, already in the 1960s and 1970s Brecht and Eco called for a liberalisation of media-assisted forms of interaction. However this has spawned 40 years later a whole field of artistic and creative discourse which finds itself confronted to an increasingly commercial environment that undermines the originally democratic potential of the network society [138].

1.3 Conclusion

To summarize we can say that the chapters presented in this volume give us a glimpse into how complex and historically relevant the issue of interface and interaction design is, as it can be analyzed not only from an engineering point of view but from a social, artistic and conceptual, and even commercial angle as well. We hope to have presented to the readers an interesting selection of articles that highlight these cultural and societal aspects of interaction and interface design, as we are convinced that these issues will become ever more relevant in the future.

Acknowledgements

We are grateful to the authors of this volume for their contributions and time during the development phase of this volume. We are also very grateful to the authors whose articles had to be delayed to the second volume of this book, due to space and content considerations. Especially we would like to thank our families and children for bearing with us in spite of all the extra time effort that went into editing this book. We are especially indebted to our teachers and collegue artists, who taught us about the wonders of interaction design and how gratifying it can be to partake in the design of an interactive future. We are especially thankful to Springer Verlag that they entrusted us with the wonderful challenge to edit this book on the ever expanding issue of artistic interface and interaction design.

References and Further Reading (to be Completed)

[1] Reichardt, J. (ed.): Cybernetic Serendipity. Praeger, New York (1968)
[2] Goodman, C.: Digital Visions – Computers and Art. Harry N. Abrams, Inc. Publishers, New York (1987); Everson Museum of Art
[3] Weibel, P.: Action Lecture: intermedium. In: Protokolle, vol. 2, p. 88 (1982)
[4] Weibel, P.: Momente der Interaktivität. In: Kunstforum, vol. 103, p. 87 (1989)
[5] Sakane, I.: The Expanding Perceptual World – "A Museum of Fun Part II". The Asahi Shimbun, Tokyo (1984)
[6] Sakane, I.: Introduction to Interactive Art. In: Wonderland of Science Art – Invitation to Interactive Art, Kanagawa International Art and Science Exhibition, pp. 3–8, 38–42 (1989)
[7] Laurel, B. (ed.): The Art of Human Computer Interface Design. Addison-Wesley, Reading (1990)
[8] Leopoldseder, H. (ed.): Der Prix Ars Electronica 1990 - International Compendium of the Computer Arts, pp. 152–224. Veritas Verlag, Linz (1990)

[9] Leopoldseder, H. (ed.): Der Prix Ars Electronica 1991 – International Compendium of the Computer Arts, pp. 118–156. Veritas Verlag, Linz (1991)
[10] Weibel, P.: Transformationen der Techno-Ästhetik. In: Rötzer, F.R. (ed.) Digitaler Schein - Ästhetik der Elektronischen Medien, pp. 205–248. Edition Suhrkamp, Frankfurt am Main (1991)
[11] Leopoldseder, H. (ed.): Der Prix Ars Electronica 1992 – International Compendium of the Computer Arts. Veritas Verlag, Linz (1992)
[12] Cornwell, R.: Interactive Art. Touching the "Body in the Mind". In: Discourse, Nr. 14.2, pp. 203–221 (1992)
[13] Popper, F.: Art of the Electronic Age. Thames and Hudson Ltd., London (1993)
[14] Rötzer, F., Weibel, P. (eds.): Cyberspace: Zum medialen Gesamtkunstwerk. Boer Verlag, Munich (1993)
[15] Leopoldseder, H. (ed.): Der Prix Ars Electronica 1993 – International Compendium of the Computer Arts, pp. 90–136. Veritas Verlag, Linz (1993)
[16] Daniels, D.: Von der Mail-Art zur E-mail: Kunst der Kommunikation. In: Neue Bildende Kunst, vol. 5, pp. 14–18 (1994)
[17] Leopoldseder, H. (ed.): Prix Ars Electronica 1994 – International Compendium of the Computer Arts, pp. 100–134. Veritas Verlag, Linz (1994)
[18] Wilson, S.: Information Arts – Intersections of Art, Science and Technology. MIT Press, Cambridge (1995)
[19] Sakane, I.: The Interaction 1995, Dialogue with Media Art – Introduction to Interactive Installations. Gifu Prefecture Government, Gifu (1995)
[20] Hong-hee, K., Goodman, C. (eds.): Info Art – Kwangju Biennale. Kwangju Biennale Foundation, Seoul (1995)
[21] Leopoldseder, H., Schöpf, C. (eds.): Prix Ars Electronica 1995 – International Compendium of the Computer Arts, pp. 94–134. Osterreichischer Rundfunk ORF Landessstudio Oberösterreich, Linz (1995)
[22] Druckrey, T.: Electronic Culture – Technology and Visual Representation. Aperture, Denville (1996)
[23] Moser, M., MacLeod, D. (eds.): Immersed in Technology – Art and Virtual Environments. MIT Press, Cambridge (1996)
[24] Klotz, H.: Media Art Perspectives, Edition ZKM. Cantz Verlag, Osfildern (1996)
[25] Leopoldseder, H., Schöpf, C. (eds.): Prix Ars Electronica 1996 – International Compendium for the Computer Arts, pp. 118–156. Springer, Wien/New York (1996)
[26] Sakane, I.: The Interaction 1997, Toward the Expansion of Media Art. Gifu Prefecture Government, Gifu (1997)
[27] Dinkla, S.: Pioniere Interaktiver Kunst von 1970 bis heute. Cantz Verlag, Ostfildern (1997)
[28] Schwarz, H.-P. (ed.): Media-Art-History. Media Museum, ZKM - Center for Art and Media Karlsruhe. München (1997)
[29] ICC Concept Book. NTT Publishing Co. Ltd., Tokyo (1997)
[30] Hünnekens, A.: Der Bewegte Betrachter. Wienand Medien, Köln (1997)
[31] Lovejoy, M.: Postmodern Currents – Art and Artists in the Age of Electronic Media. Prentice-Hall, New Jersey (1997); (first edition 1989, 1992, Simon & Schuster/A Viacom Company)
[32] Leopoldseder, H., Schöpf, C. (eds.): Cyber Arts 1997 – International Compendium Prix Ars Electronica, pp. 106–150. Springer, Wien/New York (1997)
[33] Wirths, A.: Der Elektronische Raum – 15 Positionen zur Medienkunst. Kunst- und Ausstellungshalle der Bundesrepublik Deutschland. Cantz Verlag, Ostfildern (1998)

[34] Sommerer, C., Mignonneau, L.: Art@Science. Springer, Wien/New York (1998)
[35] Leopoldseder, H., Schöpf, C. (eds.): Cyber Arts 1998 – International Compendium Prix Ars Electronica, pp. 66–106. Springer, Wien/New York (1998)
[36] Sakane, I.: The Interaction 1999 Expanding the Human Interface. World Forum for Media and Culture Committee, IAMAS, Gifu (1999)
[37] Leopoldseder, H., Schöpf, C. (eds.): Cyber Arts 1999 – International Compendium Prix Ars Electronica, pp. 58–92. Springer, Wien/New York (1999)
[38] Druckrey, T. (ed.): Ars Electronica – Facing the Future. The MIT Press, Cambridge (1999)
[39] Leopoldseder, H., Schöpf, C. (eds.): Cyber Arts 2000 – International Compendium Prix Ars Electronica, pp. 66–92. Springer, Wien/New York (2000)
[40] Frieling, R., Daniels, D.: Media Art Interaction: The 1980s and 1990s. Goethe Institute, Munic/Karlsruhe (2000)
[41] Sakane, I.: The Interaction 2001, Dialogue with Expanded Images. World Forum for Media and Culture Committee, IAMAS, Gifu (2001)
[42] Gendolla, P., Schmitz, N.M., Schneider, I., Spangenberg, P.M. (eds.): Formen Interaktiver Medienkunst. Suhrkamp Verlag, Frankfurt/Main (2001)
[43] Manovich, L.: The Language of New Media. The MIT Press, Cambridge (2001)
[44] Leopoldseder, H., Schöpf, C. (eds.): Cyber Arts 2001 – International Compendium Prix Ars Electronica, pp. 80–118. Springer, Wien/New York (2001)
[45] Wilson, S.: Information Arts: Intersections of Art, Science and Technology. The MIT Press, Cambridge (2002)
[46] Rieser, M., Zapp, A.: New Screen Media, Cinema/Art/Narrative. British Film Institute, London (2002)
[47] Leopoldseder, H., Schöpf, C. (eds.): Cyber Arts 2002 – International Compendium – Prix Ars Electronica, pp. 70–122. Hantje Cantz Verlag, Ostfildern-Ruit (2002)
[48] Grau, O.: Virtual Art: From Illusion to Immersion. The MIT Press, Massachusetts (2003)
[49] Paul, C.: Digital Art. Thames & Hudson (world-of-art), London (2003)
[50] Meredieu, F.: Arts et Nouvelles Technologies, Art Video, Art Numerique. Larousse, Paris (2003)
[51] Lister, M., Dovey, J., Giddings, S., Grant, I., Kelly, K. (eds.): New Media: A Critical Introduction. Routledge, New York (2003)
[52] Ascott, R.: Telematic Embrace – Visionary Theories of Art, Technology, and Consciousness. Shanken, E.A. (ed.), University of California Press, Berkeley/LosAngeles/London (2003)
[53] Leopoldseder, H., Schöpf, C., Stocker, G. (eds.): Cyber Arts 2003 – International Compendium – Prix Ars Electronica 2003, pp. 82–122. Hantje Cantz Verlag, Ostfildern-Ruit (2003)
[54] Lovejoy, M.: Digital Currents: Art in the Electronic Age. Routledge, New York (2004)
[55] Leopoldseder, H., Schöpf, C., Stocker, G. (eds.): Ars Electronica – 1979-2004. Hantje Cantz Verlag, Ostfildern-Ruit (2004)
[56] Leopoldseder, H., Schöpf, C., Stocker, G. (eds.): Cyber Arts 2004 – International Compendium – Prix Ars Electronica 2004, pp. 102–148. Hantje Cantz Verlag, Ostfildern-Ruit (2004)
[57] Leopoldseder, H., Schöpf, C., Stocker, G. (eds.): Cyber Arts 2005 – International Compendium – Prix Ars Electronica 2005, pp. 92–138. Hantje Cantz Verlag, Ostfildern-Ruit (2005)

[58] Leopoldseder, H., Schöpf, C., Stocker, G. (eds.): Cyber Arts 2006 – International Compendium – Prix Ars Electronica 2006, pp. 102–142. Hantje Cantz Verlag, Ostfildern-Ruit (2006)
[59] Zielinski, S.: Deep Time of the Media. The MIT Press, Cambridge (2006)
[60] Kyong Chun, W.H., Keenan, T.: New Media, Old Media. Routledge, New York (2006)
[61] Munster, A.: Materialzing New Media. Dartmouth College Press, Hanover (2006)
[62] Ascott, R.: Engineering Nature - Art and Consciousness in the Post-Biological Era. Intellect Ltd., London (2006)
[63] Leopoldseder, H., Schöpf, C., Stocker, G. (eds.): Cyber Arts 2007 – International Compendium – Prix Ars Electronica 2007, pp. 152–188. Hantje Cantz Verlag, Ostfildern-Ruit (2007)
[64] Grau, O.: Media Art Histories. The MIT Press, Cambridge (2007)
[65] Wands, B.: Art of the Digital Age. Thames & Hudson, London (2007)
[66] Popper, F.: From Technological to Virtual Art. The MIT Press, Cambridge, MA/London (2007)
[67] Pias, C.: Chimäre Interaktivität: Wohin gehen wir, wenn wir drin sind? Texte zur Kunst 58, 92–103 (2005)
[68] Römer, S.: Interaktivität ist die größte Lüge. Texte zur Kunst 32, 70–73 (1998)
[69] Engelbart, D.C.: Augmenting Human Intellect: Experiments, Concepts and Possibilities. Summary Report to the Airforce Office of Scientific Research, Menlo Park (1965)
[70] Kay, A.C.: A Dynamic Medium for Creative Thought. In: Proceedings of the 1972 Minnesota NCTE Seminars in Research in English Education, Minneapolis (1972)
[71] Papert, S.: Mindstorms: Children, Computers and Powerful Ideas. Basic Books, Inc., New York (1980)
[72] Reingold, H.: Tools for Thought: The People and Ideas behind the Next Computer Revolution. Simon & Schuster, New York (1985)
[73] Kittler, F.: Grammophon, Film, Typewriter. Stanford University Press, Berlin (1986) (new edition 1999)
[74] Kay, A.C.: User Interface: A Personal View. In: Laurel, B. (ed.) The Art of Human-Computer Interface Design. Addison-Wesley, Reading (1990)
[75] Weiser, M.: The Computer for the 21st Century. Scientific American 265(3), 94–104 (1991)
[76] Laurel, B.: Computers as Theater. Addison- Wesley, Reading (1993)
[77] Bolz, N., Kittler, F., Tholen, G.C. (eds.): Computer als Medium. Fink, Paderborn, München (1994)
[78] Coyne, R.: Designing Information Technology. The MIT Press, Cambridge (1995)
[79] Weiser, M., Brown, J.S.: The Coming Age of Calm Technology. Xerox PARC (1996)
[80] Negroponte, N.: Being Digital, Coronet Books (1997)
[81] St. Johnson: Interface Culture: How New Technology Transforms the Way we Create and Communicate. Perseus Books Group (1999)
[82] Raskin, J.: The Human Interface: New Directions for Designing Interactive Systems. Addison-Wesley, Reading (2000)
[83] Carroll, J.M.: Human-Computer Interaction in the New Millenium. Addison-Wesley, Reading (2002)
[84] Carroll, J.M.: HCI Models, Theories, and Frameworks. Morgan Kaufmann, New York (2003)
[85] Lidwell, W., Holden, K., Butler, J.: Universal Principles of Design: 100 Ways to Enhance Usability, Influence Perception, Increase Appeal, Make Better Design Decisions, and Teach Through Design. Rockport Publishers (2003)

[86] Association MetaWorx (ed.): Approaches to Interactivity. Birkhäuser Verlag, Basel (2003)
[87] Stary, C., Stephanidis, C. (eds.): User-Centered Interaction Paradigms for the Universal Access in the Information Society – 8th ERCIM International Workshop on User Interfaces for All, Vienna, Austria, June 2004. Springer, Heidelberg (2004)
[88] Shneiderman, B., Plaisant, C.: Designing the User Interface: Strategies for Effective Human-Computer Interaction, 4th edn. Addison-Wesley, Reading (2004)
[89] Stanton, N.A., Salmon, P.M., Walker, G.H., Baber, C., Jenkins, P.D.: Human Factors Methods: A Practical Guide for Engineering And Design. Ashgate Publishing (2005)
[90] Buurman, G.M. (ed.): Total Interaction. Birkhäuser Verlag, Basel (2005)
[91] Greenfield, A.: Everyware: The Dawning Age of Ubiquitous Computing. New Riders Press, Berkeley (2006)
[92] Fishwick, P. (ed.): Aesthetic Computing. The MIT Press, Cambridge (2006)
[93] Moggridge, B.: Designing Interactions. The MIT Press, Cambridge (2007)
[94] Saffer, D.: Designing for Interaction. New Riders Press, Berkeley (2007)
[95] Cornelis, J., Wynants, M.: Brave New Interfaces, crosstalks (2007)
[96] Buxton, B.: Sketching User Experiences: Getting the Design Right and the Right Design (Interactive Technologies). Morgan Kaufmann, New York (2007)
[97] Sears, A., Jacko, J.A.: The Human-Computer Interaction Handbook: Fundamentals, Evolving Technologies, and Emerging Applications. CRC Press, Boca Raton (2007)
[98] Sharp, H., Rogers, Y., Preece, J.: Interaction Design: Beyond Human-Computer Interaction. Wiley, New York (2007)
[99] Richter, H.: Dada: Art and Anti-Art. Thames & Hudson (world-of-art), London (1997)
[100] de Maggio, G.: Ubi Fluxus Ibi Motus 1990-1962. Nuove Edizione Gabriele Mazzotta La Biennale di Venezia, Milano (1990)
[101] Marzona, D., Grosenick, U.: Conceptual Art. Taschen Verlag, Cologne (2005)
[102] Gere, C.: Art as Feedback. In: Feedback: Art Responsive to Instructions, Input, or Its Environment, pp. 62–78. LABoral Centro de Arte y Creacion Industrial, Gijon (2007)
[103] Botelle Diez del Corral, A.: Feedback: Art Responsive to Instructions, Input, or Its Environment. LABoral Centro de Arte y Creacion Industrial, Gijon (2007)
[104] Krueger, M.: Artificial Reality. Addison-Wesley, Reading (1983)
[105] Malina, R.: The Beginning of a New Art Form. In: Der Prix Ars Electronica 1990 - International Compendium of the Computer Arts, pp. 152–160. Veritas Verlag, Linz (1990)
[106] Krueger, M.: Golden Nica Myron Krueger – Videoplace 1970-1990. In: Der Prix Ars Electronica 1990 - International Compendium of the Computer Arts, pp. 167–178. Veritas Verlag, Linz (1990)
[107] Sutherland, I.: Sketchpad, a man-machine graphical communication system. In: Proceedings of the Spring Joint Computer Conference, IFIPS, pp. 329–345 (1963)
[108] Sandin, D.: IP (Analog Image Processor) 1972. In: Dunn, D., Vasulka, W., Vasulka, S., Weibel, P. (eds.) Eigenwelt der Apparate Welt – Pioneers of Electronic Art. Ars Electronica 1992, pp. 132–135. Österreichisches Landesmuseum, Linz (1992)
[109] Duchamp, M.: Writings of Marcel Duchamp. In: Sanouillet, M., Peterson, E. (eds.) Da Capo (1989)
[110] Cage, J.: Silence: Lectures and Writings. The MIT Press, Cambridge (1966)
[111] Eco, U.: Das offene Kunstwerk. Suhrkamp Verlag, Frankfurt am Main (1973)
[112] Brecht, B.: Gesammelte Werke (20 Volumes). Suhrkamp Verlag, Frankfurt am Main (1967)
[113] Art+Com: `http://www.artcom.de` (retrieved on December 10, 2007)

[114] http://earth.google.com/ (retrieved on December 10, 2007)
[115] Duchowski, A.T. (ed.): Eye Tracking Methodology: Theory and Practice, 2nd edn. Springer, London (2007)
[116] Levi, D.M. (ed.): Vision Research - An International Journal for Functional Aspects of Vision. Elsevier, Amsterdam
[117] Grueneis, G.: "T-Vision". In: Siggraph 1995 Visual Proceedings, p. 134. ACM Siggraph, New York (1995)
[118] Energie Passagen: http://energie-passagen.de (retrieved on December 10, 2007)
[119] Sommerer, C., Mignonneau, L.: Interactive Plant Growing. In: Siggraph 1993 Visual Proceedings, pp. 164–165. ACM Siggraph, New York (1993)
[120] Kuribayashi, S., Sakamoto, Y., Tanaka, H.: I/O plant: a tool kit for designing augmented human-plant interactions. In: CHI 2007 Conference on Human Factors in Computing Systems, San Jose, CA, USA, pp. 2537–2542 (2007)
[121] Sommerer, C., Mignonneau, L., Lopez-Gulliver, R.: Riding the Net. In: Siggraph 2001 Conference Abstracts and Applications, p. 133. ACM Siggraph, New York (2001)
[122] Minority Report (December 2002) Actors: Tom Cruise, Colin Farrell, Arye Gross, Jessica Harper, Patrick Kilpatrick, Caroline Lagerfelt, Studio: Dreamworks Video; based on: Dick, P.K.: The Minority Report and Other Classic Stories, Citadel (May 2002)
[123] Tokuhisa, S., Inakage, M.: SUIRIN. ACM Siggraph 2005 Emerging Technologies. Article No.22 (2005)
[124] Tokuhisa, S., Niwa, Y., Iguchi, K., Okubo, S., Nezu, T., Inakage, M.: Ototonari: mobile ad hoc pervasive game that develops a regional difference. In: ACM International Conference Proceeding Series, vol. 223, pp. 155–162 (2006)
[125] Benford, S., et al.: Uncle Roy All Around You: Implicating the City in a Location-Based Performance. In: Proceedings of the International Conference on Advances in Computer Entertainment Technology 2004 (ACE 2004), Singapore (2004)
[126] Correia, N., et al.: InStory: A System for Mobile Information Access, Storytelling and Gaming Activities in Physical Spaces. In: Proceedings of the ACM SIGCHI International Conference on Advances in Computer Entertainment Technology 2005 (ACE 2005), Valencia, Spain (2005)
[127] Martins, T., Heidecker, C., Sommerer, C., Correia, N.: Designing a Wearable Interface for Pervasive Games. In: Proceedings of the Fourth International Symposium on Pervasive Gaming Applications (PerGames 2007), Salzburg, Austria (2007)
[128] Seymour, S. (ed.): Fashionable Technology. Springer, Heidelberg (forthcoming)
[129] Mann, S.: Fundamental Issues in Mediated Reality, WearComp, and Camera-Based Augmented Reality. In: Woodrow, B., Caudell, T. (eds.) Fundamentals of Wearable Computers and Augmented Reality. Lawrence Erlbaum Associates, Inc., Mahwah (2001)
[130] Milgram, P., Takemura, H., Utsumi, A., Kishino, F.: Augmented Reality: A class of displays on the reality-virtuality continuum. In: Proc. Telemanipulator and Telepresence Technologies, vol. 2351, pp. 42–48 (1994)
[131] Sutherland, I.E.: The Ultimate Display. In: Proceedings of IFIP 1965, vol. 2, pp. 506–508 (1965)
[132] http://wikimap.sourceforge.net/Main_Pag (retrieved on December 10, 2007)
[133] http://www.stickymap.com/ (retrieved on December 10, 2007)
[134] http://www.openstreetmap.org/ (retrieved on December 10, 2007)
[135] Ascott, R.: Art and Telematics–Towards a Network Consciousness. In: Grundmann, H. (ed.) Art and Telecommunication, Vienna, p. 33 (1984)

[136] Adrain, X.R.: The World in 24 Hours. In: Druckrey, T. (ed.) Ars Electronica: Facing the Future, pp. 346–351. The MIT Press, Cambridge (1999)
[137] Loeffler, C.E.: Performance Anthology: Source Book of California Performance Art. In: Tong, D. (ed.) Last Gasp (1990)
[138] Gillespie, T.: Wired Shut: Copyright and the Shape of Digital Culture. The MIT Press, Cambridge (2007)

2

Interactivity – A Word in Process

Katja Kwastek

Ludwig Boltzmann Institute Media. Art. Research.
Kollegiumgasse 2
4010 Linz, Austria
kwastek@media.lbg.ac.at
http://media.lbg.ac.at

Abstract. This essay investigates the concept of interactivity by means of a historical analysis of the term itself and its classification. It should be understood as complementary to the essay by Dieter Daniels ('Strategies of Interactivity') in this volume, elaborating on the related artistic and societal contexts and discourses. Understanding the history of a term and its application to the various scientific fields helps to contextualize its denotations and interpretations. Therefore the first part of this essay sources the roots of the term interactivity, adopted by scientific fields as heterogeneous as physiology and sociology, cybernetics and computer science. The second part investigates recent attempts to go beyond a mere definition by further describing and classifying the various processes that are subsumed under the umbrella term of interactivity. The compilation of exemplary studies introduces different approaches of classification, from an ideological or technical, epistemological or aesthetic perspective. As a further step towards new ways of describing and analyzing interactive art, the final paragraph of the essay presents a research project of the Ludwig Boltzmann Institute Media.Art.Research. evaluating a taxonomy for interactive art.

2.1 'Interactivity' and 'Interactive Art'

The notion of interactivity has come to be a universal catchword of new media and the information society alike. Nevertheless the significance and the value of the term are more than controversial. In a provocative statement, Claus Pias argues: "*'Interactivity' is sometimes just as embarrassing as the narrow yellow leather tie that was worn when colorful fractals were modern and chrome balls populated the screens.*"[1] Whereas Pias is referring to the inflationary use of the notion of interactivity in general, the following quotation refers more specifically to Interactive Art. Wolfgang Kemp, the leading scholar of reception aesthetics in Germany, argues: "*The suspicion already expressed in 1984 that 'interactivity aims more to optimize the human-machine relationship than to place technology in the service of communication between people' has not yet been dispelled.[...] In other words, the first bond of this art that seeks to liberate the viewer is the bond to the program. I think that even expert systems, which [...] explicitly seek to promote dialogue and communication cannot simply delete the fact with one key that freedom of choice can only be simulated, not programmed. What is programmed is the illusion of alternatives.*"[2]

C. Sommerer et al. (Eds.): The Art & Sci. of Interface & Inter. Des., SCI 141, pp. 15–26, 2008.
springerlink.com © Springer-Verlag Berlin Heidelberg 2008

It is not the intention of this essay to comment on these discussions of the value of the notions of interactivity and interactive art. Instead, its first part sources the roots of the term itself. Understanding the history of the term and its application to the various scientific fields helps to contextualize its denotations and interpretations. The second part of the essay investigates attempts to go beyond a mere definition by further describing and classifying the various processes that are subsumed under the umbrella term of interactivity.[3]

2.2 The Track Record of a Term

2.2.1 The Starting Point: Interaction as Reciprocity

In general usage the term 'interaction' conventionally denoted 'mutual or reciprocal action or influence'. In the 1901 'Dictionary of Philosophy and Psychology' interaction is defined as: "*The relation between two or more relatively independent things or systems of change which advance, hinder, limit, or otherwise affect one another*", citing as examples both the body-mind relationship and the interaction of objects in and with the environment, which is frequently also termed reciprocity, according to this dictionary.[4]

2.2.2 Social Psychology: Interaction as Social Relation

With the institutionalization of sociology as a science in the early 20th century, the idea of interaction was applied to social and societal processes. In Germany Georg Simmel first used the term interaction ('Wechselwirkung') to characterize interpersonal relationships.[5] In Anglo-American discourses, George Herbert Mead and Edward Alsworth Ross were discussing "*social interaction*" or the "*interaction of human beings*".[6] Mead's student Herbert Blumer systematized his research under the term of symbolic interactionism, comparing this in 1937 with the stimulus-response theory. For the proponents of this theory interpersonal interaction consisted of a complex process of causes and effects of the various sensory organs and muscle groups.[7] It was therefore primarily explained physiologically and investigated statistically. The symbolic interactionists, on the other hand, regarded "*social interaction as primarily a communicative process in which people share experience, rather than a mere play back and forth of stimulation and response.*" [8] Whereas the former principally investigated reactions, the latter were more interested in actions.[9]

2.2.3 Cybernetics: Interaction as Process of Feedback

A new perspective of processes of interaction opened up around the mid-20th century with the emergence of cybernetic theories: Norbert Wiener, who coined the term cybernetics in 1947, was less interested in the interactions between human beings than in analogies between the self-organization of the human organism and cybernetics. In a book published in 1950, however, he explained how society could also be investigated through analyzing messages and communication processes.[10] Although he focused on processes that could be statistically analyzed, such as the stimulus-response theory that Blumer criticized, his theory of feedback processes went beyond the stimulus-response theory in distinguishing between different types of feedback, from reflex-like reactions to systems capable of learning.[11]

2.2.4 HCI – Interaction as Man-Machine Communication

It was not until the beginning of the 1960s that computer science had developed to a stage that allowed for the idea of real-time interaction between men and computers: In 1960 J.C.R. Licklider's groundbreaking essay about 'man-computer symbiosis' attempted to "*foster the development of man-computer symbiosis by analyzing some problems of interaction between man and computing machines.*"[12] After the publication of his visionary theories, it took only a few years until the first devices actually enabling real-time interaction between man and computer were built. In 1963, at the Massachusetts Institute of Technology, computer pioneer Ivan Sutherland developed 'Sketchpad', a graphical interface that made it possible to manipulate graphics on a display using a light pen. He explains: "*The Sketchpad system makes it possible for a man and a computer to converse rapidly through the medium of line drawings. Heretofore, most interaction between men and computers has been slowed down by the need to reduce all communication to written statements that can be typed; in the past, we have been writing letters to rather than conferring with our computers.[...It] opens up a new area of man-machine communication.*"[13]

A year before that, Douglas Engelbart, founder of the 'Augmentation Research Center' at the Stanford Research Labs, had already published his program for 'Augmenting Human Intellect'. The most famous outcome of this program, patented in 1968 but already developed around 1965, was the "*X-Y position indicator for a display system*", now known as the computer-mouse. With the principal concept of the graphical user interface developed by Sutherland, and Engelbart's mouse replacing the light-pen, basic elements of the human-computer interface were available. From then on, human-computer interaction was established as a highly specialized and interdisciplinary field within computer science.[14]

This overview shows that by the early 1960s the concept of interaction had developed from an idea of reciprocity in biological, chemical and physiological processes into elaborate theories of social interaction (sociology), into a whole new science trying to establish the idea of feedback processes as a basic theorem of life and technology (cybernetics), and into a field of research and development in the computer sciences (HCI). But when did it enter the arts?

2.3 Towards 'Interactive Art'

Whereas it is relatively easy to answer the question of when the notion of Interactive Art first appeared (see below), the history of the concept of interaction within the arts is more complex. My suggestion is that the emergence of each of the three fields of knowledge identified above (sociology, cybernetics and computer science), with their respective concepts of interactivity, had a parallel in the arts.

2.3.1 Participatory Art Projects

The consideration of social interaction as a possible element of artistic projects arose more or less parallel to its story of success in the social sciences. Initial attempts to involve the public can be traced back to the classical avant-garde, although the breakthrough of these new artistic concepts did not occur until after WWII. Though these projects are extremely important for the development of the arts up to the present,[15] they are not commonly called 'interactive', but participatory or collaborative. Nevertheless the concepts of the participatory and collaborative works have always been considered an important point of reference for media based art and taken into consideration for comparative judgments about artworks using technical feedback-processes. Yet even in the 1960s, the concepts concerning the relationship between art and technology were heterogeneous. As Dieter Daniels elaborates in his essay ('Strategies of Interactivity') in this volume, artists like John Cage, Nam June Paik and Valie Export investigated the manifold interrelations of participatory ideas and the power of mass media. Meanwhile other artists were more interested in the technological feedback processes themselves, often referring to visions of Artificial Intelligence. The latter have to be seen in close relation to cybernetic theories.

2.3.2 Cybernetic Art

The second field of knowledge introducing concepts of interaction into the arts – even if to a lesser extent – were the ideas of cybernetics. Already in the 1950s, the Hungarian sculptor Nicholas Schoeffer built his 'Cybernetic Spatiodynamic Sculptures'. He used the cybernetic concept of the homeostat to organize the reaction of these works to the environment via sensors. He was followed in the 1960s by artists like James Seawright, Edward Ihnatowicz and Tony Martin. They built devices that would interact with their environment in one way or another , mostly via light and sound sensors. Or they installed environments that reacted to the audience, emitting light and/or sound. Yet they did not call their works interactive either. Instead, they were called cybernetic, responsive or reactive.[16]

In 1968 Jack Burnham, author of numerous books and essays attempting to contextualize cybernetic art within art history, observes the growing differences between cybernetic artworks and the achievements of computer science: "*The spectacle of an artifact adjusting to its environment through a series of visible maneuvers has a certain anthropomorphic fascination, but it remains hardly an efficient way of handling immense amounts of information. It is well to mention this because the gap between the romantic prototype robot – surely a leftover from the first age of machines – and the modern theory of automata is an ever-widening one.*" [17]

2.3.3 Interactive Art

Though the challenge of implementing computers was already discussed within the realm of cybernetic art, only very few works were actually based on algorithmic processes.[18] Even though computer science had successfully developed the possibilities of human-computer interaction in the 1960s, the adoption of these technologies in the arts happened very slowly.

In 1969 a group of artists and scientists set up an installation entitled 'Glowflow' in the Union Main Gallery of the University of Wisconsin. Inside a dark room, phosphorescent particles were circulating in tubes. The tubes ran through columns with integrated lights, which were illuminated by the visitors through touch-sensitive floor pads. These lights in turn caused the phosphors to glow. The subtitle of the work still followed the conventions of cybernetic artworks. It was called: "*Glowflow, a computer-controlled, light sound viewer responsive environment*". But the flyer accompanying the exhibition introduced the term Interactive Art: *"Glowflow is not an exhibit in the traditional sense, but a continuous experimentation in interactive art. Its basic elements – lights, sounds and viewers – interact through control devices which are programmed by the artist-researcher to explore a variety of relationships. While many exhibits in the past have established predetermined relationships of viewer to environment, Glowflow is capable of going beyond fixed interactions. With a computer as a control device, it is possible to explore much more dynamic viewer-environment relationships."* [19]

It is in this text that the notion of Interactive Art is first coined and at the same time related to the implementation of the computer as control device.

One of the co-creators of this project was Myron Krueger, who is now regarded as the pioneer of Interactive Art. His concept of interactivity was developed from experiences with Glowflow.[20] Rather than focusing on the creation of sculptures or robot-like creatures, he started with the idea of the responsive environment, augmented its sensorial capacities installing video cameras, elaborated its operative options by using computers as control devices, and extended its reactive capacities by projecting computer graphics onto the walls.

Although it was not until the 1990s that 'Interactive Art' became the catchword of new media art,[21] the origins of the term can be traced back to 1969. The tendency to restrict the term to computer controlled interactions also goes back to the same period:

The activities of communication arts of the 1970s and 1980s, such as artworks conducted by Robert Adrian X, Kit Galloway and Sherrie Rabinowitz, Douglas Davis and others, were not commonly referred to as Interactive Art. Nevertheless they received far more attention than the contemporaneous, scarcely known pioneers of computer-controlled installations such as Myron Krueger, Lynn Hershman, Jeffrey Shaw or Grahame Weinbren.[22] It was only with the advent of the Word Wide Web, when communication arts were identified as predecessors of net art, i.e. computer controlled communication art, that their interrelations started to become more evident.

As more and more artists created various kinds of 'interactive artworks' in the 1990s, it became increasingly obvious that there was an ambiguity between the term in the narrow sense of computer controlled interaction and its denotations resulting from the various concepts of interactivity that had emerged throughout the previous nearly 100 years.

The different discourses that accompanied these interpretations and developed along the various implementations of interactive technologies over the years are elaborated in detail by Dieter Daniels. The focus of the present essay, however, is on the fact that the awareness of the ambiguity of the term led to a growing number of attempts to investigate the different strategies, technical processes and intentions it stands for.

2.4 The Necessity of Distinctions

The second part of this essay summarizes attempts to describe and classify the processes understood as interactive. There are various reasons why such a compilation can never be exhaustive. One is the still sketchy state of research concerning classifications of interaction processes in the various disciplines; another is the interdisciplinary nature of the field, where entirely different perspectives on the same phenomenon impede a comprehensive synopsis. But most of all, the object of research itself is constantly changing and therefore subject to ongoing revisions of the related discourses.

2.4.1 Ideological Versus Instrumental Views of Interactivity

In their 'Critical Introduction' to New Media Martin Lister et al. distinguish between 'ideological' and 'instrumental' views of interactivity.[23] This is comparable to the distinction suggested by Dieter Daniels, who identifies interactivity as both an ideology and a technology. Nevertheless, Lister's compilation of ideological views is interesting in this context, because he focuses on the interrelation of the differing perspectives and the various disciplines they stem from. While information technology within the scope of research on HCI understands interactivity as a possibility for controlling and intervening in computer processes, communication studies gauge interactivity according to face-to-face situations.[24] Media studies, on the other hand, assume a latent interaction between the receivers and the objects of their interest consisting of the processes of selection and interpretation.[25] Like Daniels, Lister et al. cite as a further important perspective on aspects of interactivity the idea of "*grassroots democratic exchange*" as a counter-pole to the one-to-many strategy of the mass media, which understands interactivity as the co-determination and exchange possibilities of the members of one or more societies.[26]

2.4.2 Degrees of Interactivity

Categorizations that Lister describes as 'instrumental' are often still closely related to 'ideology', i.e. the different political, philosophical and economic goals associated with concepts of interactivity. This becomes apparent when the categorization of interactive processes attempts to create scales ranging from a low to a high degree of interactivity.

Beryl Graham, who provides a valuable compilation of categorization attempts, bases her classifications upon the division proposed already in 1977 by Cornock and Edmonds.[27] They distinguish between static and dynamic art systems,[28] with the latter divided into

- dynamic systems, that are based on an organizational dependence on environmental variables.
- reciprocal systems that treat the spectators as environment, with responses through time.
- participatory systems with a focus on the interpersonal reactions of a group of participants to a situation specified as a matrix.
- interactive systems that offer a mutual exchange between man and machine, elaborately related on either side of an interface.

Graham modifies this taxonomy using a metaphor of conversation, but sticking to the idea of the 'real conversation' as the highest degree of interaction, "*a category which is a possibly unobtainable end point but remains as a possible future aim.*"[29]

Lutz Goertz also qualifies degrees of interactivity, but he proposes determining them based on the interplay of various factors.[30] He suggests a consideration of the degree of selection options, the degree of modification options, the quantitative size of selection and modification possibilities, and the degree of linearity or non-linearity. In this way, he acknowledges that there are qualitative as well as quantitative differences within possibilities of selection or modification and that they should be related to the question of timing. With this question he refers to the possibilities of co-authoring the course and order of the interactive processes enabled.

While breaking down interactivity processes into the categories of selection, modification and (non-)linearity is an interesting approach, the general tendency to draw up scales according to degrees of interactivity is questionable, at least in the field of the arts. Judging the quality of media-based interaction by a comparison with direct communication disregards the fact that the decision to use media for an artwork is a deliberate one. The objective of interactive media art is precisely to uncover and reflect the specific forms of interaction enabled by new media.[31]

Categories describing interactive artworks should therefore attempt to identify and denominate the different processes taken into consideration by the artists and analyze their relation to the discursive backgrounds and concepts of interaction they refer to, irrespective of a validation of an assumed gradation.

Although Lister et al. also adhere to some extent to the idea of degrees of interactivity, in their delineation of the 'instrumental view' they attempt a description of specific characteristics of interaction processes. They distinguish between hypertextual versus immersive navigation as well as between registrational interactivity and interactive communication Hypertextual navigation, according to Lister, is based on choices available from a pool of data to construct an individual 'text'. They equate this kind of interaction with Peter Lunefeld's definition of 'extractive' interaction.[32] Immersive navigation, on the other hand, focuses on the investigation of spatially organized information: "*We might say that the navigation of immersive media environments is similar to hypertextual navigation, but with additional qualities [...] Instead of a text-based experience aimed at finding and connecting bits of information, the goals of the immersed user will include the visual and the sensory pleasures of spatial exploration.*"[33]

With registrational interactivity they describe the possibility of storing one's own data, which then become part of the 'text'. Interactive communications define a computer-aided human-to-human communication, which they again relate to the idea of gradation:

"*When email and chat sites are considered from the point of view of human communication, ideas about the degree of reciprocity between participants in an exchange are brought into play. So, from a Communication Studies point of view, degrees of interactivity are further broken down on the basis of the kinds of communication that occur within CMC. Communicative behaviors are classified according to their similarity to, or difference from, face-to-face dialogue, which is frequently taken as the exemplary communicative situation which all forms of 'mediated' communication have to emulate.*"[34]

While their study gives a valuable description of different types of interactivity, it clearly focuses on text-based interactions and is therefore far from being complete.[35] It is hardly possible to analyze all the different kinds of interaction that employ tactile devices or serve as a tool or instrument for performing specific tasks with these categories. In recent years there has been a growing desire to transcend the focus on the graphical interface, in media arts as well as in the realm of interface design. A considerable amount of research has already been conducted, for example, on notions of mediated embodiment [36] or the idea of device art.[37]

2.4.3 Epistemological and Aesthetic Categories

On a more general level it is doubtful whether the aforementioned ideological and instrumental views of interactivity would constitute a sufficient base for a thorough research of artistic projects in the realm of media-based interaction, even if they were broken down into a comprehensive compilation of the specialized studies. The classifications summarized so far do not investigate artworks specifically, but are applicable to all kinds of interactive media. Within the arts, epistemological and aesthetic concepts play a seminal role in the artist's realization of their ideas and in what the participant may gain from the artwork. Often the insights and aesthetic processes encouraged by different interactive works do not correspond to the instrumental categorizations available. For example, a hyperlink system may offer a non-linear narration of a story, but it may also involve the viewer in a question and answer session. The storage of user-inputs may serve as a surveillance device, but also as a means of co-creating a multimedia environment.

Therefore, there is a need for approaches that denominate the epistemological intentions of the works and identify the processes that enable their mediation. Two examples may suffice to illustrate possible steps in this direction:

In his extensive historical overview of 'Closed Circuit Video Installations' Slavko Kacunko does not deal exclusively with works that allow for digital feedback processes.[38] On the contrary, his explicit decision to neglect the distinction of algorithmic versus non algorithmic feedback processes is based on the realization that – from an epistemological point of view – the works conventionally called 'closed circuit installations' and the works conventionally called 'interactive installations' have common characteristics. He breaks down the various artistic concepts distinguishing different strategies of subject/object relations, reality constructions, system models and behavior patterns, and game concepts and learning processes. In relation to reality constructions, he distinguishes between the creation of reality models, post-technological visions and their psychological effects, computer-aided media reflections and interlocking levels of reality and virtuality. Kacunko thus concentrates on the philosophical views and models implied within the different interactive strategies.

A second example for the various possibilities of going beyond instrumental classifications is the approach of Jean-Louis Boissier, media artist from Paris. He focuses on the structural or aesthetic issues in the interaction process. He describes how, in the course of his own artistic work, he was able to identify certain 'figures of interaction' that had emerged. He distinguishes between figures that are generated from internal structures and those that describe attitudes on the part of recipients. The internal figures he identifies are forking ("*bifurcation*"), break/interruption ("*suspension*"), change ("*mutation*") and transfer ("*réversion*"). The recipients, according to Boissier, may conduct actions of comparison ("*comparaison*"), description ("*désignation*"), distancing ("*distanciation*") and empowerment ("*procuration*").[39]

Unlike Kacunko's suggestion, Boissier's terms do not try to embrace the philosophical backgrounds of the works, but constitute a first attempt to describe the single (also partial) processes that together constitute their structural and receptional characteristics.

2.4.4 Case Study Prix Ars Electronica

As yet another step within the attempts to identify suitable descriptive models for interactive art, the Ludwig Boltzmann Institute Media.Art.Research. is conducting a case study based on the entries for the annual competition of the Prix Ars Electronica. On the basis of research into former approaches to categorizations (summarized above) we developed a tentative taxonomy that we applied to the more than 350 entries of the 2007 competition. The taxonomy was adapted in process, thus reacting to the works that were actually entered by the artists in this category and thereby defined as interactive by them. While the project also investigates questions of the form and range of the works, the media applied and the topics covered, the focus here is on the more specific categories of interaction.[40] With the classification of interaction processes we soon realized that describing them with nouns is problematic. Gerhard Dirmoser, an Austrian system analyst, suggested testing descriptions in verb form instead. The advantages are obvious: whereas a noun may be suitable for describing a process, it leaves open the question of the direction/perspective of the process. A verb, on the other hand, urges the user to specify the subject of the process. When the word 'observation' is used, for example, it is unclear whether the work is meant to observe the viewer or vice-versa. If one instead formulates the categorization as 'the viewer can – observe,' then the direction of interaction is clear.

Our suggestion is to first identify the partners involved in the interaction process. Once the interaction partners have been specified, there should be a further definition of the type of interaction that is enabled. We suggest describing whether the visitor / performer is encouraged to observe, to explore, to activate, to control, to select, to navigate, to participate, to leave traces or store something or to exchange information – and of course further activities should be added.

But there is also another part of the interaction process that originates from the work or mediating device. The work may tell or narrate something, document or inform, visualize or sonify, it may be built to enhance perception or to offer a game, to monitor something or to serve as an instrument, to transform, to collect and store, to process or mediate.

The compilation of a rich variety of verbs describing the various activities and strategies relevant to interaction processes should improve the possibilities for describing and analyzing interactive media art.

To further evaluate this approach, we are providing this taxonomy for the artists to be applied by them during the 2008 competition entry process. We hope that this project will contribute to the attempts to develop a suitable vocabulary for the manifold field of Interactive Art.

The scope of these suggestions, as well as the scope of the present essay, is not to determine a fixed set of normative categories. Rather they provide an extendible but still comprehensive vocabulary that helps to differentiate between the various types, strategies and intentions of interactive processes and artworks. They will hopefully encourage further discussion and new interesting concepts of interactivity. As media artist Michael Naimark states, referring to the Boltzmann Institute's taxonomy project: *"lists like these serve two different functions: 1) they help organizers organize and 2) they encourage artists to do something 'unclassifiable'. Both are noble goals."*[41]

References

1. Pias, C.: Chimäre Interaktivität: Wohin gehen wir, wenn wir drin sind? In: Texte zur Kunst 58, pp. 92–103 (2005); here p. 1 (transl. Aileen Derieg)
2. Kemp, W.: Zeitgenössische Kunst und ihre Betrachter. Positionen und Positionszuschreibungen. In: ibid (ed.) Zeitgenössische Kunst und ihre Betrachter. Jahresring 43, pp. 13–43 (1996); here p. 19 (transl. Aileen Derieg)
3. They attempt to cover more than 100 years of a term's history and breakdown can of course only offer a very rough and exemplary summary of the relevant objectives and theories. The references given in the footnotes try to compensate this deficit by suggesting further readings
4. Baldwin, J.M. (ed.) Dictionary of Philosophy and Psychology. The German translation Baldwin offers for 'interaction' is 'Wechselwirkung', London, vol. 1, p. 561 (1901) (in German)
5. Cf. Abels, H.: Einführung in die Soziologie. In: Die Individuen in ihrer Gesellschaft, Wiesbaden, 2nd edn., vol. 2, pp. 204–206 (2004)
6. Ross, E.A.: Social Psychology. An Outline and Source Book, New York, p. I (1909,1st edition 1908); Mead, G.H.: Social Psychology as Counterpart to Physiological Psychology. The Psychological Bulletin VI(12), 401–408 (1909); Cf. The history of the term in Hans Dieter Huber: Der Traum vom Interaktiven Kunstwerk (2006), http://www.hgb-leipzig.de/artnine/ huber/aufsaetze/nt.html (October 24, 2007); I would like to thank Gunther Reisinger calling my attention to this essay
7. Cf. Blumer, H.: Social Psychology. In: Schmidt, E.P. (ed.) Man and Society. A Substantive Introduction to the Social Sciences, New York, pp. 144–198, here p. 170 (1937)
8. Ibid., p. 171
9. Ibid., p. 191
10. Wiener, N.: The Human Use of Human Beings. Cybernetics and Society, New York, p. 16 (1954,1st edition 1950)
11. Ibid., p. 33
12. Licklider, J.C.R.: Man-Computer Symbiosis. IRE Transactions on Human Factors in Electronics, HFE-1, 4–11 (1960), http://groups.csail.mit.edu/medg/people/psz/Licklider.html (October 24, 2007)
13. Sutherland, I.E.: Sketchpad: A Man-Machine Graphical Communication System, PhD thesis, Massachusetts Institute of Technology, reprinted as 'Electronic Edition', University of Cambridge, Computer Laboratory, Technical Report 574 (2003), here p. 17, http://www.cl.cam.ac.uk/techreports/UCAM-CL-TR-574.pdf; See Preface to 'Electronic Edition' by Alan Blackwell and Kerry Rodden for further contextualization

14. A good overview of early developments in this field is provided by Brad. Myers, A.: A Brief History of Human Computer Interaction Technology. ACM interactions 5(2), 44–54 (1998), An extensive recent compilation is offered by Bill Moggridge: Designing Interactions, Cambridge, MA (2007), http://www.cs.cmu.edu/~amulet/papers/uihistory.tr.html (October 24, 2007); Thanks to my colleague Guenther Kolar for reviewing this this paragraph
15. A good compilation of texts relating to participatory art and a comprehensive literature survey can be found in Bishop, C., Participation, London/Cambridge, MA (2006)
16. The first attempt to contextualise cybernetic arts within art history was conducted by Jack Burnham. Burnham, J.: Beyond Modern Sculpture. The Effects of Science and Technology on the Sculpture of the Century, New York (1968) See also my more detailed investigation into the objectives of cybernetic arts: Katja Kwastek: The Invention of Interactivity In: Daniels, D., Schmidt, B.U. (eds.): Artists as Inventors, Inventors as Artists, Ostfildern (forthcoming, 2008)
17. Burnham, p. 338 (1968) (see note 16)
18. Cf. exhb. cat.: Software. Information technology: its new meaning for art. Burnham, J. (ed.), New York, p. 11 (1970)
19. Exhibition flyer. Archive of Myron Krueger
20. Cf. Krueger, M.: Artificial Reality II, Reading (1991)
21. Even though he used the term 'Interactive Art' in his publications, Krueger himself preferred the term 'artificial reality' to generally characterize his activities
22. For a detailed investigation of these 'pioneers'. Dinkla, S.: Pioniere Interaktiver Kunst, Ostfildern (1997)
23. Lister, M., et al.: New Media. A Critical Introduction. London (2003)
24. The more computer-aided interactivity approaches real interaction between two or more persons, the more successful it is according to this view. Lister, et al., p. 43 (2003) (see note 23)
25. There is a perspective on interactivity, based in literary and media studies, that argues that nothing much has changed in principle. We are just offered more opportunities for more complex relationships with texts, but these relationships are essentially the same. Lister, et al., p. 43 (2003) (see note 23)
26. Lister, et al., p. 44 (2003) (see note 23)
27. Graham, B.: A Study of Audience Relationships with Interactive Computer-Based Visual Artworks in Gallery Settings, through Observation, Art Practice, and Curation, Ph. D. University of Sutherland (July 1997), http://www.sunderland.ac.uk/~as0bgr/cv/sub/thesis.pdf (July 3, 2007)
28. Cf. The more recent publication, which however still adheres to the same categories: Candy, L., Edmonds, E.: Interaction in Art and Technology. Crossings. eJournal of Art and Technology 2.1 (2002), http://crossings.tcd.ie/issues/2.1/Candy/ (July 3, 2007)
29. Graham, p. 137 (1997) (see note 27)
30. Goertz, L.: Wie interaktiv sind Medien (1995) In: Bieber, C., Leggewie, C. (eds.) Interaktivität. Ein transdisziplinärer Schlüsselbegriff, Frankfurt, pp. 97–117 (2004)
31. See for example: Huhtamo, E.: Seeking Deeper Contact. Interactive Art as Metacommentary. Convergence 1(2), 81–104 (Autumn 1995)
32. Lunenfeld, P.: Digital Dialectics: a hybrid theory of computer media. Afterimage (April 21, 1993)
33. Lister, et al., p. 21 (2003) (see note 23)
34. Lister, et al., p. 22 (2003) (see note 23)

35. Thanks to the students of Interface Culture at the University of Art and Design, Linz for a revealing discussion of Lister's publication
36. Cf. Dourish, P.: Where the Action is. The Foundation of Embodied Interaction, Cambridge, MA (2004)
37. Cf. Kusahara, M.: Device Art: A New Form of Media Art from a Japanese Perspective. Intelligent agent 6(6) (2006), http://www.intelligentagent.com/archive/Vol6_No2_pacific_rim_kusahara.htm (October 24, 2007)
38. Cf. Kacunko, S.: Closed Circuit Videoinstallationen. Ein Leitfaden zur Geschichte und Theorie der Medienkunst mit Bausteinen eines Künstlerlexikons, Berlin (2004); also Kacunko's discussion of concepts of Interactive Art, ibid., pp. 54–62
39. Boissier, J.-L.: La rélation comme forme. L'interactivité en art, Geneva, p. 289 (2004)
40. The extensive report of the research project, http://gams.media.lbg.ac.at:8080/fedora/get/o:ia-079-2/bdef:PDF/get/
41. E-Mail to the author (December 13, 2006)

3

Strategies of Interactivity

Dieter Daniels

Ludwig Boltzmann Institute
Media. Art. Research.
Kollegiumgasse 2
4010 Linz, Austria
director@media.lbg.ac.at
http://media.lbg.ac.at

Abstract. The ideological and technological frames of reference for the changing paradigms of interactivity are presented in an overview. The topics range from the early days of media and modernism to a typology of interactive art in the 1980s and 1990s and include the mass media interactivity models of the last decade.

3.1 Reception as Participation—A Leitmotif of Modernism

In one of his apparently timeless sentences, Marcel Duchamp writes: "The personal 'art coefficient' is like an arithmetical relation between the unexpressed but intended and the unintentionally expressed."[1] Accordingly, no work of art can communicate to the viewer exactly what the artist intends. To put it more candidly, one might say that the greater the misunderstanding, the higher the personal coefficient. Duchamp therefore establishes that in every aesthetic experience, the viewer is assigned a constitutive role and that he "thus adds his contribution to the creative act."[2] On another occasion he even radicalizes his statement, claiming that "a work is made entirely by those who look at it or read it and who make it survive by their accolades or even their condemnation."[3]

That the reception of a work of art requires the viewer's participation proves to be a leitmotif of Modernism that emerged as early as in Charles Baudelaire's writings. Faced with the images in his *Painter of Modern Life,* "the spectator becomes the translator, so to speak, of a translation...."[4] In his examination of Wagner he goes even further: "In music, as in painting, and even in the written word, which, nevertheless, is the most positive among the arts, there is always a gap (a lacuna), bridged by the imagination of the listener."[5] Stéphane Mallarmé formulates the obvious conclusion in his concept of creative reading. As early as the end of the nineteenth century, he anticipated the idea of processual art with permutative, aleatory elements, which in the mid-twentieth century then became a platform of the avant-garde as an "open work of art."

The attack on the ideal of everlasting, unchangeable beauty carried out under the colors of Modernism therefore had a centuries-old history before it was ultimately conferred with a new technological basis within the concept of interactive media art.

C. Sommerer et al. (Eds.): The Art & Sci. of Interface & Inter. Des., SCI 141, pp. 27–62, 2008.
springerlink.com © Springer-Verlag Berlin Heidelberg 2008

The reason for Baudelaire's rejection of photography also lies in the knowledge that the reception of a work of art always requires an element of participation in its constitution, because he understands photography as a purely technical reproduction of reality that allows no space for the imagination. What might the function of technical media be for the participation of the viewer?

Baudelaire fails to recognize that in their reproduction of reality, technical media produce a side-effect that can be regarded as analogous to Duchamp's "art coefficient." Even if it is used in as flawless a way possible, any device constructed or operated by a human being to record or transmit images or sounds never registers only what his or her intention is directed toward at the moment of its use. Who is not familiar from one's own experience with the photograph that shows much more or something completely different than intended? Or the voices on the video that one did not really want to record? For the recipient of a technical media channel, this "excess" information allows an interpretation that may more or less deviate from the intention of the person who produced it. The side-effect heightens as the channels multiply. For the TV zapper or the Internet surfer, it becomes the actual "content" of his or her experience.[6] Putting it more concisely, one might assert that the "art coefficient," which Duchamp describes as a psychical phenomenon, finds its equivalent in the technical principle of all media devices that are not—such as the drawing or writing hand—directly linked to the intention of those who use them. This effect, which becomes so noticeable with technical media that in the extreme case it ends up in noise, already begins with the pencil, which also reveals something about its material property that points beyond its purely instrumental character.

At the outset of the lecture in which Duchamp introduces his concept of the "art coefficient," he describes the artist as a "medium" or a "mediumistic being."[7] He is therefore applying the term as it was used in the nineteenth century to describe a person with paranormal or telepathic abilities, in much the same way it was taken up by the Surrealists. However, when Duchamp used the term in the United States in 1957, it had already taken on different connotations than those it bore in Paris of the 1920s.

Without wanting to accuse Duchamp of making reference to technical media, which he certainly did not have in mind, it turns out, however, that even a text like *The Creative Act* is not quite as timeless as it at first seems to be. At the time, John Cage, who was a close friend of Duchamp, was already vigorously working on making the side-effects of technical media described above the theme of his music. The focus of his considerations was the very same problem regarding the intentionality of art that Duchamp examines with the "art coefficient." Cage's approach is often understood as an attack on the Old European *Geniekult* (cult of the genius) surrounding the creative, yet he had no inhibitions about time and again drawing attention to the role Duchamp played in his own introduction of random processes into the "creative act."

At the beginning of the 1950s, with his compositions for radio Cage achieved perhaps the first completely "open work of art" employing technical media—even before Umberto Eco coined the term in 1958.[8] In *Imaginary Landscape No. 4,* a piece composed in 1951 for twelve radios and twenty-four performers, Cage uses randomness in a twofold way: firstly, he determined the parameters of the score, which consists solely of instructions on how to operate the radios, using *the I Ching,* and secondly, the sounds coming from the received transmitters produce a relationship to the here and now of each performance that is just as random as it is coeval. Thus the side-effect of technical media mentioned above, which is to always transport more information than intended by their users, becomes the primary working means of an art that makes the participative and constitutive role of the recipient the principle of a new form of creativity and a new concept of an artistic work.

This digression to the history of Modernism elucidates the fact that the issue of viewer participation arises even before technical media are used in art. However, technology lends it a new dimension, as an interference occurs in media art between the two forms of the non-intentional emergence of information mentioned above. Here, the psychic role of the artist as a "medium" and the technical function of media devices join together. For about the last three decades, the ambivalent and ever-changing meaning of the concept of interactivity, which constitutes the main theme of this essay, has been evolving in this intermediate area.

3.2 From Participation to Interaction

The above considerations with respect to the participation of the viewer, the listener or the reader start out from the assumption that modern art has changed the role of the recipient. They make reference to the aesthetic experience and assessment of art, but the material existence of the work of art remains unchanged. This is why the modern role of the recipient can also make reference to works from history. Duchamp cited the rediscovery of El Greco during the era of Expressionism as an example for contemporary art "making the picture" by changing the view of history.

Yet if the work of art itself is aimed at the active role of the recipient, the second step toward a new form of artistic production, which is decisive for the following, takes place, allowing interaction between the recipient and the work by intervening in its visual, acoustic or textual form. In the process, the work of art becomes a kind of collaborative process participated in by the artist and various recipients. The issue with respect to the intention of art thereby changes direction in that the viewer—literally, so to speak, along the lines of Duchamp's dictum—now makes the pictures. Interaction can take place in a large variety of ways, for example in the form of an object, within the context of a situation or by means of a technical medium. In the simplest case, recipients can modify an object that has been created by the artist, which occurred in 1960s' Kinetic Art. More complex structures of interaction develop through the combination of pieces of text or sounds, which in object-like form or as a score demand the recipient's active participation in order for the work to even be produced in the first place.

This transition from participation to interaction can be exemplified by John Cage as a precursor of the Happening. Cage's use of random factors makes every performance of a piece a debut that sounds different than any other previous performance. The performers and the audience therefore do not have any expectations directed toward a perfect reproduction; rather, they are open for a new experience. Since the end of the 1950s, John Cage's compositions have been giving performers more and more freedom. The graphic scores ultimately only specify the method and the material for a kind of "do-it-yourself" music, such as the sheets of transparent film in *Fontana Mix* (1958) and their multiple use for different pieces. From here it is only a single step to Allan Kaprow's concept of the Happening. The Happenings are also based on scores with a frame of action; however, these now address all the participants, who are no longer separated from the audience but who create their own aesthetic experiences. Thus the completed work is replaced by an open field of action that is first engendered by the participants. These interact among themselves as well as with the specified frame of action, so that communication becomes the central factor of the aesthetic experience. This may even result in the complete removal of the boundary between author, participants, and the audience.

When interacting via a technical medium, various modes are possible, whose range from human-machine to human-medium-human communication constitutes the principle theme in the following. The role of media technology goes far beyond the previously examined side-effect of producing "excess" information, which enables a kind of "creative reading" beyond the producer's intention. When the two basic media-technological functions of "storing" and "transmitting" are implemented for an open work of art, they allow overlying production with reception, which—as in the case of the Happening—can remove the boundary between the author and the audience.

These different forms of interaction demonstrate the broad range of the meaning of this concept even in the area of art. However, the meanings in general language usage are even more diverse. Since the end of the 1980s, two usages have received more and more attention: on the one hand, the theory stemming from the social sciences of reciprocal actions by humans, and on the other hand, the technological category of human-machine communication, which is largely referred to as interactivity.[9] Because in a media society, people communicate with other people by means of machines, the overlapping of the two fields is evident. For this reason, in the following the concept of interactivity will stand for all forms of media-based communication and interaction that occur between human and machine as well as between humans.

3.3 Ideology or Technology—Brecht or Turing

In the current discussion on interactivity, the issues regarding the social ideology of a media-based human-human relationship overlap those regarding the technological feasibility of the human-machine connection. The roots of these two fields of meaning go back to a period way before the emergence of today's concept of interactivity. They can be traced back to the 1930s and may be illustrated using two positions that could not be any more contradictory: Bertolt Brecht's and Alan Turing's. In 1932, Brecht called for the following: "Change this apparatus [the radio] over from distribution to communication.... By submitting ever persistent, incessant suggestions for the improved use of the apparatus for the general public, we have to rock the social foundations of this apparatus, to discredit its use for the benefit of a few."[10] While Cage's composition only changes the reception form of the radio without intervening in the system of the mass medium, two decades prior to that, Brecht's approach goes all out and maps out an active role by the listeners as a political utopia that also includes the transmitter side of the medium. In 1929, with his

Fig. 3.1. Bertolt Brecht, *The Lindbergh Flight*, stage performance of the radio play with demonstration of the audience participation, Baden-Baden 1929

radio play *Der Flug der Lindberghs* (The Flight of the Lindberghs) (Fig. 3.1) Brecht attempted to translate this idea into practice. But because German radio was unreceptive to the concept of listener participation, his idea was not realized in a radio broadcast but only demonstrated in a stage production by Brecht.

From 1935 onward, Alan Turing worked on his theory of a universal machine that later culminated in the famous question: "Can machines think?" This included the problem of possible ways of establishing a connection between artificial intelligence and human consciousness: "We may hope that machines will eventually compete with men in all purely intellectual fields. But which are the best ones to start with? Even this is a difficult decision. Many people think that a very abstract activity, like the playing of chess, would be best. It can also be maintained that it is best to provide the machine with the best sense organs that money can buy ... I think both approaches should be tried."[11]

Both of these theses stem from completely divergent discourses. On the basis of pure mathematics, Turing developed the scientific foundation for the feasibility of human-machine communication up to a level where it would be impossible to distinguish one from the other. Brecht transferred his theory of the theatre to media and acknowledges the social and political effects of human-human communication characterized by evermore perfect media machines. Yet despite these extremely different starting points, today, the extrapolations of these theses meet in the form of information sciences and cultural media theory—for instance in the discussion on the connection between the political and the technological function of the Internet. Against this background, in the following the concept of interactivity and its relevance in media art will be examined as a field of interference between ideology and technology.

3.4 Open or Closed Systems—John Cage or Bill Gates

Despite the allegedly depersonalizing force of new communications technologies, individual names today stand more than ever for ideas and agendas—in politics as much as in business and the arts. This is why two names also stake out the territory for a closer examination of the ideology and technology of interactivity. The mottoes pointing to their common ground might read 'programs instead of instruments' or 'software, not hardware.' This aesthetic stance made John Cage the precursor of the New Music and intermedia art of the 1960s. Bill Gates, in contrast, realized the economic potential of this perception and in the course of the 1990s multimedia boom became the richest man on the planet. Both men no doubt attach a different meaning to these statements, as becomes obvious from their radically different concepts of 'interactivity.'

Most of Cage's compositions do not define a precise musical human-instrument interaction, but open up a field of possibilities to be interpreted by the performer of his composition, each time producing differing results through elements of chance and variation.[12] Some pieces modify the instruments (prepared piano) or leave the choice of instruments up to the performers. Through the performance process, the individual's freedom to modify the structure results in social interaction among the group of musicians. This non-hierarchical form of creativity can be compared with the 'bottom-up' structure by which open-source software such as Linux is constantly enhanced by its users. In either case, it is possible to vary and reinterpret a specified code with the result that the boundary between author and user becomes fluid. The opposite model would be a 'top-down' structure as represented by the precise notation of a classical composition as well as the proprietary software developed by Bill Gates' Microsoft Corporation, for which the secrecy of the source code is the basis of a capitalist monopoly. Program users work in line with the patterns of interaction decreed by the software industry, just as the classical musical composition specifies the manner in which musical instruments are used in the most precise way possible.

As Cage saw it, the purpose of composition was not to deliver an optimum 'operating system' for musical instruments but to initiate an individual and social creative process which successively detaches itself from the intentions of its author. By contrast, the software of Bill Gates and other proprietary systems keeps users in the dark about the structures 'inscribed' by its writers. A model derived from the time-honored, idealistic notion of art—that of the deep mystery inherent to all creativity—is being kept alive solely by artificial secrecy. Instead of serving the sacred goals of the genius, it panders to the mammon of monopolists. Cage's concept of interactivity stems from an aesthetic and ideology leading to the dissolution of the boundary between author, performance, and audience. That was why he deployed media technologies like radio, record, tape and, later, computer— the interference of musical production and reception became possible through the information structures of such devices. Technology could not only replace human labor, but also open up a creative sphere.[13] For Gates, by contrast, interactivity is an economically and technologically determined pattern according to whose specifications millions structure their workflow—a view he pinpointed in an in-house paper stating that Microsoft treats human users like it does computers: it programs them.[14]

While the computer is indisputably replacing the piano as the most frequently used keyboard instrument in the home, liberation from the often tortuous obligation to practice has not reached young people in an open, Cagean form but instead in the voluntary self-conditioning of interaction with industrial software such as computer games. This admittedly bold comparison serves to bridge the gap between Cage's art and Gates' technology in order to show that their conflicting models of interactivity ultimately stand for two different blueprints of society. The respective principles of openness and closedness could act as a leitmotif for the changing meaning of the term 'interactivity' from the 1960s to the 1990s.[15]

3.5 Shifting Paradigms of Interactivity from the 1960s to the 1990s

In the 1960s, interaction among audience, artwork, and artist became a defining element of an aesthetic aspiring to the ideal of a new art form that would leave behind established genres, categories, and institutions. This artistic field is most aptly described by the term 'intermedia.' The origins of intermedia art as inspired by John Cage and molded by Fluxus and Happening lie in the decision to replace an autonomous, finished work with an invitation to the audience to essentially self-determine how they experience the artwork and in doing so, lift the boundaries between artists and audience and those separating the genres. The suspension of the difference between production and reception in the arts has much in common with the demand made by political activists in the late-1960s for consumers to take over the means of production.

Despite the more than century-old history of active reception in the modern age, the classical, bourgeois concept of culture concedes a low ranking to the participation of viewer, reader or listener, demanding that paintings, books or concerts be enjoyed with a kindred understanding of an original work that has been tampered with as little as possible. Forms of popular culture such as vaudeville, circus or, more recently, the techno DJ, on the other hand, enter into an intense exchange with the audience. The attempts to make interaction a means of avant-garde art in the 1960s show the desire to depart from the confines of a bourgeois culture felt to be elitist and instead influence mass culture. Further ideals can be circumscribed with Umberto Eco's notion of the 'open work of art' mentioned above as well as the 'domination-free discourse' first expounded by Jürgen Habermas. The common enemy of all these artistic and theoretical approaches is the passive cultural consumerism felt to be a product of the mass media in general, and of television in particular.[16]

Models of open interaction similar to those in the arts were therefore developed and with a view to changing the role of the media. Drawing on Brecht, in 1970 Hans Magnus Enzensberger proffered the theory that the electronic technologies harbored the potential to emancipate by means of non-hierarchical communication. For the same reason, he saw the media, were they to be liberated from their perverted usage by the agents of capitalism, as potential stimulii to and instruments of social upheaval. "The open secret of the electronic media, the decisive political factor, which has been waiting, suppressed or crippled, for its moment to come, is their mobilizing power." And this power would enable people to become "as free as dancers, as quick-witted as football players, as surprising as guerrillas."[17] Comparable ideals are to be found in the anti-industrial media criticism given a forum in publications such as, from 1970 onward, *Radical Software*.

The computer hacker may personify a synthesis of these utopias; the origins of the hacker movement, however, are completely apolitical and provide a drastic example for technology as a weltanschauung. Its nucleus was formed around 1960 at MIT when the military lent the institute a computer free of charge that belonged to the first generation of computers with a screen. While programmers had previously had hardly any direct contact with the computer, developing their programs on paper and having operators feed the computers with punch cards, a group of student computer maniacs developed a "free wheeling, interactive, hands-on-über-alles style" in a direct dialogue with the machine, which today has long since come to be taken for granted, without the strictly hierarchical usage order for expensive running time.[18] The symbiotic relationship with the computer, which had virtually become the sole purpose in life for this group of people who referred to themselves as hackers, anticipates what has in the meantime become the everyday proximity to the machine, turning the calculating machine into a kind of digital partner. The social consequences of this attitude are much more far-reaching than the rudimentary ideology, which the hacker historiographer Steven Levy summed up in the so-called hacker ethic, reveals. Here, Levy writes, among other things: "You can create art and beauty on a computer. Computers can change your life for the better."[19]

In their detachment from the world and their complete immersion in computer programs, the pioneering hackers correspond with the ideal of art as an end in itself, as "art for the sake of art," which the intermedia art movement of the 1960s renounced in order to propagate interaction between art and life. This is why without exception, the Fluxus and Happening as well as the political movements of the 1960s took a critical approach to technology. However, around 1970 the realization was dawning on activists in art and politics that an unadulterated rejection of media amounted to nothing less than self-incapacitation. The emergence of the phenomenon today known as 'media art' is rooted in this interference of social theory and mass-media technologies.

By combining ideological strategies with technological means, the movements of the 1960s aimed to link the influence of art with that of the media. The social and cultural utopias supplied the objective of a hoped-for role of media in the future triggering a macro-change in society. This relation was turned on its head in the 1990s: media technology is now often seen as the leitmotif from which all social, cultural, and economic changes emanate. Today, for instance, the meaning of 'interactivity' is essentially defined through the electronic media. Interface and software designs specify the framework of this technologically determined interaction from human to human via a machine, or solely between human and machine. The 'mobilizing power' of the media, in which Enzensberger was still able to discern potential for attacking the dominance of industry, has long since become fuel for advertisements plugging telecommunications shares or cellphones and deploying the same heroic images of dancers, footballers or guerrillas. Since the 1980s, the original hacker ethic, which was committed purely to the thing itself, has likewise been marginalized by a partially criminal, partially commercial twilight zone.

The same is true of the concept of interactivity through interdisciplinarity, a cultural paradigm redefined in the 1990s to become one of technology. In the digital realm, the difference between text, sound, and image is apparently reduced to varying data storage-space requirements. To combine various media in a single multimedia program is in line with the basic principle of digital technology, and therefore requires no aesthetic legitimization, as propagated by intermedia art. Admittedly, there is a connection between the subdivision of artistic genres and the specific media deployed, but the belief that a shared media platform alone could facilitate or even implement a cultural exchange has proved, by and large, to be an illusion encouraged by the superficial resemblance of various interfaces.[20]

The 1990s attitude that grasps social and cultural transformation as an effect of the media, contrasting with the calls in the 1960s for media to be the instruments of such change, is not without historical roots. These extend from the Italian Futurists' fascination with technology to Marshall McLuhan, who as early as in 1964 described the media as being the de facto realization (that solely artists refused to accept) of the dreams of a new perceptual form first devised in the arts.[21] The contemporary scientific follow-ups are the media-theoretical approaches such as as that of Friedrich Kittler, according to whom it is only possible to "continue mistaking for art the output of media because the design and nuts and bolts of technical devices ensure they remain black boxes." In Kittler's view, artists are forbidden from opening up the covers of the devices, this privilege being reserved, "as the warning signs make very clear, for qualified specialists. What goes on beneath the covers, in the actual circuitry, is not art but the end of the same in data processing that takes its leave of humanity."[22]

Measured against such a view, some of the interactive forms tried out and developed in 1990s media art may indeed seem naive and, above all, wholly dependent on technological specifications. Yet a look back at the pioneering forms of artistic, media-based interaction reveals that in many cases they penetrated far below the equipment cover plates. Indeed, long before the prefabricated media were packed inside casing and became commodities, artists were drafting new models of perception and action that would decades later become part and parcel of mass-media routine. Walter Benjamin already saw the Dadaists' montage of language and images as anticipating the media effects used in films.[23] Ever since the Futurist movement, avant-garde art has envied technology for its influence on the masses, while at the same time displaying vast far-sightedness with regard to technological effects and evolutions. For the same reason, the debate surrounding interactive art in the 1990s can be truly understood only against the backdrop of the preceding developments, especially those in the 1960s.

3.6 Examples of Media-Assisted Interaction in Intermedia Art of the 1960s and 1970s

Compared with the visual arts, music requires considerably lower data volumes and storage capacity for its electronic processing. That is why radio came before television, and the tape recorder before the video cassette recorder. For the same reason, numerous artistic approaches to media clearly first emerged in work with music.[24] This is also true of the notion of 'interactivity,' as demonstrated by John Cage's pioneering role. Yet Cage's point of departure was not technologically defined; on the contrary, he started with silence. His piece *4'33"* (1952) can be seen as the ideal 'open work,' precisely due to the absence of instruments. In it, nothing is fixed; everything depends on the conditions of the respective performance. The sounds made by the audience and coming from the environment are the content of four-and-a-half minutes of heightened sensibility. During the same period, Cage began to devise pieces that transferred the same principle of open interaction to the deployment of electronic media, for instance in his composition for twelve radios of 1951, which allowed experiencing the mass-media variety of the broadcasting station as raw aesthetic material at the very moment the composition was being performed. TV zapping, another form of 'interactive' media perception likewise produced in real time by individual selection, namely by 'assembling' a new 'film' from the TV programs being broadcast at any given time, is based on the same synchronism and redundancy of available channels as Cage's *Imaginary Landscape No. 4.* This analogy between experimental composition in the 1950s and day-to-day reception in the 1990s would serve as a good example of the way artistic models prefigured media effects.

Cage's approach was indeed seminal for the entire field of intermedia art in the 1960s, but Happening and Fluxus scarcely picked up the media-related issues he addressed. A major exception is Nam June Paik, who in the title of his 1963 show "Exposition of Music—Electronic Television" in Wuppertal already indicated his crossover from New Music to the electronic image. The various "Participation TV" models presented there were the first blueprint for viewer interaction with TV pictures. By manipulating the electronic circuits of normal TV sets, Paik was able to achieve complex visual structures that viewers could alter and which anticipated by decades the industry-marketed video and multimedia devices serving similar ends.[25] Since only one TV channel existed in Germany at the time until (by coincidence also in 1963) a second public broadcaster, the ZDF, went on the air, actuating the ON/OFF switch had until then been the TV viewers' only possibility of interaction.

The uncompromising openness and infinite indeterminacy of the Happening and Fluxus ideal of an art that has no creator/viewer hierarchy proved to be a transitory phase. Although these movements made a vital contribution toward changing the static concept of a work in the visual arts, they supplied no sustainable model for tangible results. Above all, such an ideal was unable to satisfy the recipients' need for symbols and fictions. As the Happening of the 1960s progressed to become the Performance of the 1970s, audience interaction was either no longer desired or underwent severe ritualization and formalization. Bruce Nauman expressed this change unmistakably: "I mistrust audience participation."[26] This attitude is evident in Nauman's closed-circuit installation *Live-Taped Video Corridor* from 1970, which by irritating viewers through their presence or absence in the video image, makes them more test objects than participants.[27] This is why Nauman can be regarded as a precursor of an attitude producing the very opposite of creative participation, namely the radical conditioning of a viewer through a work that forces him or her to fall back on their own experience of body and image. Artists including Dan Graham, Peter Campus, and Peter Weibel used video technology in similar fashion in the 1970s in order to confront viewers with their own image by means of closed-circuit installations. Such works were, together with Nauman's corridor, without doubt the first interactive installations that were suitable as art exhibits. They were no longer designed to solicit the spectator participation aspired to in the 1960s, however, but constructed situations reflecting upon the relationship between viewer and medium. At the same time, they marked an attitude of resignation towards video's potential for mass-media broadcasting, perpetuating instead, in almost symbolic fashion, the 'closed circuit' of the art system.

Valie Export's *Tap and Touch Cinema* from 1968 provides the antithesis to this self-reflection through the aesthetics of the media. Describing her outdoor action as an 'expanded movie,' she strapped a box to her chest and allowed passers-by to poke their hands through the curtain covering the front of the box in order to feel her breasts. "As always, the screening takes place in the dark. Only the picture-house is a bit smaller. There's only room for two hands," wrote Export.[28] This was an even more drastic conditioning of the viewer than in Nauman's *Corridor*, and again, doubt was cast on the boundary between public and private space.[29] Now that the relation to the film medium had been placed on a metaphorical plane, the sensory deprivation of the seat-bound passive cinema or TV viewer was all the more distinct. As a direct sensory experience in Export's action, 'interactivity' was the countermodel to one-sided, mediated perception. The 'tactilism' the Futurists demanded back in 1921 as a way of expanding the spectrum of arts was transformed by Export into a critique of the social role of the media. That her street action took place in 1968 was certainly no coincidence, but expressed an aspect of the calls for a "structural transformation in the public sphere" (to cite the title of a book by Habermas) that led to the 1968 movement.

3.7 Fiction and Function of Multimedia Technology and Cyberspace

Most of the examples presented so far involved interactive reapplications of media primarily serving the purposes of distribution and reproduction (video, film, TV, radio). The underlying artistic approaches deliberately worked against mass-media consumerism by modifying, to a more or less subversive degree, the consumption of the media in which the works were produced. From Brecht to Paik, such approaches demanded the alteration of the one-way structure of such mass media. In computer-based multimedia technology, by contrast, the interaction of user and device is integrated into the medium itself. Networking makes the computer an interpersonal communication medium in which all previously separated media converge. The current technological development of networked virtual reality merges the two formerly separate development strands of computer-based simulation and communication. These spaces for a new experience, as virtual as they are real, were becoming tangible in the late 1990s. That their roots stretch back to the 1960s is evident in the way present-day ideas regarding Cyberspace were anticipated in the technological blueprints of that period, but even more strikingly in the theories with respect to the potential social, aesthetic, and political implications.

Nearly all of the technical requirements for the current state of human-machine communication were created in the course of military developments. As has already been mentioned, until well into the 1960s, most computers were abstract computing machines used to process columns of figures and punched cards. With the introduction of the monitor, the first step was taken toward visual display. The first possibility of graphic interaction in real time had emerged in the 1950s with the linkage of a visual display unit and a light pen on the Whirlwind computer developed for air-defense purposes in view of the nuclear threat.[30] The dissemination of similar principles in the 1960s and 1970s opened up the option of visual, intuitive, instantaneous man-machine dialogue. When in 1966 Ivan E. Sutherland linked up the head-mounted display developed for military purposes with the simple computer-generated wireframe representation of a three-dimensional space, his combination already contained the essential elements of virtual reality technology, lacking merely faster computing speed and sufficient storage capacities. What would later turn into the Internet also began in the 1960s on the basis of the decentralized ARPA Net installed in 1968 in order to safeguard military communications in the case of a nuclear attack. Both components of Cyberspace today enabling the networking of virtual spaces are therefore products of a Cold War anti-nuclear defense strategy.

There was an astonishing synchronism between such technological blueprints and the artistic theories regarding their potential. Ivan E. Sutherland's first description of an 'ultimate display,' written in Harvard in 1965, shows considerable similarity to a concept for a 'bioadapter' drafted in the same year by the author Oswald Wiener in Vienna.[31] According to Peter Weibel, this concept was the "linguistic draft of a data suit." Interestingly, Sutherland was working on the technical implementation of a man-machine interface, while Wiener, wholly independently, was investigating the cultural consequences of a synthesis of this kind. The difference between technological practice and theoretical analysis did not lie in the belief in feasibility, but in the expectations this feasibility roused.

In this respect, Oswald Wiener began with the following finding: "The new branches of science known by the collective term cybernetics have produced sentences that can be applied virtually unchanged to sociological matters within so short a time that it is reasonable to suspect the formulators had in mind the establishment of fundamental correlations between the requirements of technology and those of the state."[32] The logical conclusion Wiener drew from all this was the "liberation of philosophy through technology," by means of the "bio adapter" that "for the first time fulfils the healthy-heroic ideal of a Homo sapiens who governs the universe, namely by drying out the cosmos on the one hand, and by liquidating the Homo sapiens on the other hand."[33] The consequences of one such scenario are depicted in the film *Matrix* from 1999.

Even if affirmative in tone, Wiener's well-nigh nihilistic skepticism contrasted with the naive enthusiasm for technology of many other artists, an attitude which was presumably closer to the utopia visions fostered by the developers of the technologies. Nicolas Schöffer's 1968 manifesto *The Future of Art,* for instance, reads: "The information networks must be opened up for the true aesthetic products. This however requires a new art-technology and a complete transformation of the relationship between the producing artist and consuming audience.... Today we can envision with certainty for the future a room that replaces the small screen and wholly envelops the consumer. In this room the consumer will be surrounded by audio-visual, (olfactory, tactile) programs, will bathe in a truly, consistently aesthetic climate he is able to dose, re-assemble and program according to his own wishes. This bath will put him in a position to continuously advance and perfect himself, to sensitize, concentrate, and express himself; it will lead to a new notion of human hygiene. This aesthetic hygiene is likewise indispensable for the those communities, or social groups, living in urban areas of various size."[34] What Schöffer chose to ignore (in best Futurist fashion) was the marginal role art and artists would play in the de facto development of the world-model he outlined. Indubitably, the contemporary ear detects a sinister undertone to the technology-based "aesthetic hygiene" he propagated.

It is tempting to place the technical enhancements of man-machine interaction in relation to the lifting of boundaries of 1960s art. Toward the end of the same decade, this synthesis was promoted by the first Art & Technology events.[35] There are undeniable similarities between the technological futures mapped out by Schöffer,

Sutherland, and Wiener, but their theories regarding the social, psychic, and political effects were radically different. We are reminded once again that interactivity always stands both for a technology and for an ideology. Either field has continued to overlap with the other up to the present day. The term 'Cyberspace' was coined 1981 by science-fiction author William Gibson in his short story *Burning Chrome* and becomes common with his novel *Neuromancer* 1984. The emphatic, sometimes even ecstatic, books by scholars like Donna Haraway or popular authors such as Howard Rheingold were more conducive elements of 1990s cyber euphoria than the general public's hands-on experiences with technology of this kind. Yet this hype, for its part, stimulated technological developments and, above all, the need for them. In this area, it follows, there is a very close reciprocal relationship between fictive visions (be they expressed in literature, science or art) and the creation of the technological functions these visions describe. The fiction and function of Cyberspace evolved in a process of constant feedback.

Nevertheless, it is possible to plainly state the real motives for the creation of Virtual Reality technology, and with them the ideological background. From the 1960s onward, the practical implementation of such blueprints was financed almost exclusively by military budgets. Regardless of whether the philosophical and aesthetic designs originated from scientists, writers or visual artists, their ideological basis clearly differed from that of their practical implementation. One ideology was trying to remove the aesthetic boundaries between individual and collective, or between producer and recipient, while the other—wholly contrarily—was aiming at the military transgression of a frontier shielding an enemy defined by this very ideology. Since the aesthetic ideal of removing boundaries was dependent on the device developed for other purposes, art was now suspected, not without justification, of recycling or even pseudo-legitimizing, military technology. Computer games, as one example, represent the broadest worldwide usage of these technologies. While as games their combinatorics give them some relation to the arts, most of them have an ideological and psychological basis making them notorious illustrations of the military origins of their technologies.[36] If artists are unaware of the inherent contradiction of using means developed for military purposes to advance their aesthetic aspirations towards lifting boundaries, then they are naive at best, opportunists at worst.[37]

3.8 Examples of Media-Assisted Forms of Interaction of the 1980s and 1990s

"Virtual Reality and Cyberspace are 1960s ideas, even if their technology was first implemented in the late 1980s," asserts Peter Weibel.[38] Like Jeffrey Shaw and Valie Export, he counts among those artists whose work with different forms of interaction spans the divide between the approaches of the 1960s and 1990s. With the same statement, however, he joins Shaw and Export in ignoring the paradigm shift between the removal of aesthetic and social boundaries in the 1960s and the technological interactivity of three decades later. This might be partially explained by the fact that the notion of interactivity only reemerged in the 1990s as a result of technological development, after being almost entirely absent from the Conceptual and Minimal art dominant in the 1970s as well as from the postmodern retrospection of the 1980s.

Toward the end of the 1980s, realistic 3-D animation in real time became possible thanks to higher computing speeds and storage capacities. Interfaces like data gloves and cyber helmets could now be used for physical immersion in data space, and presented the basis on which in the following decade various models were developed for the interaction of human and machine, of real space and data space. The capacity of elaborate technology was the hallmark of most of the models produced in collaboration with media institutions, universities or business enterprises. Commentaries accompanying the art-related projects all emphasized the aspects of technical-aesthetic innovation and of the joint research conducted by engineers and artists. The emancipationist or media-critical approaches that were obligatory in the video art of the 1960s and 1970s now almost disappeared. Several typical models of human-machine interaction are briefly outlined below and placed in relation to parallel developments outside the field of art.[39]

i. Interaction with a video story through multiple options

Counting among the first successful examples of technology based interactivity, these works of the 1980s are strictly speaking not part of the Cyberspace domain. They connect video and computer technology in order to enable a plot with several variants and loops that, unlike linear narration, offer the viewer options for the further progression of the story.

Fig. 3.2. Lynn Hershman, *Lorna,* (1983/84) interactive videodisk installation left: the viewer controls Lorna's actions, the installation and Lorna's room in the video contain the same objects right: the on-screen menu offers several options for Lorna's actions.

In her pioneering work *Lorna* (1983/84) (Fig. 3.2), Lynn Hershman places the viewer in the same space as the actress, whose fate he or she determines by way of a remote control. In her later installations, from a playfully feminist stance Hershman thematicizes primarily the sexual and erotic dimension of interaction, turning the viewer into participant or voyeur *(Deep Contact,* 1989/90; *A Room of One's Own,* 1992). Grahame Weinbren's installations develop complex relations between several plot levels, pointing toward the interactive cinema of the future *(The Erl King,* 1986; *Sonata,* 1991/93). The potential to expand such approaches for collective productions was demonstrated by the *Videolabyrinth* jointly developed in 1988 by video filmmakers Rike Anders, Ilka Lauchstädt, Mari Cantu, and programmer Martin Potthoff. Their labyrinth contains three interactive plots subject to interruption by questions, quiz assignments or scoreboard readings. As a West German production, it still had to rely on computer-controlled videotapes that entailed long waits between the sequences. Weinbren and Hershman, by contrast, deployed videodisk technology that was already available in the USA but failed to succeed on the mass market. With the launch of the CD-ROM in the early 1990s, the first interactive medium became commercially available, but its storage capacity was insufficient for longer video stories.

Attempts by the entertainment and TV industries to make interactive film and TV a mass medium have not succeeded so far.[40] This may be partially due to the still complicated operating conditions, but the commercial failure of all interactive mass media models to date (from CD-I to interactive TV) might be taken as an indication that audiences prefer linear narration.[41] The dramatic structures of interactive narration likewise remain in the early stages. Zapping, which remains the most popular form of interaction with linear programs, is an anarchic form of personal montage that eludes all control or structuring. Oliver Hirschbiegel's TV thriller *Mörderische Entscheidung* (Murderous Decision) was an attempt to transform the destructive principle of zapping into a constructive method of interaction. Its two plot strands were transmitted concurrently on ARD and ZDF, the first and second German channels, in 1991. Similarly, the non-linear film *nomad* (1998) by Petra Epperlein und Michael Tucker uses DVD technology to offer the viewer a choice between three versions running parallel to each other but allowing no interaction.

ii. Interaction with a closed data world through which the viewer can navigate

This is the classical model of basic 3-D interaction, such as presented in Jeffrey Shaw's installations *The Legible City* (1988) and *The Virtual Museum* (1991) (Fig. 3.3). The viewer passes through an unchangeable data landscape, which is not unlike being on a tour around a town or through a museum. Decisive in Shaw's works is the quality of the interface that, moving away from the keyboard or mouse, places the viewer on a bicycle or into a reclining chair and so enables everyday physical movements to be intuitively transposed into the data world. These installations have a certain resemblance to information systems such as museum guides on CD-ROM or interactive maps that navigate a driver through the streets of an unfamiliar city. The forerunner of all such systems was the *Aspen Movie Map* developed in the late 1970s by the Architecture Machine Group at the Massachusetts Institute of Technology. The notion of a

Fig. 3.3. Jeffrey Shaw, *The Virtual Museum*, 1991, installation at Ars Electronica Linz 1992

'virtual museum' has become particularly popular, with museum visits being deployed as metaphors in products ranging from demo programs to science CD-ROMs for a general audience. With digital technology, it would seem, the ideal medium for the 'museum without walls' ('musée imaginaire') envisioned by André Malraux is made available for the first time. One way of compensating for the lack of communicative processes with the 'museum-like' data world is to heighten the illusionist quality and in this way extend the sensations to bring about a complete immersion of the viewer. Such hi-tech installations are situated on the fringes of the art context, however, and often find more appropriate appreciation as scientific visualizations.[42] These products slot into a tradition stretching back to the nineteenth-century panorama.[43] At the same time they make it clear that, ultimately, illusion and interaction are mutually exclusive.

iii. Interaction between body and data world

All virtual-reality techniques constitute an expansion of perception and establish a connection between data structure and body. The development of such interfaces represents one of the most creative intersections of artistic and technological approaches in the 1990s. The classical set composed of data glove and VR headset proves impracticable for exhibitions, conflicting as it does with the habits of museum visitors and only able to be used by one visitor at a time. ART + COM delivered a paraphrased version of the museum situation with *Zerseher* (1990–91) in which a Renaissance painting *(Boy Holding a Child's Drawing* by Giovanni Francesco Caroto) is dissolved as a digital reproduction through the gaze of the viewer. An eye tracker developed for medical and military purposes records the motions of the eye within the field of vision, and in this way makes possible, at least in a symbolic destruction, the technical implementation of the active role of the art observer. Peter Weibel's installation *The Tangible Image* (1991) permits direct, haptic access to a digital image. This work can also be viewed as the human-machine version of the human-human interface provided by Valie Export's *Tap and Touch Cinema;* again, the paradigm shift from the 1960s to the 1990s becomes evident.[44] Weibel carries forward this fusion of image and spectator in *The Curtain of Lascaux* (1993), embedding it in a philosophical concept taking in the history of human perception from prehistoric paintings through Plato's cave to Cyberspace.[45] A feedback between body and data takes place in Ulrike Gabriel's installation *Breath* (1992/93). Via a sensor belt, the viewer's breath influences the computer-generated projection of crystalline-amorphous visual structures and the soundtrack. The visuals and sound are designed to affect the viewer and initiate a biofeedback between user and machine. Experiments with interactive choreography took the first step toward connecting human and technical action. *Electro Clips* (1994) by Christian Möller and Stephen Galloway as well as *Binary Ballistic Ballet* (1994) by Michael Saup and William Forsythe were both the products of collaboration between a media artist and a choreographer.

David Rokeby's sound installation *Very Nervous System,* whose reaction to body language he continued to develop between 1982 and 1995, is a pioneer work in the area of body-computer interaction. There are different versions for different areas of implementation: it exists as an exhibition object for interaction with visitors and as an interactive instrument for performances with musicians and dancers. There is even a version for medical use that enables a completely paralyzed woman to communicate with the outside world by blinking her eyes. In this respect it is not only an installation, but above all a tool for multiple applications which its users furnish with content. The media-assisted body performances by Stelarc, who in countless self-experiments since the end of the 1980s has temporarily integrated media technology into his body, are even more spectacular. External sensors, for example, control the movements of his arm, which can be animated alternatively by Internet users or by an echo in the transmission times in the Net *(Fractal Flesh* 1995, *Ping Body* 1996). McLuhan's dictum that media are "extensions of man" is taken so literally in this case that for its part, the body becomes an extension of media. Two diametrically opposed approaches: Stelarc works on incorporation of the machine in the body, and Rokeby works on the body-like reactivity of the machine. In doing so, both of them work continuously on new aspects of an almost alchemical, unfinishable opus of a lifetime in the undefined space between art, physiology, and software. This also applies to other pioneers of interactive art, such as Myron Kruger, who as early as 1974 began tinkering with perfecting his installation *Videoplace* in order to make the countless application modes for interaction between the body and the electronic image appear more and more human.

iv. A data system with momentum that is enhanced through interaction

Since Turing, the ability of machines to learn was always considered to be an essential condition of Artificial Intelligence. Even on the low-tech level, a number of models were created that assigned a 'work' a life of its own in interaction with the viewer. Peter Dittmer's installation *The Wet Nurse* (*Die Amme*) (1992–ongoing) (Fig. 3.4) is another live time development project. Based solely on language, this apparatus involves the user via the keyboard in a complex dialogue. If the computer becomes finally agitated, this results in the symbolic spilling of milk into a large glass cabinet. The software's conversational skills are continuously expanded and enriched through usage. Thanks to the modest storage requirements of text, no more than a PC is required back in 1992 for the user interaction. The sculptural appeal of the whole is as important and has since been significantly expanded and modified up to the last version of 2007 which filled a whole museum space. The basic set up is a clear demonstration of the principle of the Turing Test with its distinction between rational and libidinal function—while it is

Fig. 3.4. Peter Dittmer, *Die Amme (The wet nurse)* since 1992, installation at Minima Media Leipzig 1994

possible to mistake the machine for a human being in the conversation based on written language, no such confusion is likely to occur with the milk served by the 'wet nurse.'

Daniela Plewe's installation *Muser's Service* (1994–95) is likewise based on linguistic exchange. Unlike the impertinent answers of the *Wet Nurse*, however, the PC in this case provides assistance to daydreamers, or musers, by freely associating between two keywords entered by the user. That computers take over human chores is commonplace—but what about daydreams or even fundamental decisions? The latter are served by the model Daniela Plewe presents in *Ultima Ratio,* whose various modes range between 'cascades of doubts' and 'war of convictions.' The artist states: "In contrast to classical logic (but in unison with other AI systems), the (modified) decision-support system of Ultima Ratio tolerates contradictions and exceptions to rules.... The visitors are required to explain their intuition, and in doing so possibly feel the desire to continue refining, ad finitum, an ultima ratio that slips out of control again and again. It was not primarily a question of practicing AI, then, but of using software and its syntactical units (rules, exceptions, contradictions) to vary and comment upon something of the culture that surrounds us."[46]

When artists decide to incorporate self-developing dynamics into graphic-spatial displays, the technical requirements escalate. The installation *A-Volve* (1994) by Christa Sommerer and Laurent Mignonneau invites visitors to sketch on a monitor the outline of small, artificial beings, whose subsequent brief digital lives in a virtual aquarium are guaranteed only by the pseudocaresses of their creators. Instead of the linkage of 'art and life' propagated in the 1960s, the concern is now to overlap technology with biology in order to simulate artificial life. Yet the entertainment aspect partially invalidates the intended character of scientific visualization.

The degree to which dynamic processes in computers can be considered 'creative' was the subject of partly serious, partly ironical, debate. As early as 1985, Richard Kriesche presented the following radical theory: "As long as natural and artificial intelligence are two separate properties, art will remain a mystery," and their synthesis might however be achieved "thus rendering art unnecessary."[47] Following this line of thinking, Turing's question, "Can machines think?" would now have to read, "Can machines make art?"

v. Dialogue-based models

In these models the human-medium-human interaction is more important than human-machine interaction. The simplest case are telecommunication pieces with live video or TV links between two exhibition venues (on different sides of the world or across the street). In the 1970s Douglas Davis began to use television for live art actions. Some of them allow a real dialogue *(Talk Out!,* 1972) others only metaphorically, if not to say metaphysically, by staging a pseudo-telepathic connection *(The Austrian Tapes,* 1974). With the proper presentation, technically flawless pieces like Paul Sermon's *Telematic Dreaming* (1992) and *Telematic Vision* (1993) attract maximum audience participation today. While the telematic expansion of everyday situations such as sitting on a sofa or lying in bed is unlikely to rouse technological inhibitions, the dialogue mostly remains on the playful communicative level of 'hello there.' Agnes Hegedüs' installation *Between the Words* (1995) places directly opposite each other two partners in conversation separated only by a wall housing the interface in which virtual gestures are superimposed over real physical expressions. This lyrical approximation of virtual and face-to-face encounter became a drastic message in the 'cybersex suits' presented by the artists Kirk Woolford and Stahl Stenslie in 1994.[48] More effort, it would seem, is being put into similar models for practical application outside the art world. The press coverage granted to such experiments seems to point to a virulent area of the collective subconscious. All these dialogue-based approaches tend to exaggerate the symbolism of media connectivity. Douglas Davis, for instance, in 1975 described the feedback resulting from his work as, "Knowing that I am involved in the evolution of a deeper, more diversified system of communication, between myself and the world and back. It has nothing to do with specific response."[49] A fresh illustration of McLuhan's statement "the medium is the message," but nothing more.

vi. The "exemplary viewer"

In the installations described so far, the visitor takes on a new role: not just viewer, but also performer. Yet this self-evident explanation of the term interactivity disregards a second, equally (if not more) important change in the viewer's role. Due to the fact that most interactive installations allow only one viewer to act, he or she occupies a specific position and is a part of the work's completion. They become 'exemplary viewers,' not just one viewer among many and no longer part of a group assembled in front of a work and walking around it at individual leisure.

In the hi-tech simulations of the 1990s, the exemplary viewer acts as the link between data space and the real world. The meeting between visitor and mediated image in the closed-circuit video installations of the 1970s was comparable. In the Cyberspace installations, that element of self-duplication termed "video narcissism" by Rosalind Krauss in her analysis of 1976 produces the symbolic loneliness of the viewer in virtual space.[50] The same applies to telecommunications projects in which two viewers are placed in relation to each other but the actual fascination is due precisely to the insurmountable spatial and physical separation that accompanies the intensive connection. Along these lines, Paul Sermon's telematic linkage of two people in two beds for the purpose of televisual pseudophysical contact *(Telematic Dreaming)* is also a rejoinder to the media role shown in Valie Export's *Tap and Touch Cinema* of 1968.

At an exhibition, the actual situation of the exemplary viewer is of course often anything but lonely. Other visitors perhaps observe the interaction, offer advice, laugh, or wait impatiently for their turn to come—long queues are frequently a problem at popular shows in the 1990s. Jeffrey Shaw's account of an experience with his *Legible City* during a show with late-night opening demonstrates that isolation in front of the apparatus is one of the central experiences with this form of interactivity. Suddenly, Shaw saw his own installation on its actual plane of experience—that of cycling by night through a deserted city.[51] Loneliness, then, extends into the visual realm of the works. In none of the numerous virtual museums developed during the 1990s, will a visitor bump into other visitors.[52]

Almost all of the models of interaction described so far are implemented in installations that remain bound to real space. This physical relation enables the works to be placed in art contexts with their site-specific valuation criteria of being more, or less, prominent exhibition venues. Their technical complexity, however, makes the installations considerably more difficult to transport than pictures or objects.

Ironically, in the 1990s the price of the 3-D animation hard- and software generally surpasses by far the potential market value of the artwork generated with the aid of the same technology. The paradoxical relation of media and market is, that in the 1990s virtual-reality pieces cannot be sold because their technology is too expensive for the average collector or museum budget, whereas artists' videotapes in the 1980s and early 1990s were still too low-priced to be taken seriously as collectible art. Even more crucial is the fact that illustrations or documentations fail to produce essential aspects of the user interaction in hi-tech installations compared to traditional, static artworks. This is why books and press reviews or even TV features can only convey a fraction of the whole. The most elaborate media inventions are precisely the ones which exceed the capacities of the mass media and are for this reason often neglected by media coverage. Ironically, the anachronistic result is that the viewer wishing to experience the actual interactive quality must travel to festivals and media-art exhibitions, just as in the past people traveled for the sake of art. The stationary interactive installation has proved to be a dead end due to these distribution problems and the limitedness of its interaction potential. While the availability of high-tech equipment was a financial question in the 1990s, this is getting even more complex with the problems of maintenance and preservation of the hard- and software.

vii. Approaches toward collectivity in media space

Through the interconnection of several users as part of a collectively developed structure, the electronic realm can be transformed into a social and to some degree public domain. Complex communications structures began to emerge, mainly in the form of text-based systems, even before the Internet boom. Long before then, the *cadavre exquis* of the Surrealists had already demonstrated the poetic potential of collective authorship. Roy Ascott's *La Plissure du Texte* (1983) and the project with twenty-six authors initiated by Jean-François Lyotard on the occasion of his exhibition "Les Immateriaux" in 1985, or John Cage's *The First Meeting of the Satie Society* (1986) are initial efforts for a networked authorship among artists and writers. Hypertext concepts followed in Germany from 1988 onward by the project "PooL-Processing" of Heiko Idensen and Matthias Krohn.[53] The postmodern thesis of the "death of the author" thus finds its contemporary technological form, because already in Lyotard's project, it was possible to modify the texts written by the other participants in the collective.[54] An examination of the relation between ideology and technology is therefore also possible in the context of the postmodern discourse, in particular with respect to the concept of the rhizome, which Deleuze and Guattari coined for an interlinked text structure as early as 1976.

As a collective form of communication, networked writing has now become an everyday form of discourse on the Internet. In the 1990s textual worlds of the MUD (Multi-User Dungeon) and MOO (MUD Object-Orientated), which were originally created as networked games, are becoming part of the Internet way of life together with chat rooms and mailing lists in the tradition of the once famous Californian "The Well." [55] These playful communities were for a long time areas of non-compromised creativity and self expression, but did not surface in the context of media art. Exceptions are Evelyn Teutsch's "FOOGUE" (1996 onward) and the installation *[DPsNtN] = DISPLACED_PERSONS say NOTHING to NOBODY* (1997–99) by Christin Lahr. Lahr conducted research in the LambdaMOO on the

Fig. 3.5. Van Gogh TV, *Piazza Virtuale*, 1992, interactive TV broadcast for Documenta IX left: viewer interaction by telephone and touchtone right: interface for collective on-screen painting of the TV spectators

debates surrounding "presence and absence, truth and falsity, gender, appearance, identity and location." The findings are transferred to the art context by an installation in which only the static visitor can experience the overlayering of virtual and real space, making the encounter contemplative rather than interactive.

Even before the Internet boom, the project "Van Gogh TV" (Fig. 3.5) at the documenta 9 in 1992 created a computer-operated link between television and the telephone. This enabled viewers who had managed to acquire one of the few dial-in numbers to collectively make music, paint, or chat live on the television screen by means of an interface operated via the telephone keypad. However, despite—or even due to—the widespread acceptance by the public, the lack of a thematic parameter and the primitive interface caused the project to end in one hundred days of irrelevance. In retrospect, one could interpret "Van Gogh TV" in line with Benjamin—as a magnificently failed attempt to anticipate the effects of the World Wide Web using television and the telephone.

Physical interaction in simulated 3-D space can be combined with information data downloaded from the Internet in elaborate installations of the type implemented by the Knowbotic Research group (*Simulationsraum mobiler Datenklänge,* 1993; *Dialogue with the knowbotic south,* 1994). The viewers do not enter not a predefined data space but a digital environment which is continuously developed through the participants. It represents an attempt to find new forms of visualization for complex scientific procedures such as those used in Antarctic research. By implementing the means of associative, spatial, and physical experience, the artistic concern of the group is to make imaginable scientific and technological correlations that, due to their vast complexity, might seem to surpass our imaginative capacity. Since 1997, this field has been expanded to the analysis of urban structures. These images of 'computer-aided nature' produced at the crossover between art and scientific visualization are often seductively aesthetic, possibly even too beautiful to be true.

3.9 Interactivity and the Internet [56]

Although scientists around the world have been using the Internet as a matter of course since the 1980s, the art world hit upon a new vision only with the hype that surrounded the Internet boom a decade later. Artistic interest in the Internet from around 1994 onward was due mainly to the introduction of new software making the

World Wide Web multimedia-capable and opening it up for visuals and sound in addition to written communication. At the same time, interactive data carriers in the form of the CD-ROM and later DVD-ROM appeared on the mass market. The most important effect of these new technologies is that interaction becomes an option for the mass media. Interactivity is due to leave the laboratory and announced as the bright future of the media industry. Restrictions on access to local interactive hi-tech installations belong to the past as the interactive data can be delivered to everybody's home. The viewer, relieved of the necessity to make extensive journeys in quest of the interactive art, is turned into a data traveler on the Net. The shift of concepts is evident in the changing meaning of the central terms. 'Cyberspace' is no longer understood primarily as a virtual extension of real space into an immersive data environment, but instead as a meta-network of communication structures. 'Interactivity' is leaving behind human-machine interaction to again become interpersonal interaction whose structures are molded by the supra-machine of the Internet.

An overabundance of connectivity has replaced the symbolic loneliness experienced by the viewer in the Cyberspace of the early 1990s or on meeting his own video image in the 1970s closed-circuit installations. As a point of convergence for all media and genres, the Internet appears to supply the technical means to fulfill the utopias of intermedia art. The idea of a 'Net' is older than the technical reality; as early as in the 1960s, it was a central motif of alternative culture and aspirations to political and social influence. These ideals are being rediscovered in view of the new technologies in the first wave of Net utopias of the early 1990s. An 'open work' that is generated through the communication of participants and the 'domination-free' discourse of all Net users are basic forms of this Internet ideology and aesthetics. This attitude was anticipated in projects like the "Electronic Cafe" of Kit Galloway and Sherrie Rabinowitz, which joined up various districts of Los Angeles in a multimedia network for the 1984 Olympic Games. Without offering any form of content, the makers were determined to show that merely the opening up of communication channels possesses an ethical and democratic dimension. "Electronic Cafe" is thus the precursor of all Net utopias that cast a social model in a technological mold.

The most successful projects in the grey zone between politics and culture are the 'digital' and 'international' cities created from 1994 onward, initially in Amsterdam and then in many other European cities. One programmatic statement read: "New interpersonal relationships are initiated by the 'International City' and influence everyday life in the real city. In contrast to other media, new information will be created through social exchange."[57] The "global village" propagated by McLuhan in the 1960s was now scaled down to a regional electronic neighborhood but with a potential exchange between the interconnected digital municipalities. Many of these projects soon faced the question of whether they wanted to remain within the self-determined free space of alternative-artistic media work or, like the rest of this booming commercial environment, become professional service providers. This conflicting role led to the self-dissolution in 1997 of the prominent "Internationale Stadt Berlin," while its digital counterpart in Bremen became an Internet service provider.[58] Such scenarios echo—but at a much faster pace—the way the video scene split up in the 1980s into those who did commercial work for TV and those who continued to produce art and had no further association with the TV networks.

A number of Internet projects can be related to the ‘context art’ of the 1990s, even if in the museum and gallery context it was never generally accepted that such a thing as ‘Net art’ even exists. The first such project was “The Thing,” founded in New York in 1991, which opened at least temporary nodes in Berlin, Frankfurt, Hamburg, Düsseldorf, Cologne, London, Stockholm, and Vienna. Wolfgang Staehle, the founder of what began as a purely text-oriented discussion forum with its own BBS network (Bulletin Board System) outside the Internet, cited big names: “Beuys was concerned with social sculpture, with art production made collectively by a group or community. ‘The Thing’ is a sculpture of that kind—it realizes Beuys’ idea of direct democracy, of the political community as a social structure. At the same time, it represents an expansion of the concept of art.”[59] Can the problem of expanding the notion of art be solved using the appropriate medium? Or does this statement imply that artists too are now proclaiming the paradigm shift from the 1960s ideology to 1990s technology? To its users and creators, “The Thing” offered a preview of the time, when network communications have become a commodity and part of everyday live. Intended at first as a temporary project, it became a permanent structure that moved to the Internet and offered a discussion platform and provided web-space for artists projects. The internationality of the first years soon split up into more or less autonomous locations which led a live of there own during most of the 1990s – most of them disappearing from visibility as commercial providers offer the same services.

In the early 1990s, all arts-related Net projects were still determined to pursue the parallel goals of creating public access to the Internet and installing a new platform for discourse and dissemination whose content would develop along the lines of its members’ activities. The Internet boom, however, soon rendered superfluous (or outmoded) this coupling of content to technology. Since access to the Internet is supplied on a commercial basis, the demise of projects like “The Thing” or “Internationale Stadt” perhaps represents the commodification of the last twentieth-century vision of combining technical and artistic progress. The historical relevance of these projects, as models for a general shift in the public access to and awareness of the Net, is not acknowledged in art history neither in media history. There is no institution that has per definition a responsibility for the documentation and preservation of this part of the history of digital culture. This is why the initiative has been taken up by the Ludwig Boltzmann Institute Media. Art. Research for the exemplary republication and scholarly documentation of these early 1990s net-based artists’ collaborations to prevent a total loss of this part of the cultural digital heritage. [60]

Certainly, art is a minor player on the sidelines of the 1990s Internet boom. Artists, however, were anything but slow to grasp the central problems raised by the medium, as is demonstrated by “The File Room,” a project initiated by Antoni Muntadas in 1994 and still in progress. Acting as an open archive of current and historical cases of censorship, it is continuously being expanded by a worldwide body of users. Although Muntadas, who began exploring the political function of mass media in the 1970s, launched the project without any thought of the Internet as its medium, it became acutely relevant due to the Internet. Because the repeated calls for Internet censorship and the Communications Decency Act that narrowly failed to become law in the USA in 1996, “The File Room” directly synthesizes medium with message. Early Internet statistics placed the project’s Web site among the most frequently visited addresses, ranked closely after the Microsoft site.

Ingo Günther's "Refugee Republic" is another project that coincidentally meets with the emergence of the Internet as a conceptual model in the early 1990s. As a project aiming to make the twenty million refugees worldwide a potent capital asset rather than an economic burden, it might well appear to be a typical Net utopia. In fact, like "The File Room," it was not conceived (in 1993) as an Internet project, but the Net proved to be an appropriate medium. Günther does not base his work on the assumption that since we've got the Internet, artists are supposed to do something with it. Rather he starts to map out a new political function of the medium in relation to an international dispersed population, that has no central government. In 1516, Thomas Moore had to use an undiscovered island state as the pretext for his 'Utopia,' although he was attacking conditions in his own country. Günther's non-territorial state, by contrast, can exist only on the Internet, meaning it is a perfected utopia whose medium also designates the means of its realization—even if realization is not yet in sight.[61]

The political utopias from high-ranking sources that marked the beginning of the Net boom—for instance, the "new, Athenian age of democracy" conjured up by Al Gore for Bill Clinton's information-superhighway election campaign in 1992—have remained unfulfilled. Their influence, however, cannot be denied, even if the results were just the opposite to those intended. Remembering the Internet community's organized mass e-mail protest in reaction to the first senders of spam mail in 1993, is like looking back on a distant, bygone Net era. Since the mid-1990s, the value-free vehicle of scientific discourse, chat rooms, and newsgroups that was the Internet has been undergoing a transformation into a commercial mass medium. Thanks to the multimedia capabilities of the Word Wide Web, communication and interaction is giving way to new models of broadcasting and consumerism. The Internet has since turned upside down the market for most cultural products – the most prominent example is the ongoing crisis of the music industry since Napster.

Ironically, the art market is the one place that has not been seriously affected by the Net economy. The ubiquity of the Net contradicts the socially and spatially defined context of art and the necessarily elitist discourse of those inside the art world. This led in the mid-1990s to the promising development of new structures like the Äda-Web and Rhizome in New York or the Public Netbase in Vienna, which encompassed the potential of both the art world and network communications.[62] That these projects have remained largely without impact inside the art world is due to the reciprocally exclusive discursive processes of the art and network cultures. Either side seems to have little real knowledge of the other, but the mutual reproaches sound alike: commercial dependency, pseudo-progressiveness, superficial openness or blank arrogance. In consequence, Netbased art is merely the lowest common denominator in two discourses that fail to engage with each other, and as such a marginal category caught between two fringe groups.[63] It represents the pinnacle of a paradox that has accompanied media art from the beginning: mass media, and above all the Internet, dissolve all contextual relationships. In the twentieth century, by contrast, the art of Modernism has become ever more context-specific and, accordingly, ever more context-dependent in regard to evaluation, even to perceptibility. Netbased art in the 1990s thus faces the dilemma of addressing everybody through its medium, but nobody through its context.

The fast-developing Internet economy has passed by the field of art, which is possibly more immune to the dangers of commercialization than its exponents might want it to be. The 1990s attempts at 'Net-art galleries' where based mostly on the model of conventional galleries and shared the fate of Gerry Schum's *TV gallery*, which was an abortive attempt in 1970 to transport the art context into a mass medium.[64] Further evidence is offered by the unsuccessful Internet auction of the New York "The Thing" Web site in 1999: bids only reached around five percent of the limit of $45,000.

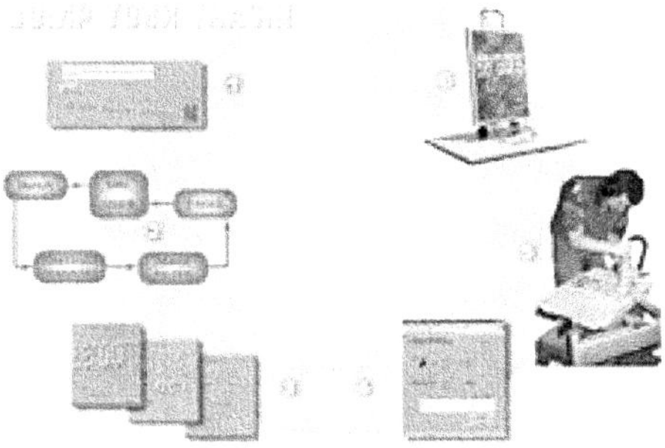

Fig. 3.6. Blank / Jeron, *Dump Your Trash!*, 1998, diagram of the project

That slow access to the congested World Wide Web has brought the World Wide Wait instead of the promised land of freedom for modem-users in the 1990s is the topic of a project entitled "www.antworten.de" (1997) by Holger Friese and Max Kossatz. An Internet project such as "Dump Your Trash!" (1998) (Fig. 3.6) by Joachim Blank and Karlheinz Jeron is the symbolic gravestone marking the drowning, in an ocean of data trash, of faith in liberation through communications. On their server, named sero.org in tribute to the garbage recycling operation in the former GDR, Blank and Jeron also offer a "re-m@il" service for the public disposal of unanswered e-mails. In times when users may easily find over 1,000 messages waiting when they return from a week's holiday, a service of this nature is a realistic satire on the self-blockage of the communications explosion. These concepts, which can comfortably be termed anti-interactive and anti-communicative, show the transition from Net utopia to Net critique. This shift is equally evident in the changing attitude emergent in the series of publications by Agentur Bilwet (Geert Lovink, Arjen Mulder, and others) from 1991 to 1997, as well as in the debates conducted on the Nettime mailing list since 1995.[65] As formulated in art and also in theory, this Net critique has two targets: the false promises of the telecommunications industry and the lost utopias of the critics' own past. Geert Lovink, Joachim Blank and Karlheinz Jeron are Net pioneers of the early 1990s whose involvement in the digital cities movement proposed cultural and communicational alternatives to mainstream media.

At the end of the 1990s a critique of interactivity and its unresolved promises is common ground of media art and media theory. Under the motto "Interactivity is the biggest lie of all!" Keith Seward and Eric Swenson condense the mixture of hard

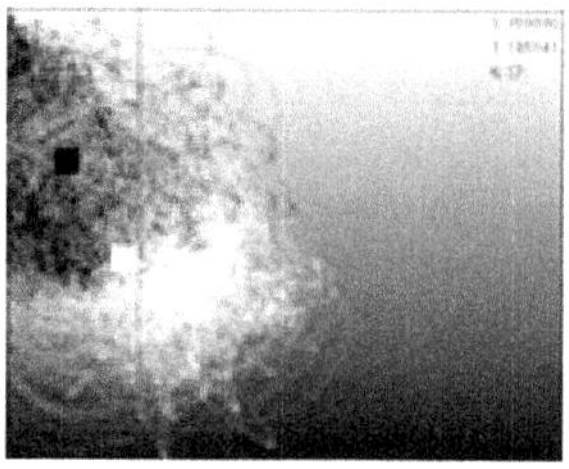

Fig. 3.7. Mark Napier, *Black and White (CNN)* 2002, application for *Carnivore* by RSG, reading each bit of cnn.com, 0 moves black horizontal, 1 moves white vertical, black and white attract each other

pornography, radical politics, advertising, and propaganda that characterizes the Net and, above all, the discussion surrounding it, on their CD-ROM *Blam! 3*. [66] A direct route leads from anti-interactivity to software subversion of the type awaiting unsuspecting viewers of the Jodi.org Web site. Confronted with constantly changing images of the final crash, the viewer is helpless until the realization dawns that these images simulate the non-simulatable end of all simulation machines.

The subversive software *Carnivore* (Fig. 3.7) by the group RSG (Radical Software Group) goes even further. The project, which was launched in October 2001 three weeks after the 9/11 terrorist attacks, makes the complete surveillance of the Internet evident by allowing one to be able to read all data traffic in plain text after it has been installed on a local network. Using freeware, anyone can defy the FBI's elaborate spy program of the same name and in this way keep up with the anticipated heightened government surveillance in the wake of 9/11. The artistic side of *Carnivore* consists in its use as a basis for various forms of the alternative visualization of data streams in real time. These "diagnostic clients," which were developed by several artists, allow transforming the consternation over being observed into an aesthetic contemplation of randomly determined structures. This conscious fatalism transformed into randomized aesthetics can be related all the way back to John Cage's pieces for radio from the early 1950s. RSG makes specific reference to the origins of the Ethernet in terms of radio technology, which is still evident in its designation.

That brings us back to the starting point of this text in two respects. Firstly, before any kind of interaction, participation is a basic principle of the modernist aesthetics—and in the RSG project it is being extended to otherwise invisible processes at the highest technical level possible. Secondly, the conflict between individual communication and mass communication reemerges, which has characterized all electronic media since the advent of the radio and ultimately to the Internet.

As early as 1972, Jean Baudrillard had refuted Enzenberger's theories regarding the emancipating, democratic function of the media: "Now, the totality of the existing architecture of the media founds itself on this latter definition: they are what always prevents response, making all processes of exchange impossible (except in various forms of response simulation)."[67] According to Jochen Gerz, it was the questions as opposed to the answers that were crucial for the political upheaval of 1968. The impossibility of providing answers in the media is shown by his Internet project "The Berkeley Oracle" (1997–99), which consists solely of questions from the general public. Just as the new departure of the 1960s ended with Bruce Nauman's declaration of mistrust in participatory art forms, so the 1990s come to a close with a skeptical revision of a concept of interactivity molded by media technologies.

The term interpassivity, coined by Robert Pfaller, provides the theoretical background for this.[68] His skepticism with respect to the general euphoria surrounding interactivity goes back to phenomena such as the canned laughter heard on TV comedy shows, which is symptomatic for art that contemplates itself, so to speak, and thus anticipates viewer reaction.[69] Pfaller also sees this tendency toward "delegated amusement" in interactive art, which allows abstaining from forming an aesthetic opinion. Thus the concept of interpassivity constitutes the updated counterpart to the pseudoactivity of media consumers demonstrated as early as 1938 by Theodor Adorno in his examination of radio and record listeners.[70]

Is it possible, at this point, to bring to full circle the ideal of aesthetic sensitization extending into interaction with the media as demonstrated in 1951 by John Cage's *Imaginary Landscape No. 4* for twelve radios and twenty-four performers? Are artists merely the 'exemplary listeners' who allow us to recognize the media-induced change of world view through a process of selection and bundling—or have precisely the new technologies restored to art the opportunity and the claim to intervene in the dynamics of the development of a media society? Conversely: how 'resistant' will the notion of art prove to be against the mediatization of all areas of life? Or, specifically in relation to the subject of this essay: does it still make sense to ponder upon the significance of interactivity from the perspective of art, or would it suffice to point out developments in the fields of software, hardware and interface design?

The fact that in 1999 the jury of the Prix Ars Electronica chose to award the main prize in the '.net' category to the operating system Linux may have something to do with the difference between the open and closed systems described above in relation to John Cage and Bill Gates. However, the underlying implication that programming is the actual art and what artists make of it will always remain secondary matches up with Friedrich Kittler's suggestion that only our ignorance makes us confuse the products of media with art. The emphatic confirmation of this theory by an art jury can be criticized, from the perspective of art, as superfluous affirmation of media art's unquestioning faith in technology and the final truncation of all links to the art context.[71] From a cultural-historical stance, on the other hand, this decision can also be seen as indicating the unquenchable yearning to return to an age in which art and technology were not separated. A festival such as Ars Electronica would then be the rightful successor to the Ars inveniendi of the Baroque age, whose attractions included military art, water art, and firework displays along with the first mechanical computing device and the android automatons that so impressed court audiences. The countless projects falling somewhere between art and media and boasting the name Leonardo in their title suggests the same need to make the painter, anatomist, master builder of forts, and inventor of flying machines a symbol of a wholeness of cultural achievement, technical innovation, and scientific research that is forever lost.

3.10 Again: Is Interactivity an Ideology or a Technology?

Even with the aid of computers, the yearning for the resurgence of the Renaissance Man is doomed to remain unfulfilled in these times of an explosion in knowledge and communications. All the same networks are producing a convergence of previously separated cultural, social and technological fields. The questions posed by Brecht and Turing regarding the social or technological significance of media-assisted interaction, which were still radically disparate in the 1930s, are now beginning to overlap. Due to the interweaving of human society and its digital back-up, it is becoming increasingly difficult to define the boundary between ideology and technology, and indeed technology forms a central part of ideology in the 1990s.[72]

The possibility of a future convergence of ideology and technology was present throughout the development of media-assisted interactivity and the surrounding debates. Even before any media artworks had been produced, the participatory forms introduced by Happening and Fluxus were attempting to remove the boundary between producer and recipient. These movements were also a reaction against mass-media consumer conditioning, as is shown in the symbolic deconstruction of radio and television by Cage, Paik, and Vostell, and also in Expanded Cinema. With its equally aesthetic, social, and political foundation, this ideology results in the belief that by means of the media it will be possible to disrupt the macrocultural inclination towards passive reception—provided these media can develop their inherent potential for interaction and communication. Therein lies the source of the proposition that the media have the power to emancipate, a thesis put forward in identical form in such disparate contexts as Enzenberger's criticism of the Left's inadequate media skills, written in 1970 with a nod to Brecht's radio theories of the 1930s, and Weibel's call in 1989 for a future dominated by interactive art.[73] The thesis of the liberating power of the media was likewise reflected in the 'Californian ideology' of the 1990s, as embodied by *Wired* magazine or the would-be alternative European projects such as the digital cities or "Nettime."[74] Bill Clinton's superhighway electoral campaign in 1992, however, already heralded a radical turnabout. In a record period of time, the idea of free network communications hatched somewhere between hackers, ex-hippies, and a small avant-garde in art and politics became the central message of the media industry leading directly into the bubble of the New Economy. Even quicker the new consumer-as-producer (prosumer) culture of the Web 2.0 has been turned around into a commercial data mining ground for the ever increasing exploitation of the attention economy. For the myspace und youtube generation the promise of the user emancipation in network media is cannibalized by the self-consumerism following the slogan "broadcast yourself". This is why, finally, the theory that media-assisted interaction and communication would overcome the hegemony of the media industry has become more utopian than ever. Instead, the good old media skepticism of the high culture representatives is loosing ground as the conflict between ideology and technology is melting down. The on-line lifestyle of the young digerati has a pragmatic view of privacy and promotes the personal data profile as part of the public personae and as a personal commodity.

Ranging from Bertolt Brecht via Happening and Fluxus to the left-wing activism of the 1960s and 1970s, and, finally, the interactive art of the 1980s and 1990s, the experiments in the laboratory of the avant-garde have all produced the same findings.

The influence of the mass media cannot be changed permanently or on a large scale. The posited liberating potential of media can be put into effect only in closely demarcated, culturally screened-off niches but—even in the allegedly post-capitalist era—will not survive against market forces. That was why Brecht deemed another usage of radio would be "impossible to implement in this social order, feasible in another."[75] Now that the issue of Capitalism vs. Communism has been decided, however, we know there will be no other social order. If consumerism is an inevitable effect of all mass media, then to have faith in the liberating potential of media amounts to much the same thing as giving an alcoholic the key to the alcohol cabinet. Indeed, in 1932 Brecht compared the radio listener's isolated passivity to that of the secret drinker, that most wretched of addicts. The interactivity euphoria of the early 1990s is coming to an end with the morning-after feeling of "electronic loneliness" which Agentur Bilwet summarizes in 1997 with the motto, "Change the world; stay at home."[76] This is still a good motto ten years later in the Web 2.0 context, where social life is more and more virtualized.

3.11 "Where Do We Go from Here?"

The depression that followed the Internet euphoria in the late 1990s can be seen alongside the attempts by the mainstream mass media to adopt forms of interactivity developed in the Net culture and in media art. Already in these failed attempts around the year 2000, the ultimate goal of activating the audience through merging the Net and broadcast media was clearly discernible. [77] It is not the emancipation of consumerism but a new edition of the economics of attention based in high-technology in which any activity by a viewer potentially becomes commercially measurable and exploitable. When AOL boss Steve Case says in 2000 that "more and more people want interactivity," what he means is that in the future, viewers will "click on Britney Spear's dress during a TV show and K-Mart will deliver it to their front door."[78] At the same time a group of researchers at MIT is developing a so-called hyper soap that will lead to the ultimate perfection of the old television principle of product placement. While the show is airing, viewers can click on any of the articles on the screen and receive information about the product or view one's ordering options. Imagine this: the car the leading man is driving, "Mercedes 300 SLK, $30,000—Link to the available models and a test-drive option." The beer he drinks, "Tuborg, a six-pack for $3.99, delivery time 30 minutes for online orders." The tissue he uses to dry his lover's tears, "Kleenex, $1.99, will be delivered with the beer." Thus the entire story of a TV series takes place in a virtual department store in which the actors are living store-window mannequins. Identification with the star becomes a guaranteed commercializable factor. By buying his clothes, furniture, etc., one apparently becomes someone like him. Any further commercials would then be superfluous, even counterproductive. The interactive, netbased mass media would then achieve a total synthesis of economic and technological structure, against the divergence of which they have been fighting since the days of the radio using methods such as rating.

Ironically, the high-tech "hyper soap" exhibits marketing strategies comparable to low-tech reality TV[79]. Reality TV was likewise anticipated by avantgarde film in the style of Andy Warhol. In both cases, a principle developed within the context of media art is adopted by the mainstream media, but the original goal is turned into its opposite. To express it in Brecht's words: "Capitalism immediately and continuously transforms the poison with which it has been injected into a drug and it takes pleasure in it."[80] Thus the artistic utopia of a participatory and later interactive art as the emancipation of the viewer from consumerism, which opposes the classic, closed concept of an artistic work that embodies an art in line with market conditions as a product, sees itself faced with the paradox that its concepts are to be converted into the engine for the total commercial penetration of everyday media consumption. This process may confirm the avant-garde status of media art. However, since the collapse of the New Economy and the reemergence of Web 2.0 economy, the ideals of the 'heroic' period of interactive art sound like historical relics from the antiquity of the 'new media' age.

Today interactivity is no longer an experiment in the media lab or an experience in a media art exhibition but part of everyday life in digital culture. Does this mean, that some of the artistic and theoretic ideas behind it, have also left the field of high culture and have been embedded in a digital folklore which no longer cares about art with a capital A? Some anticipations of this idea can be traced back to Nicholas Negroponte's prophecy of a new electronic amateur "E-xpressionist" art from 1995 in his book "Being Digital": "The Sunday painter is a symbol of a new era of opportunity and respect for creative avocations—lifelong making, doing, and expressing. … There will be a more common palette for love and duty, for self-expression and group work. … Computer hackers young and old are an excellent example. … The behavior of their computer programs has a new kind of aesthetic. These hackers are the forerunners of the new e-xpressionists." [82] The themes of media art festivals in 2000 also set the expectation for amateur Internet culture that, under the motto of "do it yourself," closely scrutinizes or reprograms the industrially predefined standards of media technology, and, under the motto "take over," dismisses or simply ignores the evaluation processes of the art business.[83] The pathetic proclamation of the "century of the consumer" in the ZKM exhibition "YOU_ser" of 2007 sounds like the final re-enactement of all the utopias that have been associated with interactivity and is again merging ideology and technology. After a long summary of the history of modernism as participatory user emancipation Peter Weibel's conclusions is: "The artist no longer has a monopoly on creativity. □Users ... become producers and program designers and thereby, competitors to television, radio, and newspapers, the historical media monopoly. Audience participation reshapes itself as consumers' emancipation. □... The new installations presented in the exhibition transfer the potential for co-designing by the user that has been developed on the Internet into the context of art and allow the visitors to emancipate themselves. □They can act as artists, curators, and producers. The exhibition visitors, as users, as emancipated consumers, are at the center of focus. YOU are the content of the exhibition! ... Is this the new cultural space for the emancipated consumer, the visitor as user who will decide the culture of the twenty-first century, just as slaves, workers, and citizens as historical subjects have done in the past?" [80]

The unfulfilled utopia of an art that no longer calls itself art stems from the inheritance of early twentieth century avant-garde. The working class art of the early soviet union ended up in a totalitarian modernism and the ironic detachment of Marcel Duchamps ready-made became the model for a significant part of what we see in museums today. At the same time the media amateurs working with photography, radio, film and video have created autonomous communities of creativity, mostly outside the field of high-culture and at least in some parts independent from the media industry. Can the rhizomatic pluralism of the Internet culture rescue the ideological legacy of modernism from its totalitarian claim and thus make the question regarding art or non-art finally superfluous?

"On the Internet, nobody knows that you're a dog," announces the dog, in a cartoon from 1993, sitting at the keyboard of a computer talking to another dog. [84] Perhaps this is the updated version of Marcel Duchamp's closing statement in his 1961 lecture "Where do we go from here?": "The great artist of tomorrow will go underground."[85]

References

[1] Duchamp in his lecture "The Creative Act" from 1957. cf. Duchamp, M. (ed.) Museum Jean Tinguely, Basel, p. 43 (2002)

[2] Ibid

[3] Duchamp in a letter dated 1956. cf. Duchamp, M., Schriften, D. (eds.) Serge Stauffer, Zurich, p. 202 (1981)

[4] Baudelaire, C.: Critique d'art, Paris, p. 358 (1992)

[5] Baudelaire, C.: Œuvres complètes, Paris, vol. 2, p. 782 (1976)

[6] cf. Daniels, D.: Kunst als Sendung: Von der Telegrafie zum Internet, Munich, p. 168, 189 (2002)

[7] Duchamp, p. 239 (1981) (see note 3)

[8] Umberto Eco points out that the stimulus for his theses stems from New Music, without, however, mentioning John Cage. cf. Eco, U.: Das offene Kunstwerk, Frankfurt am Main, p. 23 (1977)

[9] The socioscientific concept of interaction can be traced back to the theory of symbolic interactionalism developed by George Herbert Mead in the 1920s. This theory examines the reciprocal conditionality of social action and communication. For a detailed conceptual history of interaction/interactivity see the essay by Katja Kwastek in this volume

[10] Brecht, B.: Der Rundfunk als Kommunikationsapparat. In: id., Werke, Berlin and Frankfurt am Main, vol. 21, p. 553, 557 (1992) Due, among other reasons, to Hans Magnus Enzensberger's renewed treatment of Brecht's theory of radio, which was noted by Marxist theorists like Todd Gitlin and artists like Douglas Davis, Brecht was similarly a point of reference for discourse on media and art in the United States in the 1960s and 1970s

[11] Turing, A.M.: Computing Machinery and Intelligence. Mind LIX 236, 433–460

[12] cf. Cage, J.: Composition as Process: Part II; Indeterminacy. In: Frieling, R., Daniels, D. (eds.) Media Art Action, Vienna and New York, pp. 27–33 (1997)

[13] Cage wrote in 1966: Are we an audience for computer art? The answer's not No; it's Yes. What we need is a computer that isn't labor-saving but which increases the work for us to do ...turns us (my idea) not 'on' but into artists. Cage, J.: A Year from Monday, London, p. 50 (1968)

[14] Gates as cited in Friedrich Kittler's lecture at the 1999 conference Wizards of Oz 1, Offene Quellen und freie Software, in Berlin

[15] Söke Dinkla writes on this subject: "The motto 'art and life' is transformed into 'art and technology'." She disregards, however, the associated shift in ideological paradigms that far surpasses the framework of art or technology. Equally, it is impossible to equate interaction based on a score written for a Happening or a Cage composition with interaction incorporated into a computer program without addressing the basic issue of human-machine interchangeability. Dinkla, S.: Pioniere Interaktiver Kunst von 1970 bis heute, Ostfildern, p. 41 (1997)

[16] Umberto Eco, For instance, in the final chapter of The Open Work (1962) examines the openness of a live TV broadcast as the mass-media counterpart to the open structures of the avant-garde. His hope with regard to the open structures: "These digressive annotations would then jolt the viewer out of the hypnotic spell woven by the plot, and, by distancing him from it, would force him to judge, or at least to question, the persuasiveness of what he sees on the screen." Eco, U.: The Open Work, trans. A Cancogni, Cambridge, MA, p. 122 (1989)

[17] Enzensberger, H.M.: Constituents of a Theory of the Media (1970) In: Hanhardt, J. (ed.) Video Culture, Rochester, p. 97 (1986), Reprinted from The Consciousness Industry, trans. Stuart Hood, New York, pp. 95–128 (1974) cf. Jean Baudrillard's critique of this utopia, in which he objects to a view of the media merely "as the relay of an ideology" determined by the powers of capitalism, saying they must be grasped as "effectors of ideology." Baudrillard, J. "Requiem for the Media" (1972), in Hanhardt 1986, op. cit., p. 128. Reprinted from For a Critique of the Political Economy of the Sign, trans. Charles Levin, St. Louis, pp. 164–184 (1981)

[18] cf. Levy, S.: Hackers: Heroes of the Computer Revolution, p. 52, New York (1994)

[19] Ibid., This 'hacker ethic' appears on the Web site of the Chaos Computer Club to this day, p. 39

[20] Here lies also the problem of the interference between scientific visualization and media art, as is investigated by groups like Knowbotic Research

[21] In Understanding Media, for instance, Marshall McLuhan describes television as an instrument of synaesthesia (1964)

[22] Kittler, F.: Fiktion und Simulation. In: Ars Electronica (ed.) Philosophien der neuen Technologie, Berlin, p. 57 (1989)

[23] Benjamin, W.: The Work of Art in the Age of Mechanical Reproduction (1935) In: id., Arendt, H. (ed.) Illuminations, New York, p. 251, note 30 (1969) Enzensberger carries on where Benjamin left off when he writes, in regard to the 1960s: "This is where the prognostic value of otherwise inessential productions, such as happenings, fluxus, and mixed-media shows, is to be found." Enzensberger, p. 122 (1970) (see note 17)

[24] Umberto Eco, too, explicitly takes contemporary music as his point of departure and refers to Karlheinz Stockhausen, Luciano Berio, and Henri Pousseur, although John Cage is not mentioned. Eco (1962/1989) (see note 16)

[25] If an electronically modified TV set is fitted with a microphone, for instance, visitors can generate an oscillating pattern on the TV screen by making sounds and noises. cf. Frieling and Daniels 1997, p. 62 (see note 12)

[26] Naumann, B. (ed.): Joan Simmon, exh. cat. Walker Art Center, Minneapolis and Basel, p. 77 (1994)

[27] Upon entering this installation, the viewer sees him- or herself at the other end on one of the two video monitors, while the other monitor shows the empty corridor in a previously recorded video without the viewer. Attempting to convince oneself of one's own presence in the image and/or space is utterly impossible, since as soon as one moves through the corridor to the video monitors, one moves away from the camera installed at the entrance and thus disappears from the video image

[28] Export, V. (ed.): Peter Assmann, exh. cat. Oö. Landesmuseum, Linz, p. 258 (1992)
[29] Valie Export's expanded-cinema project "Ping Pong, A Film to play with/a player's film" (1968) in which the viewer is asked to aim a ball, with the aid of a ping-pong paddle, at the black dots that emerge on, and disappear from, the film screen. This, according to Export, illustrated the "relation of domination between producer and consumer," since even as a participant, the viewer remains wholly dependent on the specifications of the film
[30] For a more detailed description of the technological development, see for instance. Weibel, P.: Virtuelle Realität: Der Endo-Zugang zur Elektronik. In: Rötzer, F., Weibel, P. (eds.) Cyberspace: Zum medialen Gesamtkunstwerk, Munich, pp. 15–46 (1993), cf. the excerpt of the text in Rudolf Frieling and Dieter Daniels, Media Art Interaction: The 1980s and 1990s in Germany, ed. Goethe Institute, Munich, and ZKM Karlsruhe, Vienna and New York (2000), and the comprehensive study of interactive art by Dinkla, pp. 50–62 (1997) (see note 15)
[31] cf. Sutherland, I.: The Ultimate Display. In: Proceedings of IFIPS Congress 1965, New York, vol. 2, pp. 506–508 (1965), id. "Computer Inputs and Outputs," Scientific American (September 1966). Rötzer and Weibel, p. 18, 25 (1993) (see note 30)
[32] Wiener, O.: Die Verbesserung von Mitteleuropa, Reinbek (1969/1985), p. CXXXIX. cf. translated excerpts In: Weibel, P. (ed.) The Vienna Group, Vienna and New York, pp. 666–698 (1997)
[33] Ibid., p. CLXXV
[34] Schoeffer, N.: Die Zukunft der Kunst—die Kunst der Zukunft. In: Schoeffer, N.: exh. cat. Städtische Kunsthalle, Düsseldorf (1968)
[35] cf. the "E. A. T." (Experiments in Art and Technology) program at Los Angeles Country Museum from 1967 onward, and Cybernetic Serendipity: the computer and the arts. Reichardt, J. (ed.) exh. cat. Studio International, London, New York (1968)
[36] cf. Hartwagner, G., Iglhaut, S., Rötzer, F. (eds.): Künstliche Spiele, Munich (1993)
[37] One rare example of congruence of technology and content was one of the first computer-controlled interactive visual artworks, namely the program Random War by Charles Csuri, which on the basis of a randomly generated constellation simulates the progress of a battle between two groups of soldiers. Reichardt, p. 81 (1968) (see note 35)
[38] Rötzer and Weibel, p. 27 (1993) (see note 30)
[39] Most of the examples only briefly mentioned here are documented on, http://www.mediaartnet.org
[40] A linkage of popular culture and interactivity was introduced very early by the Austrian group Station Rose
[41] Even before interactivity boomed in the 1990s, Ann-Sargeant Wooster wrote the following in the commendable article "Reach out and touch someone—The Romance of Interactivity": "Most uses of interactivity will probably be confined to mass-market populist entertainment ...and rigidly controlled by media merchants." In: Hall, D., Fifer, S.J. (eds.) Illuminating Video, New York, p. 302 (1990); See also on this subject Regina Cornwell, "Interactive Art: Touching the 'Body in the Mind'," Discourse 14.2, p. 209 (Spring 1992)
[42] From 1993 onward, Jeffrey Shaw collaborated with engineers and computer scientists at the Kernforschungszentrum in Karlsruhe on developing the project "EVE—extended virtual environment," which corresponds to a viewer-interactive panorama. In 1997, Shaw and the Frauenhofer Institut, Stuttgart, jointly carried out the "confFIGURING the CAVE" project in a "Cave Automatic Virtual Environment" (a 3-D simulation developed for research purposes and able to be physically entered)

[43] Oliver Grau investigated this development in the Deutsche Forschungsgesellschaft research project "Kunstgeschichte und Medientheorien der Virtuellen Realität" conducted at the Kunsthistorisches Seminar, Humboldt University, Berlin, http://www.virtualart.at
[44] Peter Weibel participated as a 'crowd-warmer' in Valie Export's action
[45] cf. Weibel, P.: Der Vorhang von Lascaux. In: First Europeans: frühe Kulturen—moderne Visionen, exh. cat. Orangerie Charlottenburg, Berlin, p. 78 (1993)
[46] Plewe in an e-mail to the author
[47] Kriesche, R.: Artificial Intelligence in the Arts, Graz, p. 13 (1985); see text in Frieling and Daniels 2000 (see note 30)
[48] Stenslie, S.: Cyber SM, and Kirk Woolford, "A touch at the end of the century," both in Lab 1: Das Magazin der Kunsthochschule für Medien, Cologne, pp. 40–43, 72–75 (1994)
[49] Douglas Davis, interviewed by David Ross. In: Schneider, I., Korot, B.: KorotVideo Art, An Anthology, New York and London, p. 33 (1976)
[50] Krauss, R.: Video: The Aesthetics of Narcissism (October 1, 1976)
[51] Shaw, J.: Reisen in der virtuellen Realität: Gespräch mit Florian Rötzer. Kunstforum 117, 295 (1992)
[52] Such virtual museums are only beginning to become potential sites of communication in the late 1990s thanks to the incipient synthesis of 3-D graphics and the Internet. cf. Grassmuck, V.: Das lebende Museum im Netz. In: Schade, S., Tholen, G.C. (eds.), Konfigurationen zwischen Kunst und Medien, Munich, pp. 231–251 (1999)
[53] cf. Roy Ascott's theses on art and telematics, which although written as early as 1983, were comprehensive and concrete. Grundmann, H. (ed.) Art Telecommunication, Vienna and Vancouver, pp. 25–59 (1984)
[54] Lyotard, J.-F. (ed.): Les Immatériaux, vol. 1, Epreuves d'écriture, vol. 2, Album: Inventaire, exh. cat. Centre Georges Pompidou, Paris (1985)
[55] cf. Turkle, S.: Live on the Screen, New York (1995)
[56] The last part of the essay refers only to art works before 2003, when it was first published in German
[57] Manifesto on Web site of the "Internationale Stadt Berlin" (1994) (offline)
[58] cf. Kerscher, G., Blank, J.: "brave new city," Kritische Berichte 1, pp. 10–16 (1998); special issue on Net Art
[59] Staehle in Vera Graf, "Kunst im Informationszeitalter," Süddeutsche Zeitung, p. 11 (March 22, 1994)
[60] http://www.netzpioniere.at
[61] cf. Daniels, D.: Utopia—What For? In: Rennert, S., von Wiese, S. (ed.) Ingo Günther: Republik.com, exh. cat. Kunstmuseum Düsseldorf, Ostfildern, pp. 48–61 (1998)
[62] Äda-Web and Public Netbase both went online at the start of 1995 and in a collaboration with artists produced WWW-specific works that were then embedded in a theoretical context. After its sponsor, a telecommunications company, withdrew its support, Äda Web ceased operations in 1998 and was sold to the Walker Art Center as an archive offering access via the Internet. Public Netbase was forced to stop its activity in 2006 due to lack of funding and will be documented as part of the netzpioniere. at project by the Ludwig Boltzmann Institute Media. Art. Research
[63] cf. on this subject the debate, telling for the misconceptions on either side, conducted on Net art between Isabelle Graw and Tilman Baumgärtel: Graw, I.: "Man sieht, was man sieht: Anmerkungen zur Netzkunst," Texte zur Kunst 32, 18–31 (1998), Tilman Baumgärtel, "Das Imperium schlägt zurück!," Telepolis (on-line journal) (January 20, 1999)

[64] While Olia Lialina's Net-art gallery Art Teleportacia in Moscow has received plentiful press coverage, it has so far sold only one work of art (by the gallery owner). The online version of the New York Times did at least find worth a notice the purchase of the project, http://www.antworten.de by Holger Friese and Max Kossatz by the private collectors Hannelore and Hans-Dieter Huber

[65] Agentur Bilwet has published the following books: Bewegingsleer, 1990 (engl. Cracking the Movement: squatting beyond the media, 1994); Media-Archif, 1992 (engl. The Media Archive, 1997); Der Datendandy, 1994; Elektronische Einsamkeit, 1997; also Geert Lovink, My First Recession, 2003; nettime, Netzkritik, Bosma, J., et al. (eds.) (1997); Read Me! filtered by nettime: ASCII culture and the revenge of knowledge, New York (1999)

[66] cf. Römer, S.: Interaktivität ist die größte Lüge. Texte zur Kunst 32, 70–73 (1998)

[67] Baudrillard in Hanhardt, p. 129 (1986) (see note 17)

[68] Pfaller, R. (ed.): Interpassivität: Studien über delegiertes Geniessen, Vienna and New York (2000)

[69] Valie Export's first interactive video installation Autohypnose, likewise shows the conditioning of the viewer by means of a systematic behavioral program and his or her being rewarded with applause from the videotape (1973)

[70] Whenever they attempt to break away from the passive status of compulsory consumers and 'activate' themselves, they succumb to pseudoactivity.... Their ecstasy is without content. That it happens, that the music is listened to, this replaces the content itself. Theodor Adorno, On the Fetish-Character in Music and the Regression of Listening. In: Arato, A., Gebhardt, E. (eds.) The Essential Frankfurt School Reader, New York, pp. 270–299, p. 292 (1987)

[71] cf. on this subject Armin Medosch in the on-line journal Telepolis (June 1, 1999)

[72] cf. also Jean Baudrillard, in whose view the media produce an ideology as opposed to merely being the means of the latter (see note 17)

[73] Peter Weibel in 1989 expressed the view that modern art as a whole was undergoing a development towards the 'inter' principle, and announced his own program of concentrating "on the actual, utopian social possibilities ...offered by technology, such as participation in and interaction with the artwork as a model for emancipationist communicational forms." Peter Weibel, "Momente der Interaktivität," In: Kunstforum 103, p. 87 (1989)

[74] cf. on this subject: Richard Barbrook and Andy Cameron, Californian Ideology, first published in 1995, The authors call for a specifically European position in which, in opposition to the US enthusiasm for technology, the "hi-tech artisans" re-connect with the theory and practice of the visual arts, http://www.hrc.wmin.ac.uk/theory-californianideology.html (accessed November 10, 2007)

[75] Brecht, p. 556 (1992) (see note 10)

[76] Bilwet, A.: Elektronische Einsamkeit, Cologne, p. 11 (1997)

[77] In November 2000, the major German TV broadcasting stations RTL and ZDF launched Internet series that attempt to translate the tried-and-true television format of the soap opera into an interactive, Internet-based form. RTL's Internet soap opera Zwischen den Stunden comes from the producers of the TV series Gute Zeiten schlechte Zeiten and is shown at designated "airtimes." With etagezwo, ZDF developed a more intricate Internet-specific presentation, but the viewing audience is also unable to influence the plot. ARTE, a joint Franco-German cultural TV channel, even offered an interactive novel, where the audience was supposed to write the complex plot for actors provided by the TV station. Although each of the stations takes great pains to win over the target group of young, future-oriented audience, none of the projects are successful, and all of them are eventually discontinued

[78] Steve Case and AOL manager Myer Berlow, cited in Christian Tenbrock, "Online sucht Inhalt," Die Zeit, p. 32 (September 14, 2000)
[79] Product placement in reality TV also leads to a duplication of the medium in reality instead of to a depiction of reality in the medium. The media theoretician Douglas Rushkoff speaks of an "ossification of the interactive capabilities" of the Internet due to marketing strategies. Rushkoff, D.: Virtuelles Marketing. In: Maresch, R., Rötzer, F. (eds.) Cyberhypes, Frankfurt am Main, p. 103 (2001)
[80] Weibel, P.: http://www.zkm.de/you (accessed November 10, 2007)
[81] Brecht, p. 516 (1992) (see note 10)
[82] Negroponte, N.: Being Digital, New York, p. 221 (1995)
[83] In 2001, the theme of the Transmediale Berlin was "do it yourself," and the theme of the Ars Electronica Linz was "take over"
[84] Cartoon by Peter Steiner, The New Yorker, vol. 69 (LXIX) (20), p. 61 (July 5, 1993), The cartoon did not receive much attention at the time, but in 2000 it is the most reproduced cartoon ever from the New Yorker. The sentence ... has slipped into the public consciousness, leaving its source behind ... and the saying has become practically an industry of its own. Fleishman, G.: Cartoon Captures Spirit of the Internet, The New York Times (December 14, 2000)
[85] Duchamp, p. 242 (1981) (see note 3)

First published in German in Dieter Daniels, *Vom Readymade zum Cyberspace* (Ostfildern, 2003) updated for the English translation 2007.

4

Interfaces in Public and Semi-public Space

Joachim Sauter

[1] ART+COM, Kleiststr. 23-26, 10787 Berlin, Germany
js@artcom.de
http://www.artcom.de/js
[2] University of the Arts, Grunewaldstr. 2-5, 10823 Berlin, Germany

Abstract. In the late 1980s, a Berlin based group of designers and artists from Berlin's University of Arts teamed up with hackers and programmers coming from the ChaosComputerClub environment. Together, they founded ART+COM. Up to that time, everyone from this group had used computers only as a tool. At the same time, all of them knew that this technology was on the verge of turning from a tool to a (mass)medium used not only to process and edit information but also to spread and to convey it. The quality most important to this new medium was and still is its potential for interaction (generating a mutual dialog between the users and the application) that distinguishes it from the classic mass media like print, radio, TV and the traditional fine arts like painting and sculpture.

What are the possibilities, the strategies and the right approaches to use interactivity and interfaces to access information in public space (e.g. urban environments) and semi-public space (e.g. museums)?

4.1 Expert Interfaces vs Non-expert Interfaces

In contrast to other research facilities or design studios, ART+COM has realized only a small number of interfaces for products, services or tools. The focus lies on application based design and development of applications and installations in the context of mediating information e.g. for interactive installations in museums, trade fairs or in an artistic context. Because these applications were mostly developed for public or semi-public spaces where visitors would have to understand them quickly and where they would often be used only once, their interfaces had to be intelligent, simple and functional without giving a feeling of banality. (These interfaces are in contrast to such e.g. as can be used in experts' applications as users will have more time to get to know them and employ them on a daily basis.)

In its first decade (1988 to 1998), ART+COM developed interaction design and interfaces primarily in application-based research projects commissioned by third parties or by itself. Interfaces and interaction principles were developed on a technological as well as on a content basis and realized as working prototypes. During the second decade, the focus shifted to end-user applications that are characteristically unique copies designed for special contents and contexts (museums, trade fairs, urban environment).

C. Sommerer et al. (Eds.): The Art & Sci. of Interface & Inter. Des., SCI 141, pp. 63–73, 2008.
springerlink.com © Springer-Verlag Berlin Heidelberg 2008

4.2 1988 to ∼1998: Application-Based Research in the Realm of Interfaces and Interaction Design

Among the multitude of projects from the first ten years, four that are prototypical of this time, its attitude and approach are described.

4.2.1 Cyber City (1989-1992)

In 1989, the Wall came down and Berlin was once more one city. This motivated a research project interested in Berlin's urban planning. It was financed by Deutsche Telekom and was based on an already existing broadband test bed in Berlin. For this project, a model of Berlin was created in all its historic as well as planned future states and made navigable by an in-house developed real-time system. Users (architects, city planners, interested citizens) were to have the possibility of navigating - via the broadband net - in real-time through time and space, in order to newly plan or judge these plans, based on the knowledge of the historical situation (Fig. 4.1).

Fig. 4.1. Cyber City, 1989-1992

In exchange for the ART+COM real-time software and the content designed by ART+COM, the first Eyephone realized by and the first data glove Jeron Lanier were made available by VPL to the project. It became clear quite quickly that this interface, while being attractive and effective in gearing the media's attention, was ergonomically and functionally unusable. Because of this, only the Polhemus sensor located on the Eyephone and the data glove were used in the Cyber City project. For navigating virtual Berlin, a system was developed consisting of an aerial map of the city, a projection, and the sensor removed from the goggles. This sensor could then be moved on the aerial map by the users. The projection showed the virtual city rendered from the sensor's position.

With this radical reduction to a small orientation sensor and an aerial map, a system easily understandable and manageable for every user in a very short time was created.

4.2.2 Zerseher (1991)

The project "Zerseher / De-viewer" took a step forward in the physical reduction of interfaces. For the first time, an image based eye tracking system was used in an art installation; every visible or physical interface between user and application was eliminated. The observers find themselves in a museum environment, a framed picture is hanging on a wall. Upon coming closer, the visitors notice that exactly the spot of the picture they are looking at is changing under their gaze. The image reacts to the way a viewer looks at it; it changes according to when and how they look at it. After a while, people start interactively changing the image by purposely moving their gaze over it. They never see the same image twice (Fig. 4.2).

The motivation for this project was the fact that, at the end of the 1980s, people were still looking at the computer primarily as a tool and not as a medium. The painter exchanged his brush for the mouse, but he used it to do almost

Fig. 4.2. Zerseher, 1991

exactly the same thing he used to do on an analogue basis. This was art with computers, not the beginning of computer art.

With this installation, we have thus tried to promote, in a provocative way, interactivity in the realm of art as one of the most important qualities of the new digital media technologies.

One of the most important insights from this project was the fact that especially in public space, quick acceptance of interaction needs reactivity. A system that, in a first step, reacts to users without their intending it and thus shows the system's possibilities without any conscious action will then, in a second step, lead to conscious interaction with the system.

4.2.3 TerraVision (1994-1998)

TerraVision is a system using a typologically designed interface. The physical interface (earth tracker) represents the content to be communicated and can thus be easily read and used.

Isochronously to the vision of "the earth" in Neal Stephenson's novel "Snow Crash" (a system that allows the novel's protagonists to navigate on a virtual representation of the earth), ART+COM prototypically realized such a system and presented it later at 1995's Siggraph (Fig. 4.3).

TerraVision is a virtual representation of our Earth based on satellite images, aerial shots, altitude data and architectural data. Users can navigate seamlessly

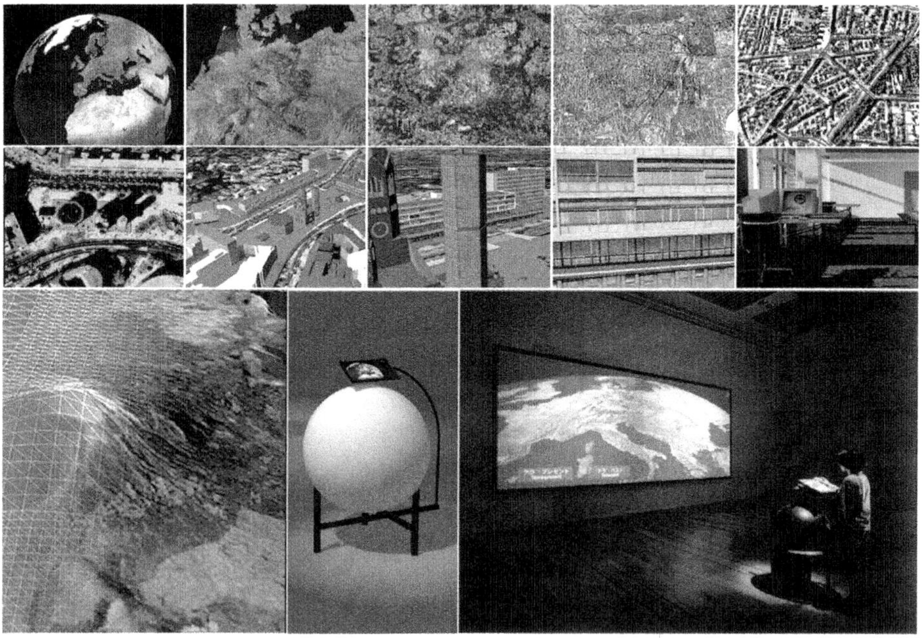

Fig. 4.3. TerraVision, 1994-1998

from overviews of the earth to extremely detailed objects in buildings. Additionally to the photorealistic representation of the earth, different kinds of spatial data were integrated (weather, geological information, tourist information, etc.). Even historical aerial shots were offered in the system. This allowed users to navigate not only spatially but also through time. All data had been distributed and networked and was streamed into the system according to the user's needs. TerraVision was the first system to provide visualization and unlimited freedom of navigation within, in theory, an infinitely large spatial data environment.

In order to navigate this data, an interface based on three components was created: A big sphere in the style of a globe to pilot the planet; a 3D mouse to fly around; and a touch screen to interact with objects on the virtual earth. Terravision is an isochronous realization of Neal Stephenson's literary idea as well as a prequel to Google Earth.

4.2.4 The Virtual Car (1996-1998)

After unsatisfying experiences with VR-goggles and gloves, ART+COM developed a new interface system in 1993: "Window into Virtuality" allows for objects to be examined in space and to interact with them. To this purpose, the Polhemus Sensor described above was mounted on a flat screen with handles, and a camera was attached to the screen's back. The camera's images were shown on the screen. Due to the sensor, the position and direction in space of the monitor

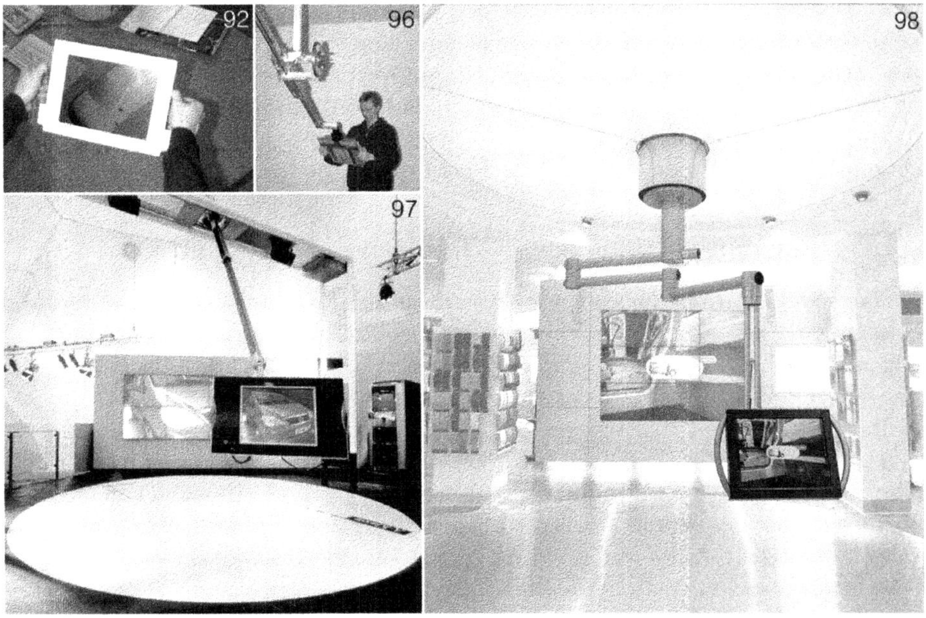

Fig. 4.4. The Virtual Car, 1996-1998

were always known; thus a virtual object could be keyed (augmented) into the camera image (Fig. 4.4).

Based on this development, the project "The Virtual Car", commissioned by Daimler Benz, was realized as a prototype of an interactive presentation system for viewing and configuring the entire model range of a given vehicle. It was designed for a future use in showrooms.

A virtual vehicle on a scale of 1:1 is standing - invisibly - in space. By means of a telescope or pivoting arm attached to a touch screen on which the vehicle is made visible, the spectators can freely move around and inside the car. If the users move one meter in real space, they will also move one meter in virtual space. The touch screen allows for direct access to the virtual model and its features: The users can configure its color, materiality and equipment. Here, a movable window was created that lets users look and reach from real space into virtual space.

4.3 1998 - 2008: Interactive Installations and Environments for Public and Semi-public Spaces

ART+COM continued its application based research from conception to prototypes in its second decade. However, we now focused mainly on design, development and realization of applications and projects for end users in public and semi-public space. As described initially, these are mostly unique installations dependent on the space and content they were designed for, where users can quickly seize the content or the designed experience. For this reason, the interfaces and interaction principles are kept on a simple level. Even if these are final products, it remains our claim to take an innovative and experimental approach on each of these works, to develop and, at best, establish new paradigms of access to information. In the following section, from again four projects have been chosen to exemplify our approach and attitude of the last ten years.

4.3.1 Bodymover (1999-2000 EXPO Hanover)

For an exhibition on the theme of interfaces for EXPO 2000 in Hanover, the brief was to design a space where visitors could create a collaborative visual and auditory experience by employing their whole body. For this project, an image recognition system was used as the interface that steered a floor projection (Fig. 4.5).

The visitors stepped on a 20 meters x 7 meters, infrared illuminated surface. They were filmed by an infrared camera under the ceiling. (Infrared cameras can only "see" moving people but not the floor projection. This way, the projection was eliminated from the analyzed image.) An image recognition system identified the outlines in real-time. Beamers also fixed under the ceiling then projected a "billowing aura" based on these outlines around the visitors. When extremities like hands or feet were extended, the aura was protracted into space and

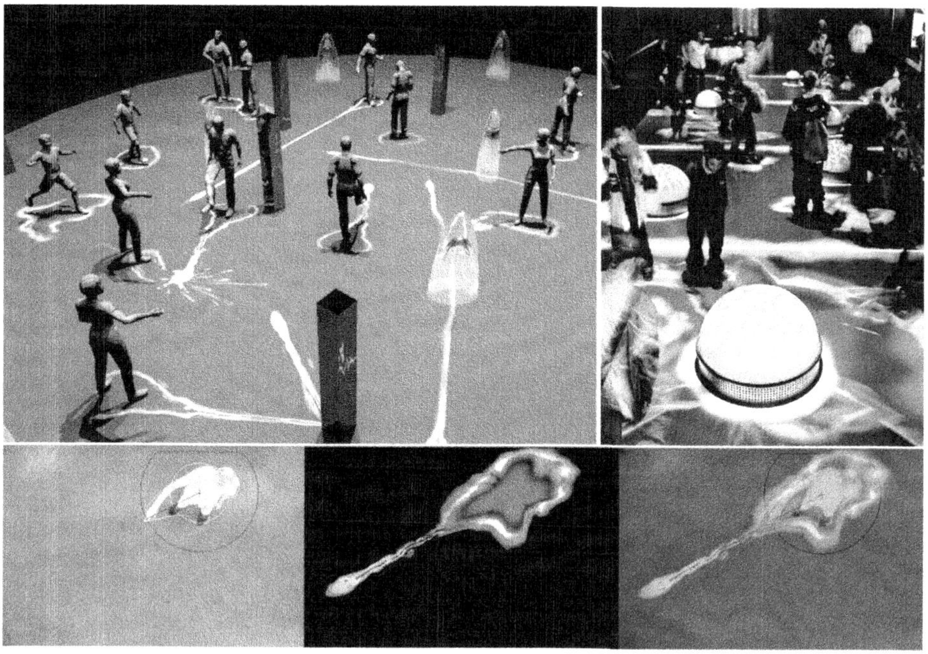

Fig. 4.5. Bodymover, 1999-2000 EXPO Hanover

produced different sounds when it hit objects positioned in the room. Visitors could thus experience their own body as an interface to graphics and sound.

This project proved that especially in public spaces image recognition systems are suitable interfaces as: On the one hand, they are quite safe from vandalism. On the other hand, if the interplay of reactivity and interactivity is appropriately designed, the visitor will quickly grasp contents and enjoy the experience.

While Bodymover was the first project working with floor projection and image recognition.

4.3.2 Interactive Medial Stage and Costume Design for the Opera "The Jew of Malta" by André Werner (2002 Opera Biennale Munich)

In 1999, André Werner was commissioned to compose an opera for the Biennale in Munich. The composer cooperated with Büro Staubach and ART+COM to design an interactive stage design backing the libretto and the composition. The ZKM, Karlsruhe supported the project. The aim was to augment the classic static stage design to a reactive, dynamic media design that could be introduced as an active part to the opera.

On the stage designed to this purpose, large planes were arranged onto which architecture, generated in real-time, was projected. The projection screens formed clipping planes through an imaginary virtual architecture positioned on

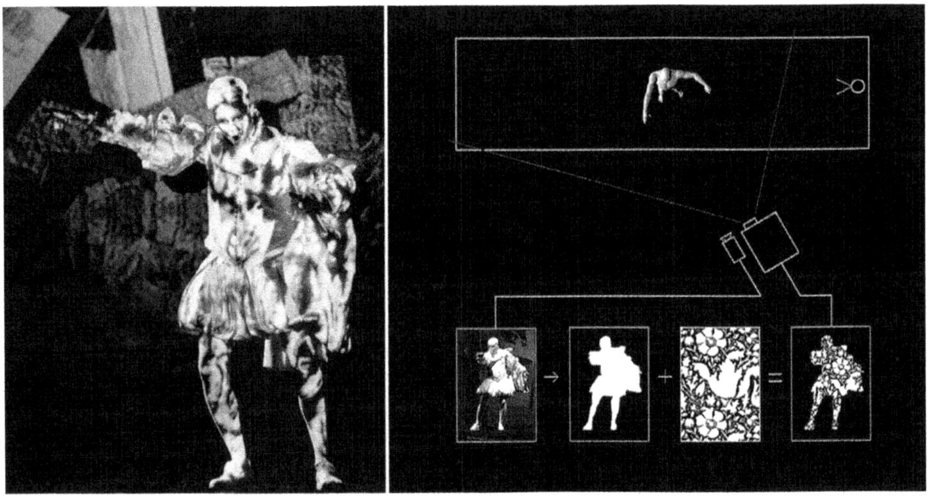

Fig. 4.6. "The Jew of Malta", 2002 Munich

stage. Machiavelli's - the opera's protagonist's - movements and gestures were camera tracked and the virtual architecture moved accordingly. This concept allowed linking the staged action and the architecture closely: Machiavelli as a powerful character in the play has power over the stage (and consequently over his co-actors) through the possibilities of interaction given to him (Fig. 4.6).

The costume was also medially augmented: Via a tracking system developed especially for this opera, digital masks were generated in real-time, and the ensuing "medial costumes" were projected exactly fitting onto the singers. This way, it was possible to depict the characters' conditions and feelings with dynamic textures on their bodies.

Even though great efforts in innovation and development were made for this project, the display of technologies was never at the forefront. The exclusive aim was to generate new ways of expression for the director and the actors. The real-time generated, mask-based body projection was the first of its kind. Other versions based on this principle are nowadays used e.g. in performances, ballets, and operas.

4.3.3 Floating.Numbers (2004 Jewish Museum Berlin; 2005 Taiwan National Museum of Fine Arts)

In 2004, the Jewish Museum Berlin housed an exhibition on the history and meaning of numbers. ART+COM realized the central installation in this exhibition as a 9 meters long interactive table with thousands of numbers floating in a continuum on its surface. Individual digits appear randomly at the surface of this stream of numbers and, once touched by a visitor, surrender their secret in text, pictures, films, and interactives. The significance of the numbers results from the various perspectives of science, religion, art or one's outlook on everyday life.

For this installation, the table was consciously employed as an in- / output interface. This everyday object is perceived as a place of communication, conversation and exchange of ideas. A common interface display of wall projection (output) and interface in front of it (input) has the visitor interacting while standing isolated, facing the wall. In contrast to this, users interacting with a table while standing around it more often start communicating and exchanging what they have only just discovered and learned (Fig. 4.7).

Fig. 4.7. Floating.numbers, 2004 Berlin

The interaction concept was deliberately kept extremely simple. The table surface is made sensitive with capacitive sensors installed below it. Simply touching one of the numbers projected onto this surface will trigger a predetermined animation.

The content projected onto the table was computationally designed and generated in real-time. It was created so as to let the visitor think that this was an autonomous, behavioral system. All numbers floating on the table are so-called typobots (type-robots) with specific behavior (move along the different currents; move forward, according to their length, like certain fish; move in the direction of attractors; etc.). The reason to install the system in this way was the assumption, later on confirmed, that visitors would rather interact with an assumedly autonomous, behavioral system than with one not behaviorally designed.

This project was the first table installation based on capacitive sensors, and its concept was re-used by ART+COM and others in all kinds of contexts and implementations. One of the advanced developments by ART+COM of this, e.g., was a space taking installation for the O2 Flagship store in Munich: A projected strip fitted with capacitive sensors goes from floor (interaction by steps), over a table (touch interaction), back over the floor to a wall and up to the ceiling (Fig. 4.8).

Fig. 4.8. O2 Flagship Store, 2005 Munich

4.3.4 Duality (2007 Tokyo)

For a new development in the centre of Tokyo, ART+COM was commissioned with an "Art in Public Space" installation in 2006. Our own claim was to design an interactive installation that responds to its location and helps build its identity. Our choice of location in the whole of the newly developed area fell on the interface between a path and a neighboring artificial pond. This interface between "liquid" (water) and "solid" (land) was thematically used and augmented by the question of "real" (water ripples) and "virtual" (artificial light waves).

On the path, a 6 x 6 meter white monochrome LED plane was installed and covered with a sand-blasted, opal glass, thus made accessible to passers-by. In the glass-tiles, there are weight sensors measuring the exact position and power of steps. They trigger corresponding virtual waves on the LED plane. When these virtual waves reach the pond's border, they are extended into the water by precisely piloted solenoid actuators (Fig. 4.9).

In contrast to the above mentioned projects, "Duality" does not act as an informational system but is a poetic and multilayered installation giving identity to its location.

Interactive installations in city space demand a different approach and entail other challenges then indoor installations. The latter are related to spaces like museums which in themselves already filter a certain audience. Exhibition visitors go there deliberately, open towards the show and often with a certain previous knowledge. In public space, however, this filter does not exist. Also

Fig. 4.9. Duality, 2007 Tokyo

medially illiterate people will find themselves facing the installation. Interfaces and interaction principles have to be adequate to this situation. It is judicious to work with the interplay between reactivity (the system simply reacts to the passer-by) and interactivity (the passer-by has incidentally understood the principle and actively plays with the system). An important additional challenge is to make the installation endurable, weather proof and safe from vandalism.

4.4 Innovation Is Substituted by Quality!

The 1990s were the decade of interfaces. Unfortunately, again and again projects were acknowledge in the media, curated, or awarded prices only because of their innovative and fascinating interface technologies. The content conveyed by them was often seen as secondary. This has changed fundamentally in the current decade: The "new" medium has matured. Today, interactive applications, installation and environments are judged because of the quality of their concept and design, the quality of experiences they evoke, the information mediated by them, and their utility.

Today, you no longer talk about a handle as an interface to a cup (and the incorporation of content this involves), but you talk about the formal and functional qualities of the cup as a whole. Nevertheless, lots remain to be discovered, researched and designed on the level of handles in the realm of new media technologies and their everyday application.

5
Interactivity as Media Reflection between Art and Science

Monika Fleischmann and Wolfgang Strauss

Fraunhofer Institute for Intelligent Analyze and Information Systems, MARS-Exploratory Media Lab / eCulture Factory, 53754 Sankt Augustin / 28199 Bremen, Germany
{Monika.Fleischmann,Wolfgang.Strauss}@iais.fraunhofer.de
http://www.iais.fraunhofer.de/mars.html,
http://netzspannung.org/about/mars/projects/

Abstract. This article describes the evolution of interactivity in the media arts ranging from immersive virtual reality to intuitive interfaces for real-time installations, to online archives and tools for knowledge discovery finally unfolding in networked environments for public space. This development is exemplified by the authors' own works and compared with selected media art works in the field.

5.1 Introduction

The importance of the so-called new media for society appears to be so all embracing, that several theoreticians already talk about "the digital" [1] in the sense of an epoch. Here, above all, they mean the decade of the 1990s. "The digital" is a revolution, which will have global consequences. It will lead to worldwide-networked communication and production structures. The media theoretician Vilém Flusser saw in computer supported communications the opportunity to free ourselves from traditional structures. "He searched for new possibilities for human cohabitation, not determined by abstract authorities, but by fast and efficient exchanges of knowledge; to form the relationship between the foreign and one's own, so that the homeless and the stranger are respected in their dignity. That can happen with the help of the new media." [1]

Our understanding of computers changed in the 20th century. We first saw them as codeable calculating machines, then as functional tools, then as interactive and "artificially intelligent" information, communication and production medium. From the end of the 1980s, media artists have worked with the phenomenon – interactivity, " Never before was it possible to operate within the thoughts of others", commented Derrick de Kerckhove, the media theoretician, on one of the first interactive virtual environments.[2] Nietzsche analysed the way "pen and ink" affect our thoughts. Kleist said that

[1] As quoted by Nils Röller, who oversaw in 2001 the "Flusser Archive" at the "Kunsthochschule für Medien" [Art College for Media Studies] in Cologne, and who wrote an essay for the 10th anniversary of Flusser's death – "Vilém Flusser: Medientheorie mit ethischen Anspruch" [Vilém Flusser: Media theory with ethical pretentions].

[2] Derrick de Kerckhove in discussion with Fleischmann & Strauss following the lecture "Virtual Walk Through Berlin – Visiting A Virtual Museum" by Monika Fleischmann at the Imagina 1992 in Monte Carlo, Monaco.

C. Sommerer et al. (Eds.): The Art & Sci. of Interface & Inter. Des., SCI 141, pp. 75–92, 2008.
springerlink.com © Springer-Verlag Berlin Heidelberg 2008

language and one's counterparts are responsible for the step-by-step manufacture of thought during speech. Similarly, de Kerkhove was convinced that our consciousness is formed by the materialising of our imagination and its translation into algorithms, and that therefore interactive media shape our thoughts.

As artists we translate this subject matter into virtual mental spaces, into reflection metaphors and interfaces, which touch a nerve because they remind us of well-known archetypes. Our projects, "Liquid Views"[3] and "Rigid Waves"[4] (1992/93), in which we made associations with the mirror stories by Ovid and Lewis Caroll, show this most clearly. Only the digital medium offers the possibility of the reversible, which assimilates human thinking and plays it back as calculated thoughts. This reflection arises through interactivity. It opens a framework of action, which changes the status of the work and allows the creation of new knowledge. The interface is the key.

5.2 The Cultural Discourse

The international competition for media art, the "Prix Ars Electronica", introduced in 1990 the category "interactive art". In that year, the first work to receive the golden Nica was "Videoplace" by Myron W. Krueger and Katrin Hinrichsen.[5] This installation highlights the relationship between people and machines in its aesthetic, rather than its technical dimensions. "Videoplace" aimed to elevate interactivity itself to an artistic medium. In the following years, interactivity became the lynchpin of Ars Electronica. However, it took nearly 10 years, before traditional arts and humanities became interested in the phenomenon – interactivity, which embraces code, network, interface design and technology.

A cross-disciplinary discourse began at the end of the 1990s. Sibylle Krämer (1998), Professor of Philosophy, proposed that we talk of interactivity instead of interaction between people and machines. "When one speaks of "new media" then we are talking about this text, image and sound simulating networked computer." [2] [...] "Digitisation, virtualisation and interactivity are therefore those phenomena, which we must study, when we see the computer within the perspective of a media. The media concept, which lets such a perspective be accentuated, moves away from the idea, that media just serves to communicate messages. [...]. Media bring across not only messages, but unfold an operating power, which influences the modalities of our thought, perception, experience, memory and communication." [3] Whilst Krämer puts this essential phenomena as the focus of her considerations, the artist and theoretician Lev Manovich, looked for further categories and proposed a code oriented definition for new media: „Rather than focusing on familiar categories such as interactivity or hypermedia, suggest a different list. This list reduces all principles of new media to five – numerical representation, modularity, automation, variability, and cultural transcoding." [4]

Since the late 1980s, teams of artists, designers, information technologists and theoreticians have established interactive media art. With self-developed participative

[3] <http://netzspannung.org/database/liquid-views/en> Rev. 2007-06-17.
[4] <http://netzspannung.org/database/rigid-waves/en> Rev. 2007-06-17.
[5] <http://www.aec.at/de/archives/center_projekt_ausgabe.asp?iProjectID=11224> Rev. 2007-06-17.

tools – the interfaces – they have tested the possibilities of interaction between people and machines.[6] However, as late as 2004, the Jury of "Ars Electronica" saw no theme focus in the concept of interactivity. In 2004 a prize was awarded to "Listening Post" by Ben Rubin and Mark Hansen. This installation made Internet communication observable and audible, but it was not interactive. The Jury statement indicated that interactivity was no longer a premise. With that, the definition of the category "interactive art" changed significantly.[7]

The media archaeologist, Erkki Huhtamo described this change as a crisis in interactive art. In his essay "Trouble at the Interface"[8], he proposed, either to do away with interactivity, or to introduce a new category for projects such as "Listening Post", such as for example "database aesthetics". Although only fragmentarily researched, interactivity no longer stood at the focus of attention. With Web 2.0 and the possibilities it offers for active participation, interactivity has now become part of everyday experience.

5.3 Interactivity as Aesthetic Experience

Notwithstanding, interactive structures remain the basic principle of digital media. Visitors should be in a position to relocate and to challenge themselves to the interactive projects to make an experience that moves over and beyond the usual contemplative observation of a work of art. Clearly, the projects should not only bring the visitors' bodies, but also their thoughts into motion. Erkki Huhtamo described the challenge "please touch" as the corner stone of the interactive art aesthetic, an echo of Marcel Duchamp's "Prière de toucher".[9] Touching an interactive work is not only allowed, but necessary. Whether with mouse, trackball, touch screen, tangible objects, video camera, responsive workbench, virtual balance, the touch less PointScreen [5] or other interfaces[10], the observer first brings the process into motion. This is very

[6] Such teams include: Art & Technology Labs such as "Art+Com", the „Future Lab" of "Ars Electronica", the Fraunhofer MARS – Exploratory Media Lab, the Dutch "V2" as well as "De Waag", the Polish "WRO", the Hungarian "C3", the Indian "Sarai Media Lab", "Videotage" in Hong Kong amongst others.

[7] Interactivity would from now on be interpreted and extended, so that 1. Computer are no longer a precondition, 2. the borders between real time and direct interaction would be relaxed and 3. the concept of passive interaction should be allowed. Therefore, active participation was no longer demanded as a necessary component of the category "interactive art". http://www.aec.at/de/archives/prix_archive/prixJuryStatement.asp?iProjectID=12807 Rev. 2007-06-17.

[8] Huhtamo, Erkki: Trouble at the Interface or the Identity Crises of Interactive Art. Revised version of an essay first published in Framework, The Finnish Art Review, 2/2004 <http://www.mediaarthistory.org/Programmatic%20key%20texts/pdfs/Huhtamo.pdf> Rev. 2007-06-17.

[9] Duchamp's text in the exhibition catalogue of the Surrealists from 1947, was designed by him, on the wrapper of a foam rubber breast. Media artist Ken Feingold, alludes to Duchamp with the installation "The Surprising Spiral (1991). The name "Pierre de Toucher" was given as author for his "book". See Huhtamo 2004.

[10] The interactive projects of the MARS Lab and its interface inventions are documented at netzspannung.org. http://netzspannung.org/about/mars/projects Rev. 2007-06-17.

different to the appreciation of a traditional work of art. Simon Penny differentiates the difference in perception as follows: "A painting is an instance of representation. A film is a sequence of representations. Interactive artworks are not instances of representation, they are virtual machines which themselves produce instances of representation based on real time inputs." [6]

Duchamp and others already discussed the active participant[11] in the first half of the 20th century, and so a new dimension of reversible works was introduced. In addition to the mental reception and bodily activities of observers, came a level relating to other visualised perceptions and processes. [7] The levels of activity and reception overlay each other. Those doing the interacting, influenced to a certain degree the appearance and therefore also the object of their aesthetic experience. Interactive art creates a situation or an environment, which the observer confronts, and through which they enjoy an experience, which arises only first out of the participatory process itself. It is this, repeatable, process, which first gives the work its distinguishable identity. Interactive art thus means, the experimental exploration of artwork and tool.[12] Interactive art is artistic research and interactivity an aesthetic experience.

Reviewing the history of media art, Söke Dinkla reflected on the concept of "knowledge arts", to which we devoted our exhibition in 2006 in the "Neues Museum Weserburg", Bremen[13]. "It's not about just one, but different forms of the arts. The plural "arts" qualifies our understanding of art, but at the same time extends it. ... "Knowledge arts" – in the 1990s we described them somewhat more categorically, but also less openly as media art – are the result of the many coups, which visual arts in the course of the 20th century had undertaken."[14] Interactive concepts are based on an altercation with forms of human interaction, with communication technologies and with the possibilities of networked activity. Artistic practice creates situations, which encourage forms of communication and interaction, and in this way it changes the modes of coming-into-contact-with-one-another. Participants can extend the possibilities of interpersonal exchanges. We find ourselves today in a culture of active participants, of interactivity, in which the digital media "become identity giving machines. Thus the current challenge is to comprehend digital media as cultural technology," as Söke Dinkla points out. [8]

---

[11] Walter Benjamin and Berthold Brecht in the 1920s, with the advent of radio, also bemoaned the poor programming, which the National Socialists, above all misused for propaganda purposes. They made proposals on the possibility of new formats being experimented with, and in Brecht's "radio theory" the participation of the listener is schematised: listeners should be made into participants.

[12] We called the results of our experiments in the EU project eRENA, [electronic areas in art and entertainment, in which the MARS Lab, the ZKM, Nottingham and Stockholm, Lausanne and Geneva Universities, as well as Illuminations London took part], „Tools for the arts of tomorrow". <http://www.erena.kth.se/> Rev. 2007-06-17.

[13] Monika Fleischmann & Wolfgang Strauss, Wissenskünste aus der eCulture Factory. An exhibi-tion in Neues Museum Weserburg Bremen. 5.October-3.November 2006 <http://eculturefactory.de/wissenskuenste> Rev. 2007-06-17.

[14] Dinkla, Dr. Söke: Von der Medienkunst zur Wissenskunst. Zur Ausstellung „Wissenskünste aus der eCulture Factory" von Fleischmann & Strauss. Shown at eCulture Factory: <http://www.eculturefactory.de/eculturetrends/download/dinkla.pdf > Rev. 2007-06-17.

5.4 Interactive Media Art as Aesthetic Laboratory

Media artists explore the culture of active participants and with it the aesthetic potential of interactive art, process-related image worlds and generative processes. The Fundamental questions of media art are: the organisation and structuring of data and information; the orientation and navigation in virtual space; questions relating to interface and interactivity processes; and additionally, telepresence and immersion. Artistic research on digital media, functions as an aesthetic laboratory for societal development. In so far as future forms of communication can be anticipated, interactive art – rather unwillingly – becomes a driver of innovation, and positions itself between everyday, scientific and artistic experience.

At the beginning of the 1990s, media artists were working often in scientific research institutes with advanced virtual reality technology and were searching for extended and networked domains, telepresence and artificial life. By means of performance, sculpture, installation or environment they stage interactive projects and communicate playfully complex topics to the public. Expansive interactive environments were created, such as "The Legible City", 1991, by Jeffrey Shaw, where cyclist-visitors pursue various narrative threads on their passage through the virtual city; "Terrain 01" 1993 by Ulrike Gabriel, where robots are equipped with photocells reacting on light that is controlled by participants' brain waves activity; and "A-Volve", 1994 by Christa Sommerer & Laurent Mignonneau, where visitors can create artificial creatures and follow the artificial live process. They prove so far unknown forms of communication.

The art theoretician Oliver Grau describes the networked and interactive media art installation "Home of the Brain"[15] 1992 by Monika Fleischmann & Wolfgang Strauss, as media theory put into practice and new mnemonic theatre that anticipated the form of communication with networks. Grau writes: "Home of the Brain" already in 1991 was an early appearance of the epistemic innovation telepresence. As a consequence, the reception of the work of art lost its fixed position. The observers do not go to the artwork, panel, panorama, cinema film etc, the work however does not come solely to them."[16]

In "Home of the Brain" the visitor navigates with a data glove through digital rooms that are made visible with data glasses. Hand movements activate the citations of four scientists, who play an important role in the theoretical formation of media culture. They are represented by their individual thought buildings: Joseph Weizenbaum has the House of Hope, Marvin Minsky, the House of Utopia, Paul Virilio, the House of Catastrophe and Vilém Flusser, the House of Adventure. The work was designed, at the beginning of the 1990s, to give new impetus to a media discourse enshrouded by technophobia. The media theoretician Claudia Giannetti endorsed the "interdisciplinary nature" of media art, "which extends far further than

---

15 "Home of the Brain" (1990-92) by Fleischmann & Strauss was developed in the context of their project: Berlin, Cyber City. It won the Golden Nica of Prix Ars Electronica 1992. <http://www.medienkunstnetz.de/works/home-of-the-brain> Rev. 2007-06-17.

16 Grau, Oliver: Immersion und Interaktion. Vom Rundfresko zum interaktiven Bildraum. 2004 <http://www.medienkunstnetz.de/themen/medienkunst_im_ueberblick/immersion> Rev. 2007-06-17.

Fig. 5.1. "Home of the Brain", Virtual Reality Installation (1992)

the already known considerations on the relationship of art to technology".[17] The multi-faceted nature of digital media, demands interdisciplinary thought and work structures that bridge the chasm between analogue and binary language – the chasm between art and ICT[18] science.

5.5 From Virtual to Real Space

The question of how digital information can translate not only the metaphoric virtual, but also the physically real into accessible and understandable domains, marked the passage to media architecture. Here we understand an architecture, which connects people, space and data with one another. It creates an extended area of activity. The multi-user installation "Murmuring Fields" (1998) adresses the overlaying of physical and data space, which embraces the body with imperceptible interfaces. The performative sound installation is a recordable field of sound, managed by movement of the body.

Digital information - sounds and figures – are located in the space as if the room were furnished with data. [9] Every movement of the body is captured with an optical body-track-technique.[19] Movement is transferred from real space into data space, and translated into a sound collage. Spoken texts are broken up into words and syllables. Movement in space creates movement in the text. Two interactors produce text samples by Joseph Weizenbaum, Marvin Minsky, Vilém Flusser and Paul Virilio. "Poli-tic-tic-tic", says Flusser's voice as a performer bows backwards and forwards and thus interprets a part of the text: "Youngsters at the terminals; they turn their backs to politics and turn to each other."[20] The dancer starts up syllables with her body and forms speech. She plays with the meaning of the concepts. Text is translated into a texture of sound and movement.

The concept of accessible knowledge space has its correspondence in the concepts of David Rokeby. With "Very Nervous System" (1982) he schematised interactivity,

[17] Claudia Giannetti: Ästhetische Paradigmen der Medienkunst. In: Medien Kunst Netz. 2004 <http://www.medienkunstnetz.de/themen/aesthetik_des_digitalen/aesthetische_paradigmen/scroll> Rev. 2007-06-17.

[18] ICT = Information and Communications Technology.

[19] Body-Track is part of the eMuse-Systems, which were developed as production system for ``Murmuring Fields". In: W. Strauss, M. Fleischmann: Imagine space fused with data. In: Cast01 Proc.2001. <http://netzspannung.org/version1/cast01/proceedings/index.html> Rev. 2007-06-17.

[20] Sentences from an interview with Vilém Flusser in 1990 in Austrian television are built into the sound collage.

Fig. 5.2. Assembling the interactive stage: "Murmuring Fields" (1998/99)

with the aim of putting computers and people into an intuitive, bodily, expansive and intimate relationship.[21] Whilst Rockeby focuses on audiovisual interaction, we deal with the communication of the participant and the acquisition of knowledge. Both activities start with the premise of sensory perception. Knowledge here is not acquired by reading, but through the body. The theorist of cognition George Lakoff, emphasized again and again the importance of the body and its entity for thought processes. Sensory experience and reflection combine together in "sensory thinking of the body", he wrote. [10] The psychoanalyst Maurice Merleau-Ponty described the body as the centre of the spatially and temporally mediated world. For him, thinking is

[21] Very Nervous System is one of the first audiovisual interaction systems, that was introduced variably into installations. In the system, a computer analysed the image created by a video camera in motion. The result is an interactive space, in which those interacting use their bodies as the active element of the interface. <http://www.medienkunstnetz.de/werke/very-nervous-system> Rev. 2007-06-17.

based on experience, which arises from bodily perception connecting to everyday activity. Oliver Grau determined, that with "Murmuring Fields" a new type of space of mind was created.[22]

5.6 Knowledge Art as Cultural Technology

Interactive knowledge structures and digital archives are current themes in media art research.[23] Under the title "Explore Information – Create Knowledge"[24] projects are presented on the media art platform netzspannung.org, which concern themselves with the structuring of unmanageable amounts of information. The online archive netzspannung.org also offers ever more amounts of information material concerning the theme digital culture.[25] To find one's way around the over 1.500 lectures, workshops, study series, scientific texts and artistic projects of the online database, innovative visualising tools have been developed for accessing the digital archives. Fundamentally, there are two different types of access to electronic data: the "precise" search and the "imprecise" browsing.

The search presupposes that users know what they are looking for, that they can formulate their interests and, where necessary, can be more precise. Browsing, on the other hand, involves the user being inspired and prodded by that, which is offered. In the article "As we may think"[26] (1945), the American scientist Vannevar Bush already bemoaned that the problematic relating to the selection of information was located in the artificiality of its indexing systems. Data in archives were filed alphabetically or numerically. Information, if at all, could only be retrieved by sifting through, index for index. He observed, "The human mind does not work that way. It operates by association. With one item in its grasp, it snaps instantly to the next that is suggested by the association of thoughts, in accordance with some intricate web of trails carried by the cells of the brain."[27] Bush called for a new relationship between the thinking person and the sum of our knowledge. He proposed mechanising the selection of information using association – and not indexing. The idea of an associative net of concepts is also fundamental to the concept of "Knowledge Discovery Tools" of the netzspannung.org.[28]

---

[22] Oliver Grau. Integrating Media Art into our Culture. Art History as Image Science. <http://www2.hu-berlin.de/grau/Grau.pdf> Rev. 2007-06-17.

[23] Thomas Goldstrasz: „Suchmaschinen – Sechs Kunstwerke und eine Suche zum Thema suchen - speichern – suchen lassen". In: <http://netzspannung.org/media-art/publications/digital-transformations/search-engines/> Rev. 2007-06-17.

[24] Wolfgang Strauss; Nina Zschocke: Explore Information / Create Knowledge. <http://netzspannung.org/media-art/explore-information/> Rev. 2007-06-17.

[25] From the beginning of 2001 netzspannung.org has recorded an ever-growing number of users. In 2007 the daily average figure is 3.300 and more that 100.000 visits per month.

[26] Vannevar Bush - As We May Think - The Atlantic Monthly,1945. Published in the Journal 'Form Diskurs', Nr. 2, I/1997, pp 136-147. <http://wwwcs.upb.de/~winkler/bush_d.html> Rev. 2007-06-17.

[27] Ibid.

[28] <http://netzspannung.org/about/tools/> Rev. 2007-06-17.

5.6.1 Navigational Map for the Data Domain

With the "Semantic Map", we developed a navigational tool for the digital data domain. It shows all the documents of the archive, organised into self-organising clusters. So, for example, a cluster with the label "virtual" includes all documents that discuss this theme. The first step is to identify the cluster labels, that is the genre terms, by comparing them with a list of keywords. This is done by an automated text analysis of the database entries. Then the relevant documents are assigned to the clusters. In the next step, semantic relationships between individual database entries are computed. According to these textual relationships, the individual database entries are sorted relationally to one another within the cluster, whereby the distance, one from the other corresponds to the relevance of their respective contents. [11] If the documents are close to one another, there is a textual relationship. On selection, a short description appears in a second window. In further zoom stages; the map is changed from a text based to a visual design.

As soon as new documents are entered into the archive, they integrate themselves according to an automatic text analysis. The archive is therefore not "stipulated" but because the documents "have knowledge of each other" they can automatically re-order themselves. With the "Semantic Map", hidden connections within the data stock are computed and visualised. A self-organising neuronal network is deployed for the computation of the data and the automatic graphic arrangement in the map, which is named "Kohonen Map"[29] after its inventor. The semantic knowledge map is prototype visual search and find machine.

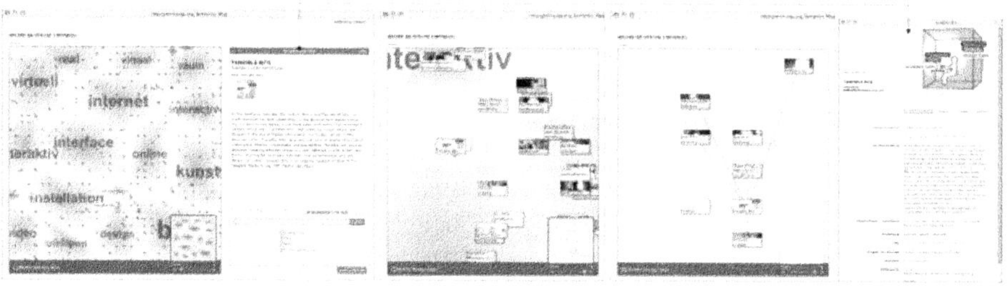

Fig. 5.3. "Semantic Map": Dynamic zooming from simultaneous overview to detail (2001-04)

5.6.2 The Archive Domain of Netzspannung.org

The difficulty of orientation in online archives is due to contents only being viewable on hundreds of individual web sites. One is always looking for new methods of exhibiting and mediating information. Since 2006 we presented the numerous database entries of netzspannung.org, as publicly accessible archive, in the context of an exhibition.[30] Two

---

[29] Teuvo Kohonen, Dr. Eng., Emeritus Professor of the Academy of Finland <http://www.cis.hut.fi/research/som-research/teuvo.html> Rev. 2007-06-17.

[30] Several installations of Fleischmann & Strauss were shown in the exhibition „Kunst-Computer-Werke" [Art-Computer-Work] in the „Zentrum für Kunst- und Medientechnologie (ZKM) [Centre for Art and Media Technology] in May 2006 and until January 2009 in "YOU_ser – the century of the consumer" <http://www02.zkm.de/youser/> .2008-01-15.

installations show how digital information can be spatially staged. "Matrix" [12] with the "PointScreen" Interface and "Medienfluss" (Media Flow), offer a complete overview of the online archive, which allows a dynamic switch of criterion, from detailed overview, down to the individual artwork. Both applications can reach back to data, which is saved in the online archive. They are exported via an XML gateway and presented audio-visually as an installation in physical space.

Interfaces were sought out, which make clicking around on websites dispensable. It is always the same problem with online archives: how can great masses of information be structured, so that everyone can easily find what they are looking for? What can one offer an audience that enjoys playing with the modern media, and in the process wants to gain experience and to learn something?

The "Medienfluss"[31] [Media Flow] is an interface, which transmits an immediate impression of the contents and number of documents in the online archive. Two parallel media flows of images and words, stream as large format date projections through the room. The flow of words shows keywords, authors and titles of the archived documents. Text-based access is complemented by visual access. The images represent respectively an archive entry. The terms are spoken out by a computer voice using text-to-speech processing. The "Medienfluss" creates an atmospheric image and sound domain. A touch screen translates the flowing images into scrollable text bands, serving as an index for specific searching. On selecting a term or image, the relevant document is visually highlighted and presented in detail in the form of text, image or video. While the active user is immersed in detail, observers on site can follow the process of selection and display of the archive. Medienfluss [Media Flow] is a living database.

With the "Matrix"[32] a browser was developed for exploring large stocks of data, which in combination with the gesture based PointScreen technology[33] can be implemented as a room installation. The interface takes up the "Matrix" theme[34] of the non-finite classification system. Each field of the "Matrix", using an image icon, represents a media project. A virtual lens is steered, contact-free, over the "Matrix". On rollover, the lens enlarges the image contents and, in addition, shows author and title of the respective project. The selected image enlarges itself and also shows a video on the project. The "Matrix" offers museums and archives the possibility, in compressed form, of accommodating a greater part, or even their complete, inventory. It also provides a tool for detailed examination. The "Matrix" supports overview browsing, whilst the lens offers a dynamic insight into the detail.

[31] <http://www.eculturefactory.de/medienfluss> Rev. 2007-06-17.

[32] <http://eculturefactory.de/digital-sparks-matrix> Rev. 2007-06-17.

[33] With PointScreen technology one accesses a novel navigational medium, which allows a contact free, gesture-based interaction. PointScreen was developed at the MARS – Exploratory Media Lab of the Fraunhofer IAIS (former IMK) by Wolfgang Strauss, Monika Fleischmann, Yinlin Li. The PointScreen Technology is based on so-called "Electric Field Sensing" (EFS) and uses the human electrostatic field to control interactive applications. US-patent number is 7,312,788. The title of the patent is: Gesture-based input for a user interface of a computer (ECCO). See References: Strauss, W. et al: Information Jukebox ... [5]

[34] In Mathematics, the matrix (plural matrices) is an arrangement of numerical values in tabular form. One speaks of columns and lines of the matrix and describes that also as (line and column) vectors. The objects, arranged in the matrix one calls components or elements of the matrix.

Fig. 5.4. "Medienfluss" [Media Flow] as a visual interface (2006)

Browsing the "Matrix" occurs by touchless interaction. The "PointScreen"[35] allows to control digital objects on a screen just by body movement and energy like the Chinese Qigong, and it represents a new interface paradigm. PointScreen provides access to any digital system just by the move of the user's hand. Unlike touch-screens there is no need to get in contact with any surface. The person in front of the interface controls or manipulates the application by natural hand gestures, even from a meter's distance. It literally seems like magic and reminds of what can be seen in science-fiction movies such as minority report. The basic principle of PointScreen technology is the sensing of electric fields. People that engage with the system enter an electrostatic field that is established by emitting antennas. The person's body modifies this field, varying with his movement. These disturbancies are measured by the PointScreen antennas and mapped onto cursor coordinates. Hand and body gestures are tracked in the three dimensional space between the user and the screen and are interpreted to control the digital system. The PointScreen research and development was inspired by the Theremin, one of the first electric musical instruments. It gave us the idea of the virtual window control by movement and gesture. If there is nothing to touch interaction gets more direct and intuitive.

Fig. 5.5. The "Matrix" with "PointScreen" technology (2005 / 2003)

[35] <http://netzspannung.org/database/pointscreen/> Rev. 2007-06-17.

5.7 Interactive Art in the Public Domain

Together with knowledge media, the public domain is a current theme of media art. Interventions in the public domain are able to make a wider public aware of the current state of research in experimental media and Internet art. Since the middle of the 1990s, activists and artists have had great hopes of the Internet as "public domain" and means of democratisation. A series of town-like communities have developed since 1994, including "Digitalen Stad" [Digital Town] Amsterdam, and, in 1995, the "International Town" [International Town] Berlin in the then new World Wide Web. In the last few years, media façades and public spaces have become popular sites for media art. One is talking here mostly about participative projects, staged temporally in public urban space. One can observe and accompany such actions in public places, but also actively take part. In this way, the digital is situated, appreciated and discussed in the public domain.

The "Chaos Computer Club" in 2001 on its 20th foundation day, supplied an enthusiastic public with online tools for creating animations, which could be sent by email and projected onto a house façade – the "Haus des Lehrers [Teachers' House] in Berlin. Using mobile phone and special software, the façade became a media skin. The installation "Blinkenlights"[36] ran for five months and has found many imitators. The project opened a series of interactive façade projects, in which participants can contribute their own contents.

Likewise in 2001, Rafael Lozano-Hemmer presented with his "Body Movies"[37], and his concept of "relational architecture", another format to involve the public and public space. This installation also used the façade of a building as screen and access to information. Here though, people and their shadows become actors in a dramatic setting. Virtual passers-by, recorded in different towns and projected onto a house wall, integrate with shadows cast by real on-site passers by. Participation in a collective process transforms public spaces into places of public action.

A similar effect is shown by the project "Energie-Passagen" [Energy-Passages].[38] The stroll though the daily news, schematises public and private interest in information. Language is understood as intellectual energy, which shapes a city. "Energie-Passagen" is a location-specific installation, which transforms the daily flow of news into an audio-visual data flow, staged as a media reading, performed in city space. Starting points are texts from a mass-media newspaper. An automatic technique converts daily RSS Feeds, and analyses some 50.000 words of the current newspaper and reduces them to 500 most common keywords.

These keywords appear as a large-screen projected flow of information in front of the "Literaturhaus" [House of Literature] in Munich. They are performed by an artificial computer voice. Passers by select terms over a microphone or touch screen. The computer voices react with a multi-voice echo to the selection. Simultaneously, the

---

[36] <http://de.wikipedia.org/wiki/Projekt_Blinkenlights> Rev. 2007-06-17.

[37] Projekte von Rafael Lozano-Hemmer <http://www.lozano-hemmer.com/eproyecto.html> Rev. 2007-06-17.

[38] In 2003, Monika Fleischmann and Wolfgang Strauss were commissioned with the implementation of their concept "Energie-Passagen" in the context of the competition for the exhibition "Ortstermine – Kunst im Öffentlichen Raum", Kulturamt [Cultural Department] of the city of Munich. <http://energie-passagen.de> Rev. 2007-06-17.

Fig. 5.6. The Energie-Passagen [Energy Passages]. Information flow shows computed keywords.

Fig. 5.7. Energie-Passagen [Energy Passages]. Electronic Reading in front of the House of Literature Munich, Germany (2004)

selected word appears in a concept grid of "friendly" words. This text movement allows connections between the terms to emerge. The associative reading of the newspaper encourages one's own thoughts. Through the selection of terms, the visitor "writes" a new newspaper. In the text montage, unexpected and new connections of meaning are created.

New meanings and relationships, which till then did not exist, are created in interaction with the text, because readers have at their command individual patterns of perception and behaviour. In the information flow of "Energie-Passagen" connected words build the site of the narrative. The process of reading becomes thought in action, through the interaction. The text space itself becomes a browser for the "search as process". The curator Christiane Paul describes the installation as public performance: "„Energie-Passagen literally re-inscribes the passages of energy that inform our daily life onto the street, allowing the passers-by to 'perform' the events of the day in multiple semantic connections."[39]

[39] <http://www.energie-passagen.de/presse2_eng.html> Rev. 2007-06-17.

A computer based tool for associative reading was developed, that became an information browser, by means of automated keywording. It also makes a contribution towards the development of accessible data archives. The sensory and cognitively experienced city space becomes, through algorithms, an electronic reading garden. The motif of the flow is embedded in city space. A connected and reactive text and sound domain, becomes through the light of the projection, a directly perceptible urban media domain: a materialised newspaper archive, which is situated directly on the way. Media art in public space reaches out to more and different visitors than in a museum. A contrast to the newspaper browser in urban space is the Internet browser 10 x10, which constantly refreshes, and presents the 100 most important news items from the whole world in an image matrix.[40] The result shows a daily updated snapshot of the images and words of the world.

5.8 Book Renaissance

The idea of the Virtual Book has its origin in artists' books. Our first example of a virtual book showed the electronic representation of the publication "Digitale Transformationen" [Digital Transformations] and described in over 50 contributions from authors, media art projects from the 1960s till the 2004. [13] The publication serves, with its individual texts in PDF-format, as entry point for a multi media presentation. The interactive work, which is described here, demands the presentation in a time-medium.

The virtual book presents video, audio and hypertext linking, and is an extension of the traditional book. It can thus be read in different ways. It can be browsed and managed, page for page, like a traditional physical book. Integrated into the virtual book are also hyper-textual navigation elements and multi-media components. The text is saved and indexed in a database. The book can be consulted, or searched chapter for chapter, according to keywords and terms, authors, images and videos, using a menu. The virtual browsing and the images, which metamorphose into video or animation, are impressing for the reader. The MP3 audio track, reads out the text using professional narrator voices, and at the same time, marks the relevant positions in the text.[41]

The "Reading Table"[42] presents another form of reading. Via a pneumatically controlled table, a so-called "swing table" – text flows can be temporally manipulated by lying on hands, and bending the projection surface. In Masaki Fujihata's "Beyond the Pages" (1995)[43], the contents of his picture book are moveable. Images of objects such as stones or printed characters make sounds or are made to ring. Touching the image of a switch turns on the writing-table lamp, which as in the real world, sits on the table next to the virtual book. Touching the image of a doorknob opens the image

---

[40] <http://www.tenbyten.org/10x10.html> Rev. 2007-06-17.

[41] The virtual book was first presented to the public at the Frankfurt Book Fair 2006 and the public reaction evaluated. <http://www.eculturefactory.de/virtual-book> Rev. 2007-06-17.

[42] Xerox Parc, Experiments in the future of reading. ACM 2001. <http://transliteracies.english.ucsb.edu/post/research-project/research-clearinghouse-individual/research-reports/tilty-tables> Rev. 2007-06-17.

[43] Masaki Fujihata, Beyond the Pages (1995) <http://on1.zkm.de/zkm/stories/storyreader$552> Rev. 2007-06-17.

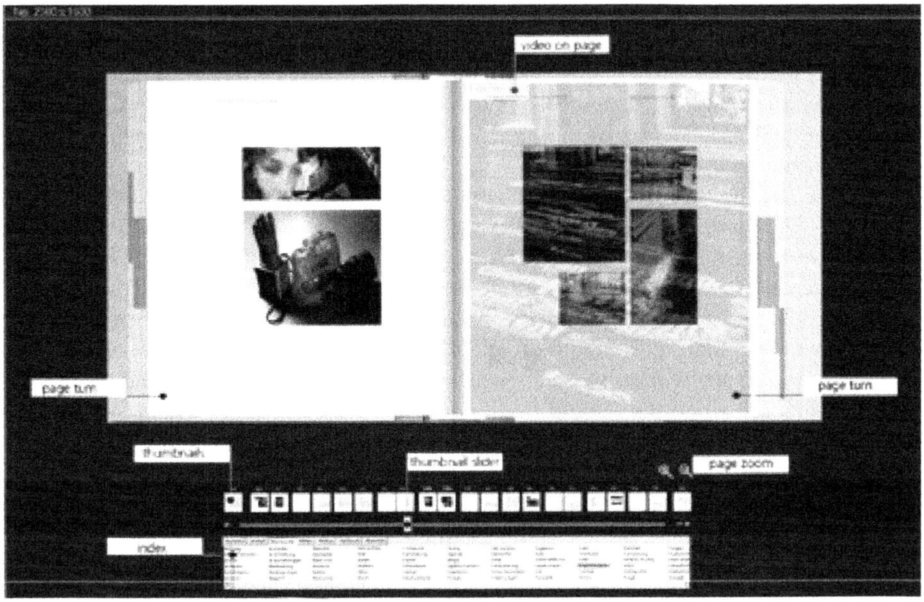

Fig. 5.8. Virtual Book (2005)

of a door, projected onto the wall, which unexpectedly opens. Fujihata tests the limits of the media book, in so far as he combines real and virtual space in a particularly amusing way. The book becomes an interface to the environment.

5.9 Deep Storage

Whilst in traditional art forms, packaging, stacking and storing were discovered as artistic form of expression, "the ever present digital date storage in the 1990s," led to, "an artistic altercation with the freeing, or rather ousting of the human memory." [14] The irritation increases over the loss of memory that is given over to machines the more knowledge can be relocated on hard drives. The machine today is not a single computer, but a global network of computers. The knowledge lost to individuals by relocation to this storage is gained through the compensations from the collective and the exchange. With the work presented, we ask questions of this digital storage, relating to knowledge, memory and recollection. Our work is an answer to the challenge of the ever-increasing mass media flood of information. Since 1990, our thematic foci regarding interactive media have shifted. [15] At first, questions of body, recollection and memory stood at centre stage. Afterwards, the increasing floods of information and the theme -knowledge as stored information-, took on a greater meaning.

With "Home of the Brain", we reflected not only on the new medium, but the media discourse itself became an object of reflection. The interactive participants were enclosed literally in the discursive environment, their field of vision filled in completely with a 360° illusionary immersion room. This isolated immersion, is extended in "Murmuring Fields" into a space of mind through dialogic forms of play with other participants. The data room of the sound archive is played like an instrument using bodily actions, and so experienced bodily. The new experience is discovered through the joint play of the participants. The theme of "Liquid Views" is the media related shaping of the body through the computer. In the mirror of this medium, the visual perception and recollection of the observer is addressed. Viewers become, at one and the same time their own observers.

We use the digital media, as Söke Dinkla established, "in order to newly structure available knowledge, to make it sensorally accessible, and so to feed it into the discourse on media culture". In much of the work of knowledge art, the contents communicated, are at least as important as the technology. Sometimes, the form and content consort together. One starts with the desire for a specific form, which bit by bit fills with content. The media art platform netzspannung.org has developed in a similar way. "Frameworks, in other words a criterion of regulations and rules, were made available, which then could be filled by the media art community with content and individual contributions. Netzspannung.org is at one and the same time a forum and an online archive".[44]

Communication and presentation formats such as "Energie-Passagen" or "Medienfluss" were developed under the umbrella of knowledge arts. With the image motif of the river, static and passive masses of information are transformed and flow out of the archive and around the visitor. In as much as the data appears animated as flowing movement, it is transformed into a time-based medium and can take up a narrative function. The "Medienfluss" as interface, embodies the psychological meaning of the term "flow".[45]

By flow, we understand a sort of intellectual elation, which leads to thought flow, uncoupled from current reality. Flow can be described as a state in which attentiveness, motivation and the environment come together in a form of productive harmony. Flow means, to forget time.[46] Knowledge maps such as the "Semantic Map", or "Matrix" widen out information, side by side. Whilst the "Semantic Map" orders the spatial closeness of documents according to textual similarities, the "Matrix" incorporates the principle of serendipity. This term describes a co-incidental observation, something not originally sought out, which proves to be a new and surprising discovery, such as, for example when surfing the Internet, one co-incidentally discovers useful information.

The "Virtual Book" allows new work techniques such as collaborative writing and participative reading. Text in the virtual book becomes hypertext. The examples presented on the future of the book, highlight a spectrum of artistic and scientific

---

[44] Söke Dinkla, Von der Medienkunst zur Wissenskunst. Einführung in die Ausstellung „Wissenskünste aus der eCulture Factory" <http://eculturefactory.de/download/dinkla.pdf> Rev. 2007-06-17.

[45] The term "flow" means the pleasure oriented feeling of a complete merging in an activity, a creative burst, or burst of activity. http://de.wikipedia.org/wiki/Flow_%28Psychologie%29.

[46] In 1975, the psychologist Mihaly Csikszentmihalyi described the „flow-experience". He is though not the first to discover the concept, the writing of the educationalist Kurt Hahn (1908) with his extensive synonym on knowing "creative passion" and the doctor and educationalist Maria Montessori with "Polarisation der Aufmerksamkeit" (1909), describe the self-forgotten, playful, explorative activity of children as separation from the environment and the concentrated turning towards a certain immediate activity.

research on the future of reading. Through the comments of the reader, furthermore, the complete reading process can be documented. Thus the text becomes starting point for online discussion, or serves as first basis for collaborative writing techniques, such as the Surrealists in the 1920s described. In Cadavre Exquis, the Surrealists developed a continuous game with folded paper, in which many people, one after the other, could create a sentence or a drawing, without anyone knowing about the previous stage. Breton argued, that in this way one could have access to an infallible means to turn off critical thinking and to create a free path for the metaphoric capabilities of the spirit.[47]

The "Virtual Book" and its search profile, functions like glasses, through which the data domain can be contextually observed. Its surface is virtual, and is a window to the temporal space of textual connections.[48] The idea of books as active knowledge structures, is inspired by Marvin Minsky's provocative vision from the 1980s: "Can you imagine that they used to have libraries where the books didn't talk to each other?" [16] Attaining such interactive structures will occupy quite a few generations of artists and scientists to come.

Acknowledgements

We would like to acknowledge the many colleagues who contributed in particular to some of the work of the MARS – Exploratory Media Lab described above, including Gabriele Blome, Adam Butler, Jochen Denzinger, Ansgar Himmel, Kai-Uwe Kunze, Yinlin Li, Lina Lubig, Jens Muuss, Andreas Muxel, Jasminko Novak, Stefan Paal, Predrag Peranovic, Kresimir Simunic, Stephen Williams, Stefan Winarzki and Stefanie Zobel. This research had been supported by the German Federal Ministry for Research and Education (ICT, Culture) in Bonn, the Department of Arts and Culture in Munich, the Ministry of Economics and Ports in Bremen, the European Commission (ICT) in Brussels and the Fraunhofer Institute for Intelligent Analysis and Informationsystems (former Institute for Media Communication).

References

1. Lunenfeld, P. (ed.): The Digital Dialectic - New Essays on New Media, Cambridge MA, London, p. VXI (2000)
2. Krämer, S.: Was haben die Medien, der Computer und die Realität miteinander zu tun? In: Medien Computer Realität. Wirklichkeitsvorstellungen und Neue Medien, Frankfurt/M, p. 9 (1998)
3. ibid., p. 14

---

[47] Definition by André Breton: Cadavre Exquis – Game with folded paper, which is about allowing a number of people to construct a sentence or a drawing, without a participant having knowledge of the previous contribution. The example, which has become a classic, which the game has given its name to, makes the first part of a sentence, created in this way: Le cadaver-exquis-boira-le-vin-nouveau. (fr="The delectable-corpse-drinks-the-new-wine").

[48] See also Michael Wetzel: Flüssige Datenströme. <http://www.freitag.de/ 1999/52/ 99522701. htm> Rev. 2007-06-17.

4. Manovich, L.: The Language of New Media, Cambridge, Mass, p. 20 (2001)
5. Strauss, W., et al.: Information Jukebox. A semi-public device for presenting multimedia information content. In: Proceedings of the 1st International Conference on Appliance Design 2003. Springer-Journal Personal and ubiquitous computing, pp. 217–220. Springer, London (2003)
6. Penny, S.: Critical Issues in Electronic Media (Suny Series, Film History & Theory). State University of New York Press (1995)
7. Hünnekens, A.: Der bewegte Betrachter. Theorien der Interaktiven Medienkunst. Wienand-Verlag (1997)
8. Dinkla, S., Leeker, M. (eds.): Tanz und Technologie / Dance and Technology - Auf dem Weg zu medialen Inszenierungen/Moving towards Media Productions, Berlin (2003)
9. Strauss, W., et al.: Staging the space of mixed reality. Reconsidering the concept of a multi-user environment. In: Proceedings of the fourth symposium on the virtual reality modelling language, VRML, Paderborn. ACM Press, New York (1999)
10. Lakoff, G.: Philosophy in the Flesh. The Embodied Mind and Its Challenge to Western Thought. New York (1999)
11. Simunic, K., et al.: Combining Visualization and Interactive Clustering for Exploring Large Document Pools. In: Proceedings of the 4th IAESTED International Conference on Visualization, Imaging and Image Processing - VIIP 2004. Marbella, Spain (2004)
12. Strauss, W., et al.: Matrix-Lupe. Browser zur Exploration multimedialer Datenbestände in Verbindung mit gestenbasierter PointScreen Technologie. In: Mensch und Computer, Konferenzband, Oldenburg (2006)
13. Fleischmann, M., Reinhard, U. (eds.): Digitale Transformationen, Heidelberg (2004)
14. Schaffner, I., Winzen, M. (eds.): Deep Storage - Arsenale der Erinnerung. Haus der Kunst, Prestel Verlag München (1997)
15. Fleischmann, M., Strauss, W.: Images of the body in the house of illusion. In: Sommerer, C., et al. (eds.) Art@Science, pp. 133–147. Springer, New York (1998)
16. Kurzweil, R.: The Age of Intelligent Machines, p. 328. MIT Press, Cambridge (1990)

6

Media Facades as Architectural Interfaces

Laurent Mignonneau and Christa Sommerer

Interface Cultures, Institute for Media, University of Art and Industrial Design
Sonnensteinstrasse 11-13, 4040 Linz, Austria
laurent.mignonneau@ufg.ac.at
christa.sommerer@ufg.ac.at
http://www.interface.ufg.ac.at

Abstract. Artists who are creating interactive systems have begun to look for new display formats for their interactive art systems. Modern architecture allows building facades to become membranes for the display of interactive digital content.

6.1 Introduction

For 15 years we have produced interactive artworks that have focused on development of human-computer interfaces for art. These projects have been mostly shown indoors in areas such as galleries, museums and other public places. Interfaces that we have developed include living plants. These were used for a work called "Interactive Plant Growing" [1] in 1992, and for "Eau de Jardin" at the House-of-Shiseido in Tokyo in 2004, where users could interact with virtual plants on a screen through touching real plants. An example of this human to plant interaction is shown in Figure 6.1. We also used light as the interface in an interactive artwork called "Phototropy" from 1995. Here users could interact and play with virtual insects by feeding or killing them using a flash light [2]. In 1999 we developed a multi-modal, multi-touch interactive web browsing system called "Riding the Net" [3]. Users could generate images streamed from the Internet based on keywords recognized from their conversations and they could also touch these images on a multi-touch window. This principle was later extended to a full room, called "The Living Room" [4]. This had 4 large interactive multi-touch walls, each of 6x5 meters. Users could touch and interact with keywords projected onto the interactive walls, and thus browse the Internet using their body gestures. The keywords were derived from users conversations detected via a speech recognition system. An overview of our various interactive installations is given in literature [5].

These interactive systems as well as systems of colleague artists [6-10] have been designed for projection walls or screen settings. As display technology has become more affordable and projection quality has become more powerful even in daylight situations, artists and designers have brought their artistic skill into public space. They have started to use large buildings as display surfaces. We will give a brief overview of some artistic media facades and then introduce our interactive media facade called "Wissensgewächs". This was developed in 2007 for Braunschweig in Germany.

C. Sommerer et al. (Eds.): The Art & Sci. of Interface & Inter. Des., SCI 141, pp. 93–104, 2008.
springerlink.com © Springer-Verlag Berlin Heidelberg 2008

Fig. 6.1. "Eau de Jardin" interactive installation at the House-of-Shiseido in Tokyo © 2004, Christa Sommerer and Laurent Mignonneau

6.2 Interactive Media Facades

One of the earliest interactive facades was built by Jean Nouvel for the Institute du Monde Arabe in Paris in 1988. Here mechanical elements, similar to zoom lenses, were arranged in geometric patterns and their opening and closing was controlled using the intensity of sunlight [11].

Another early example of media facade is the "Tower of Winds" [12] by the Japanese architect Toyo Ito. This was built in Yokohama Japan in 1986. The tower includes an air conditioning cylinder surrounded by circular neon lights that can be switched on when the ventilators in the building are activated. This produces a beautiful light pattern at night. Ito writes: "The Tower of Winds which I built a few years ago in front of Yokohama Station, Japan, embodied most efficiently the design of the winds. The tower is characterized in that it is installed admidst the neon lit downtown rather than in a museum. Although the tower which winks light similarly to other advertising neon lights is less spectacular, it is said to give an impression that the air around the tower is filtered and purified. That may be so because what I intended was not to cause a substance to emit light in the air but to make the air itself converted into the light." [13]

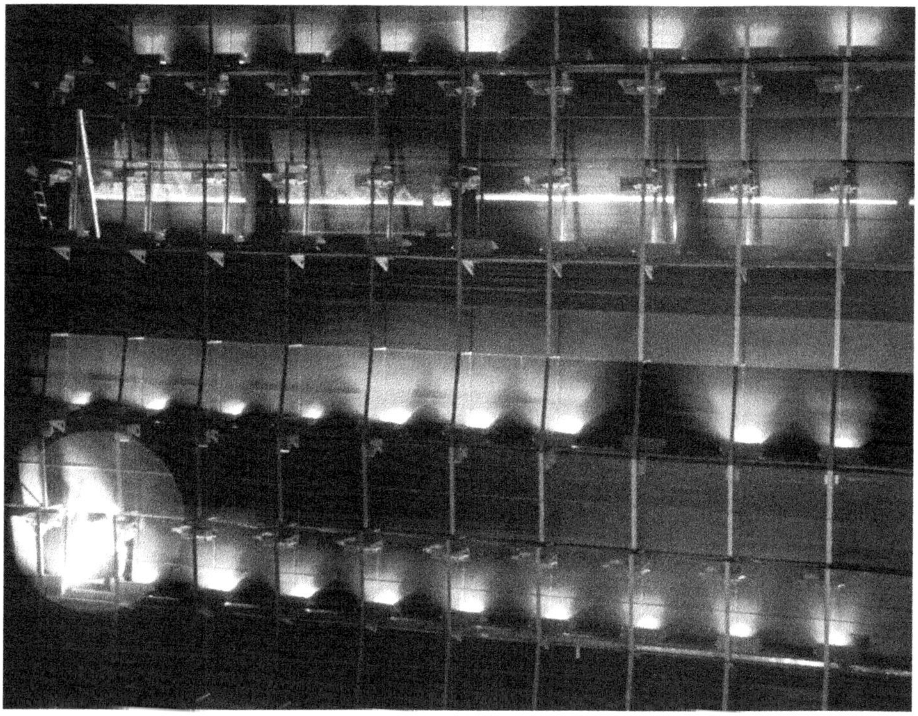

Fig. 6.2. Christian Möller's "Zeilgallery" interactive facade in Frankfurt 1992

Another beautiful example of an early interactive media facade is the "Zeilgallery" [14] done by the German artist Christian Möller, and built in 1992 in Frankfurt. Here the facade of a shopping mall in the city center was covered with blue and yellow neon lights. These changed their pattern depending on the weather conditions. That is by using the wind and temperature data. The so called "Zeilgallery" facade is shown in Figure 6.2.

During the 1998 Ars Electronica Festival in Linz Austria, Stadtwerkstadt which is an Austrian media art initiative, used the lit up windows of a large office building as pixels to create a large display. The project was called "ClickScape98" [15]. Participants could send simple grid designs through the Internet, and these images would then be transformed into a display of window lights inside the building which were switched On and Off. A result of one of these designs is shown in Figure 6.3.

Very similar to this "ClickScape98" project is the "Blinkenlights" facade by the German media art association Chaos Computer Club. This project was developed for a large building in Alexanderplatz in Berlin 2001. As in the "ClickScape98" facade, here blinking neon lights inside the windows were used as pixels which could be switched on and off, to create a large display. In a similar way to "ClickScape98" participants could send designs through the Internet or use their mobile phones to send messages [16].

Fig. 6.3. Stadtwerkstadt's "ClickScape98" in Linz 1998. Foto: Norbert Artner.

In 2001 the company Realities:United created "BIX" an interactive light- and media installation for the Kunsthaus Graz in Austria. A matrix of 930 fluorescent lamps was integrated into the acrylic glass facade of a biomorphic building structure. Each lamp's brightness could be controlled individually. This allowed the display of images, short films and animations at a speed of 20 frames per second. As a result the surface of the building was transformed into a giant low resolution computer display [17].

In 2003 German artists and researchers Monika Fleischmann and Wolfgang Strauss developed a project they called "Energie-Passagen", an interactive facade project for the Literaturhaus in Munich. Here passersby could see the daily flow of media news in the form of audio-visual data which was projected onto a large screen. It is described in Chapter 5 in this book [18].

In 2005 the Dutch artist and architect group called NOX designed a special tower, called the "D-Tower," a piece of architecture and sculpture which displays emotions by changing light patterns inside a sculptural tower. Inhabitants of the Dutch city Doetinchem, which is where the "D-Tower" is installed, could complete questionnaires that deal with daily emotions such as hate, love, happiness and fear. The tower then showed the emotions by the use of the four colors (red, green, blue and yellow). Neon lamps were used to illuminate the tower [19].

The facades described so far have mostly relied on neon lamps for illumination. An advance in display technology is the facade designed by UN Studio for the Korean Galleria Department Store in Seoul [20], 2004. Here the shopping mall's exterior was covered with 4330 separate opaque glass discs. Each functioned as a pixel which was lit from behind. The resulting surface used the whole building as a large display. Light patterns and some text could be displayed on the surface of the facade.

In addition to using neon lights or glass elements lit from behind, LEDs have also been used as display elements for media facades. In 2006 the French luxury cosmetics brand Chanel opened new head quarters in the Tokyo Ginza district [21]. The American architect Peter Marino covered the whole 56-meter high facade with a curtain wall which contained 700,000 white LEDs. The resulting surface could display prerecorded videos in black and white and in grayscale which promoted the Chanel brand and it also showed some abstract patterns.

LED based media facades have become popular in the beginning of 2000. The costs for large-scale facades using this technique are still exorbitantly high. An artistic project using LED color modulation of red, green, yellow, blue and white lights on a large building facade was designed by the renowned artist James Turrell for the Takarazuka University of Art and Design Osaka in Japan [22]. Here there is no apparent interaction with the environment.

In 2005 the Austrian architect Michael Shamiyeh proposed to use the outer facade of the "Wissensturm" as a display surface for displaying images of a bookshelf inside the public library [23]. This caption would have shown the books the citizens of Linz have recently read. The display would have been done by using a 3x4 meters movable LED video display, mounted on rails on the outer surface of the building.

In 2006, a hotel in Budapest, Hungary, known as the Lánchíd 19 Design Hotel commissioned two groups of graphic artists, photographers and fashion designers, to produce an interactive facade similar to that developed by Christian Möller in 1996. A

Fig. 6.4. Christian Möller's interactive facade on the Osaki Building Tokyo 2006

moveable accordion-like glass facade was developed, where the movement of the glass lamellas painted with tiny graphics, followed the speed of the Danube river and the general strength of the wind, measured through meteo-sensors installed on the top of the hotel [24].

In 2006 the Uniqua Insurance Company built new headquarters in Vienna which was constructed with a double layered glass facade which had LED modules installed [25]. These modules were arranged as a helix that could display simple graphics and imagery designed by the artists Mader, Stublic and Wiermann. Like many of the other systems mentioned earlier, the display looked best at dawn.

In the years 2006 and 2007 media artists who were experienced in interactive installations received commissions to produce large scale interactive facades. A well-known American media artist, Bill Seaman, designed a partly interactive video content for the "T-Tower" of SK Telecom building in Seoul Korea. This uses LCD screens to display video images on a large ribbon of screens which surround the building [27].

In 2006, Christian Möller, who already developed the "Zeilgallery" facade in Frankfurt Germany in 1992, designed an interactive facade project for a public building in Osaki Japan. Participants could see their video images projected onto the facades within large LCD screens (Figure 6.4).

6.3 "Wissensgewächs" Interactive Media Facade

In 2006 we, the authors of this paper, received a commission from the city of Braunschweig Germany to develop a special interactive facade for an exchange library in the center of Braunschweig. This was to form part of the "City of Science 2007" initiative that promoted Braunschweig's importance as city of research. To make the "City of Science" more widely known to the general public and to promote science and research in general, the city council decided to built a glass house in the very city center. This is next to the Cathedral. This glass house had a café and an open library where citizens could borrow books in exchange for their own books. The intention is to promote the concept of knowledge exchange. To encourage citicens to participate in this open book exchange library, we were asked to design a special interactive facade. This was to encourage to enter the library.

6.3.1 "Wissensgewächs" Concept

We developed the concept of a visually growing facade that would reflect the visitors attention and interest and entice them into the building. We called this interactive facade "Wissensgewächs", or the growth of knowledge. The intention is to arouse the passerby's attention and to come physically closer. The reward is a series of increasingly complex images on the screens. Images of virtual plants would grow on the screens as visitors approached. Each time the participants moved, the images would be different.

6.3.2 System Set-up

To do this work we collaborated with the architects KSP Engel und Zimmermann GmbH Braunschweig and Planning office of Assmann and Partners, Braunschweig.

Fig. 6.5. The glass house exchange library with integrated interactive facade "Wissensgewächs" © 2007, Christa Sommerer and Laurent Mignonneau. City Center Braunschweig, Adjacent the Cathedral.

The glass house was conceived as a cube of 6 x 6 x 6 meter, with one entrance door on the west side. It was made in stainless steel with 25 glass elements on each side. The height was about 1,3 meters from the ground. There are 5 large LCD screens, 1.05 x 0.75 meters each. These were integrated around the building, so as to create a ribbon of 16 screens, see Figure 6.5. The west side had only one screen due to entrance door.

Specially developed aluminum profiles and the integrated distance sensors were developed by Laurent Mignonneau. They were built into the framing of the glass elements. The sensors could detect the presence of and the distance to passersby within a range of 0.1 to 1.5 meters. A sketch of the relation between sensors and the screens is shown in Figure 6.6.

6.3.3 User Interaction

As passersby walk near the glass facade, the sensors detect their presence and distance to the screens. A special "Wissensgewächs" plant growth software developed by Mignonneau and Sommerer then interprets the distance and proximity of the passersby as growth parameters for the virtual plants on each screen. For example,

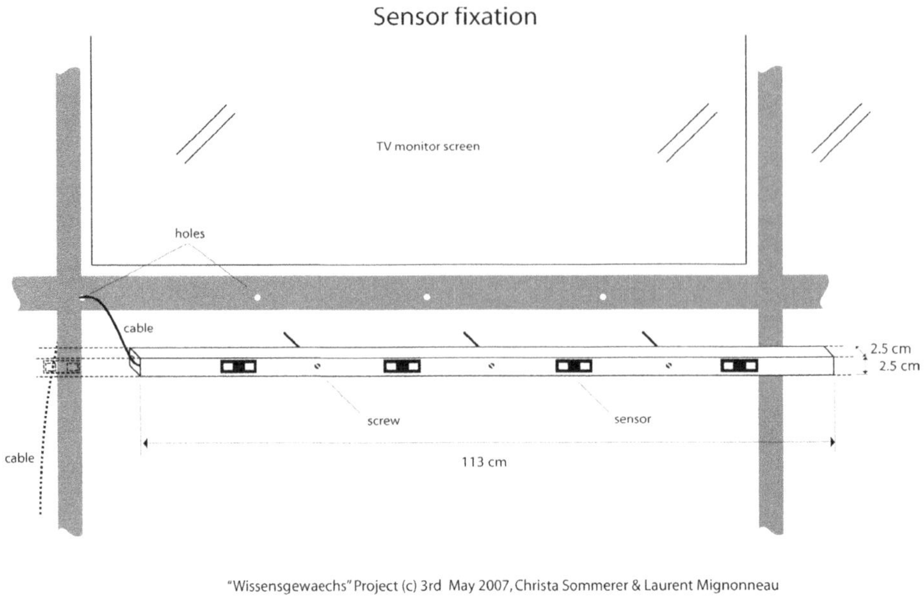

Fig. 6.6. Sensor profiles with integrated distance sensors above the LCD screen. These were developed for "Wissensgewächs" in 2007 by Laurent Mignonneau.

Fig. 6.7. A user interacting with the interactive facade of "Wissensgewächs"

standing still would create one type of virtual plant on the screen and walking slowly would make this plant follow the participant on several of the screens. To trigger the growth of new types of plants users could walk away from the screen and then come back to it at a distance of less than 1.5 metres. Altogether this user generated data would create new types of plants on the screen resulting in a perpetually growing image scenery, as shown in Figures 6.7 and 6.8.

Fig. 6.8. Screenshots of the "Wissensgewächs" plant growth triggered by the presence and movements of a participant

When several participants interacted with the facade, the multiple user interaction would become immediately visible as more plants begin to engulf the whole building. The amount of growth on the screens is thus directly linked to the degree of interaction of the participants. In return this rewarded the participants with fuller and better images. This creates a positive feedback as other passersby would become curious and would also being attracted to participate. When there is no interaction the previously generated plants slowly fade towards the background of the screens. They leave traces and shadows so as to suggest that there has been some interaction. An example image of the pattern resulting from several users interacting with the facade is shown in Figure 6.9.

Fig. 6.9. An example of multiple user interaction with the interactive facade "Wissensgewächs" © 2007, Christa Sommerer and Laurent Mignonneau at the city center of Braunschweig

6.3.4 User Feedback and Evaluation

The general acceptance of the installation by the visitors was very high. The system was designed in such a way that it does not require any previous knowledge to participate. A simple approach and presence were enough to become part of the interactive installation. Designing an easy and intuitive access proved to be the right decision as the passerby usually had only a short attention span and needed to be attracted quickly to go towards the facade. The audience profile was very diverse. The audience ranged from experts who are used to media and computers, to people with no previous contact to art or media. Being situated nearby the main shopping area of Braunschweig which is one of the busiest places in the city, the fluctuation in the number of passersby was high. This allowed us to use the feedback between the passerby and the participants so as to attract greater interaction.

After observing the participants for several days we concluded that the goal of attracting the public' attention has been achieved. We also received favorable feedback from the City of Braunschweig's marketing department. They reported that public engagement with the glass house, the artwork and the exchange library was high. The "Wissensgewächs" interactive facade installation was installed in Braunschweig for about a year and was very successful. As a result it will be moved to the university campus of the Technical University Braunschweig during 2008.

6.4 Conclusion

To conclude, the concept of creating an interactive facade in a busy public space is a successful means of not only engaging the media art experts in the creation of interactive content. It also allows the communication of art to a large and diversified audience who would not normally engage in such art forms [28].

Coming back to our initial observation that interactive art and interface design are now mature enough to be used in the public space on a large scale within cities and public buildings. We also report that the interactive facade "Wissensgewächs" opened a new space for artistic interventions. These bring art closer to the public and help eliminate borders between art and life by directly involving the audience into an interactive work of art [29].

Acknowledgements

The authors thank Dr. Anja Hesse of the City of Braunschweig for the commission of this work. Thanks are also due to the Cultural Department of the City of Braunschweig, Wolfgang Laczny, for enabling this commission and Dr. Müller-Pietralla from the Volkswagenstiftung for creating the contact. We also thank the architects KSP Engel und Zimmermann GmbH Braunschweig and the planning office of Assmann and Partners. We are grateful to the University of Art and Industrial Design Linz for supporting this project.

References

[1] Sommerer, C., Mignonneau, L.: Interactive Plant Growing. In: Siggraph 1993 Visual Proceedings, pp. 164–165. ACM Siggraph, New York (1993)
[2] Sommerer, C., Mignonneau, L.: Anthroposcope & Phototropy. In: ARTEC 1995 - The 4th international Biennale, Nagoya City Art Museum (1995)
[3] Mignonneau, L., Sommerer, C., Lopez-Gulliver, Jones, S.: Riding the Net: a Novel, Intuitive and Entertaining Tool to Browse the Internet. In: SCI 2001- 5th World Multiconference on Systemics, Cybernetics and Informatics Conference Proceedings, pp. 57–63. International Institute of Informatics and Systemics, Orland, Florida (2001)
[4] Lopez-Gulliver, R., Sommerer, C., Mignonneau, L.: Interfacing the Web: Multi-modal and Immersive Interaction with the Internet. In: VSMM 2002 Proceedings of the Eight International Conference on Virtual Systems and MultiMedia, Gyeongju, Korea, pp. 753–764 (2002)
[5] Mignonneau, L., Sommerer, C.: Designing Emotional, Metaphoric, Natural and Intuitive Interfaces for Interactive Art, Edutainment and Mobile Communications. COMPUTERS & GRAPHICS: An International Journal of Systems & Applications in Computer Graphic, 837–851 (2005)
[6] Sakane, I.: The Interaction 1995, Dialogue with Media Art – Introduction to Interactive Installations. Gifu Prefecture Government, Gifu (1995)
[7] Sakane, I.: The Interaction 1997, Toward the Expansion of Media Art. Gifu Prefecture Government, Gifu (1997)

[8] Dinkla, S.: Pioniere Interaktiver Kunst von 1970 bis heute. Cantz Verlag, Ostfildern (1997)
[9] Sakane, I.: The Interaction 1999 Expanding the Human Interface. World Forum for Media and Culture Committee, IAMAS, Gifu (1999)
[10] Gendolla, P., Schmitz, N.M., Schneider, I., Spangenberg, P.M.: Formen Interaktiver Medienkunst. Suhrkamp Verlag, Frankfurt/Main (2001)
[11] Sharp, D.: Twentieth Century Architecture: a Visual History, p. 394. Prentice-Hall, Englewood Cliffs (2003)
[12] Ito, T.: Arch+, Zeitschrift für Architecture und Städtebau, No. 111, p. 42 (1992)
[13] Ito, T.: Architecture in a Simulated City. In: Gerbel, K., Weibel, P. (eds.) Intelligent Environments. Ars Electronica 1994, vol. 1, pp. 84–91. PVS Verleger (1994)
[14] Moller, C.: Christian Moller: A Time and Place. Lars Müller Publishers (2004)
[15] Stadtwerkstadt: ClickScape 1998 Views of Linz. Clickable Public Space. In: Stocker, G., Schöpf, C. (eds.) Ars Electronica 1998, InfoWar – Information, Macht, Krieg, Part 1. Springer, Wien/New York (1998)
[16] "Blinkenlights" facade by Chaos Computer Club at: http://www.blinkenlights.de/ (accessed on December 21, 2007)
[17] "BIX" facade by Realities: United at: http://www.bix.at (accessed on December 21, 2007)
[18] Strauss, W., Fleischmann, M.: Interactivity as Media Reflection. In: Sommerer, C., Jain, L.C., Mignonneau, L. (eds.) The Art and Science of Interface and Interaction Design. Springer, Heidelberg (2008)
[19] Spuybroek, L.: Nox: Machining Architecture. Thames & Hudson (2004)
[20] Liao, A.: An Illuminated Skin designed by Arup Lighting and UN Studio transforms The Galleria in Seoul. In: Architectural Records (2005)
[21] Marino, P.: Chanel building facade in Tokyo (2004), http://www.petermarinoarchitect.com (accessed on December 21, 2007)
[22] James Turrell for the Takarazuka University of Art and Design Osaka, http://www.colorkinetics.com/showcase/installs/takarazuka/ (accessed on December 21, 2007)
[23] Shamiyeh, M.: http://www.baukultur.at/ (accessed on December 21, 2007)
[24] Lánchíd 19 Design Hotel, http://www.lanchid19hotel.hu/ (accessed on December 21, 2007)
[25] Mader, Stublic, Wiermann: http://www.webblick.de (accessed on December 21, 2007)
[26] Seaman, B.: The Thoughtbody Environment Interface interactive facade installation for SK Telecom, T-Tower. Seoul (2007), http://billseaman.com/
[27] Möller, C.: Nosy – video installation Osaki City, Tokyo, Japan (2006), http://www.christian-moeller.com (accessed on December 21, 2007)
[28] Sauter, J.: Das Vierte Format: Die Fassade als Mediale Haut der Architektur. In: Fleischmann, M., Reinhard, U. (eds.) Digitale Transformationen – Medienkunst als Schnittstelle zwischen Kunst, Wissenschaft, Wirtschaft und Gesellschaft, pp. 117–121. Whois Verlag, Heidelberg (2004)
[29] Weibel, P.: Es geht um Dinge von Öffentlichem Interesse. In: Fleischmann, M., Reinhard, U. (eds.) Digitale Transformationen – Medienkunst als Schnittstelle zwischen Kunst, Wissenschaft, Wirtschaft und Gesellschaft, pp. 45–51. Whois Verlag, Heidelberg (2004)

7

Interaction Design for Ubiquitous Content

Masa Inakage[1], Satoru Tokuhisa[2], Eri Watanabe[2], and Yu Uchida[3]

[1] Keio University, Faculty of Environment and Information Studies
5322 Endoh, Fujisawa, Kanagawa, Japan
inakage@sfc.keio.ac.jp
http://www.imgl.sfc.keio.ac.jp/English/index.html

[2] Keio University, Keio Research Institute at SFC
5322 Endoh, Fujisawa, Kanagawa, Japan
{dk,eri}@imgl.sfc.keio.ac.jp

[3] Keio University, Graduate School of Media and Governance
5322 Endoh, Fujisawa, Kanagawa, Japan
naruto@imgl.sfc.keio.ac.jp

Abstract. Ubiquitous Content is an emerging genre that uses everyday and everywhere media as a platform for creative content. This chapter covers key components for designing Ubiquitous Content to achieve emotional and entertaining experience.

7.1 Introduction

We are in the midst of the digital revolution that will transform our society from the Industrial Society to the Creative Society. In the Creative Society, personal and everywhere media will become important for interactive arts. Thus, the design for interaction of people, artifacts, and the environment contributes to the emotional and entertaining experience in everyday and everywhere media.

Ubiquitous Content is an emerging genre that uses everyday and everywhere media as a platform for creative content. Similar to the established content genres such as film, game, and music, Ubiquitous Content aims to move people's emotion through content embedded in artifacts and environment in daily life.

The vision and concept of Ambient Intelligence (or AmI in short) is built upon the environment that is responsive to the people's activities. AmI focuses on users and their experiences from natural interaction and context-aware systems.[9] Ubiquitous Content shares the vision for technology in the future society with AmI researches, designing memorable experience is the goal of creativity in Ubiquitous Content.

Ubiquitous Content is experienced in everyday life activities. For example, furniture or kitchenware may be turned into an Ubiquitous Content by embedding smart technology to add features that can respond to the context, as shown in Figure 7.1. People will be entertained for a very short moment such as 10 seconds when one is interacting with household goods. This snack sized entertaining experience in daily life opens up a new landscapes for media artists and designers as a genre for their creativities.

Creating a magic is the key to designing emotional experience. The magic becomes effective only when the trick is not obvious. In interactive media art and

C. Sommerer et al. (Eds.): The Art & Sci. of Interface & Inter. Des., SCI 141, pp. 105–113, 2008.
springerlink.com © Springer-Verlag Berlin Heidelberg 2008

design, technology creates the trick. Therefore, it is important to hide technology so that the user is not aware of the technology when experiencing the work.

This chapter guides you through some key components for designing Ubiquitous Content to achieve emotional and entertaining experience: synesthetic interaction, body interaction, place interaction and ambient interaction. With invisible computers embedded in real space, interaction design with faceless interfaces such as voice, gesture and person's presence becomes mainstream.[5]

Fig. 7.1. Kitchenware such as juice blender becomes affective and smart artifacts by embedding Ubiquitous Content platform "xtel"

7.2 Synesthetic Interaction

Synesthesia is a field of study regarding our senses. Many researches were made in Neuropsychology. The concept of synesthesia has been used by many artists. For example, a Russian composer Scriabin composed and performed a symphony that used an innovative keyboard instrument called "Luce" that Scriabin himself devised, in addition to the traditional musical instruments of orchestra. "Luce" transmits colored lights instead of sound when keys are played. The audience enjoyed the symphony by both listening to the musical performance and watching a huge screen constantly changing its color.

Synesthetic interaction comes into play when content appeals to more than a single sense. If interaction design is effectively engages our five senses, it produces a deeper experience than an experience from a single sense. When people perceive information through different senses, the information is processed in our brain to merge as a unified stimulation. Therefore, synesthetic design should be investigated to be used effectively in Ubiquitous Content.

7.2.1 "SUIRIN"

SUIRIN (the content name comes from an ancient Japanese ball-shaped artifact called "ukidama" that is made of glass) is an interactive artwork of light and sound with tactile interaction, as shown in Figure 7.2.[6] Users can interact with SUIRIN by touching and fumbling the water in the container. Through this interaction, the ukidama, or glass balls, in the container generates sound through surround speakers. The sound itself sounds like an insect's chirping, which in the Japanese culture represents a pleasing and comfortable sound – a "sound scenery". In addition, from the fog that simmers out of the container with mysterious light, a first-hand experience of a magical real space is made possible through the content of SUIRIN. The blue-green color, cold water, and the chirpy sound collectively induce the feeling of coolness and relaxing atmosphere.

SUIRIN uses 4 microphones to pick up the sound of ukidama balls hitting the containner bowl. The surround sound is processed by sophisticated FFT sound processing to produce the cricket chirping sound.

Fig. 7.2. "Suirin" integrates visual, auditory, olfactory and tactile senses to experience the work. The interaction of hand and water with ukidama glass balls creates a mysterious and soothing environment.

7.3 Body Interaction

Tangible User Interface and its interaction with everyday physical artifacts and environments have successfully moved away from computing input devices such as keyboards and mouse.[2] To further move away, the body motion becomes the input in the notion of embodied interaction.[1] It is important to design embodied interaction so that natural body movements can be used for input.

7.3.1 "Clay Tone"

Clay Tone is a playful sound generation instrument that uses clay as the input for creating sound, as shown in Figure 7.3.[10] The user physically interacts with bodies of clay, where the shapes of freely deformable lumps, and the user's physical movements against them, can affect the manner in which sounds are generated from the system.

Fig. 7.3. "Clay Tone" detects the natural motion of clay playing action to generate sound

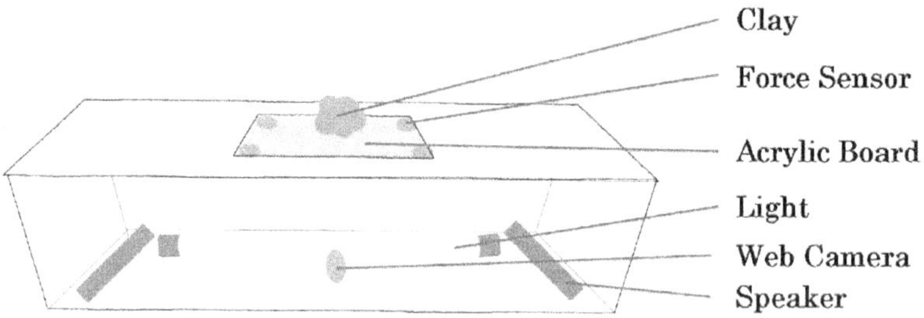

Fig. 7.4. System diagram of "Clay Tone"

By utilizing familiar materials as elements of manipulation, Clay Tone has achieved an intuitive body interaction. Expression with clay is enjoyable and visual. Users can easily share or appreciate an image of a sound with another person. Through changing sounds dependent on the form and softness of clay, the user can experience a fusion of touch, sight, and hearing.

Clay Tone senses color, shape, area and the force of kneading to control the mode of interaction, tone and pitch. Clay Tone uses pressure sensors and a web camera to sense the body interaction of clay kneading, as illustrated in Figure 7.4.

When the user plays with the green clay, the speed of kneading influences the sound. The faster the user squeezes, the higher the tone gets. Purple clay changes pitch. When the user spreads clay, pitch change. Big area of clay results in high pitch, and small clay area results in low pitch. The user can experience sound similar to the cry of a creature. Blue clay changes rhythms. When the user kneads clay, the user hears the rhythm patterns. The rhythm patterns depend on where the user kneads clay on the board and the shapes of clay.

7.4 Place Interaction

Humans are rooted in the physical world regardless of how advanced the digital technology becomes to realize many forms of metaverse activities. Interaction design needs to account for the connectivity between the physical and virtual worlds.[3] In Ubiquitous Content, the environment is used as an important component in designing the interactive work. Various sensors and pervasive computing will be embedded in the environment to have a smart, dynamic environment.[4] People, artifacts and the environment mutually influence each other to create a very complex inter-relationship.

Location-dependent interactive design can be accomplished by detecting the location of the user. The environment can therefore control what can be experienced in certain place. Localization and inter-connection of localized places create a complex interactive relationship between people, physical location, and digital content such as sound and digital video.

7.4.1 "Ototonari"

Ototonari is a pervasive game that allows the participants to create sound based on position, proximity and density of participants in real space using mobile ad hoc network as the direct collaboration among users sharing the same time and the same place, as shown in Figure 7.5.[7]

Fig. 7.5. In "Ototonari", sound can be left in a location and picked up by other people

Ototonari was exhibited at AICHI EXPO 2005, Japan using PDAs with wireless network capability. Each participant carries the PDA and walks around the park. Sound files are location-based so that unless the participant is within an area, the sound cannot be picked up to listen or save in the PDA. By walking around the park, participants will collect various sound elements that can be mixed in the PDA to create original music. It is also possible to leave sound in certain location to be picked up by other participants. Sound elements can also be shared among participants if they are nearby.

In Ototonari, location terminals are installed in each area in the park. The wireless network is set to reach only within its vicinity. Each location terminal possesses the sound elements that are left by the participants. PDA carried by the participants connects to the location terminal and other PDA carried by participants by peer-to-peer network. GPS technology was not used in this implementation.

7.5 Ambient Interaction

Interaction should be designed so that it feels invisible. Ambient interaction is a type of faceless interaction in which the natural movements of people such as gestures trigger the environment to respond. Ambient displays are used to actuate information through subtle changes in light, sound, and movement.[11] In ambient interaction, the concept of ambient display is extended to account for interaction design using natural movement of people in the environment.

If our environment becomes smart to dynamically change dependent on the context, we do not need any devices to actively interact with the environment. People always actuate sufficient information about one's emotional response such as eye movement, facial expression, voice, posture, and gesture. It is the responsibility of the environment to collect and analyze personal actuated information, similar to how we observe and understand a person. Reading the mood and adjusting the environment to match the atmosphere is the goal of ambient interaction. Entertaining flavor should be added as the hidden taste to vitalize the environment for a memorable experience.

7.5.1 "Kage no Sekai"

Interactive design work "Kage no Sekai" (Shadow World in Japanese) is designed around ambient interaction.[8] It allows the user to communicate with marvelous creatures through the shadow with a view that everyone has once felt in their childhood that something would be in the shadow, as shown in Figure 7.6. Although at first glance it looks like a regular wooden table, if one looks at the shadows on its surface one will notice the movement of mysterious life forms. When one tries to touch them by reaching the hand to the shadow area, they sense the presence and hide away. They do not emerge while human shadows are cast over the table, but the life forms hiding within a distant shadow are watching them. The ambient display of mysterious life forms entertains people nearby the table. The life forms may be treated as an adorable display, but ambient interaction occurs when one tries to reach over the shadow to touch.

In "Kage no Sekai", the technology is hidden below the table, as depicted in Figure 7.6. The wooden desk surface deceives the user that both a camera as sensor and a projector as actuator are hidden to detect the shadow of the user reaching the hand toward the shadow area and project the characters in the shadow area. The semi-transparent wooden surface creates the magic. Video input from the web camera is used for realtime detection and tracking of shadow using interframe difference image processing technique. The life forms are displayed in the static shadow area. If the system detects a shadow that is approaching toward the static shadow area, the life forms immediately run away and disappear.

Fig. 7.6. "Kage no Sekai" uses ambient interaction design to interact with creatures appearing in the shadow on the table

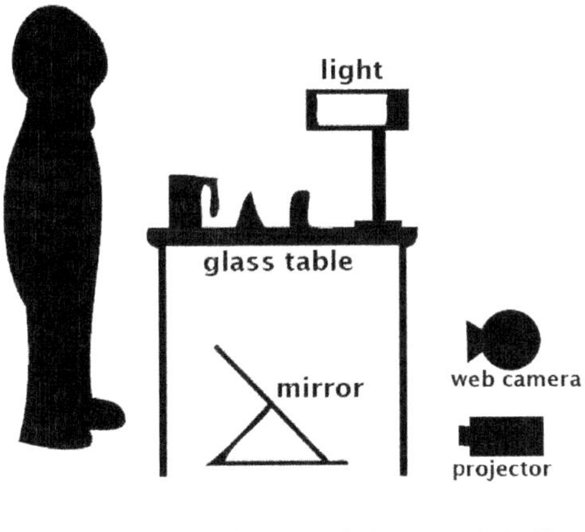

Fig. 7.7. System diagram of "Kage no Sekai"

7.6 Conclusion

In this chapter, the concept of Ubiquitous Content is presented with example prototypes. This entertaining and intuitive interaction design approaches can be deployed in various everyday products such as kitchenware, furniture, and smart toys. The techniques may also be used for designing intelligent and affective environments and architectural design.

Acknowledgements

The authors would like to thank all the members of our laboratory for the design and implementation of interaction design works in this chapter. The research was granted by CREST, JST.

References

1. Dourish, P.: Where the Action Is: The Foundations of Embodied Interaction. MIT Press, Cambridge (2001)
2. Ishii, H., Ullmer, B.: Tangible bits: towards seamless interfaces between people, bits and atoms. In: Proceedings of the SIGCHI conference on Human factors in computing systems, pp. 234–241 (1997)
3. McCullough, M.: Digital Ground. MIT Press, Cambridge (2004)
4. Philips: Vision of the Future. V+K Publishing, Blaricum (1996)
5. Saffer, D.: Designing for Interaction. New Riders, Berkeley (2007)
6. Tokuhisa, S., Inakage, M.: SUIRIN. In: ACM SIGGRAPH 2005, Emerging technologies, Article No. 22 (2005)
7. Tokuhisa, S., Niwa, Y., Iguchi, K., Okubo, S., Nezu, T., Inakage, M.: Ototonari: mobile ad hoc pervasive game that develops a regional difference. In: ACM International Conference Proceeding Series, vol. 223, pp. 155–162 (2006)
8. Uchida, Y., Naito, M., Hirayama, S., Inakage, M.: Interaction based on function of a table in real world with "Kage no Sekai". In: ACM International Conference Proceeding Series, vol. 274, pp. 184–184 (2007)
9. Vasilakos, A., Pedrycz, W. (eds.): Ambient Intelligence, Wireless Networking, and Ubiquitous Computing. Artech House, Boston (2006)
10. Watanabe, E., Hanzawa, Y., Inakage, M.: Clay tone: a music system using clay for user interaction. In: ACM SIGGRAPH 2007 posters, Article No. 156 (2007)
11. Wisneski, C., Ishii, H., Bahley, A., Gorbet, M., Braver, S., Ullmer, B., Yarin, P.: Ambient Displays: Turning Architectural Space into an Interface between People and Digital Information. In: Proceedings of the First International Workshop on Cooperative Buildings, Integrating Information, Organization, and Architecture, pp. 22–32. Springer, Heidelberg (1998)

Bibliography

Aarts, E., Marzano, S.: The New Everyday. 010 Publishers, Rotterdam, The Netherlands (2003)
Benford, S., Magerkurth, C., Ljungstrand, P.: Bridging the Physical and Digital in Pervasive Gaming. Communication of the ACM 48(3), 54–57 (2005)

Brumitt, B.L., Meyers, B., Krumm, J., Kern, A., Shafer, S.: EasyLiving: Technologies for Intelligent Environments. In: 2nd Intl. Symposium Handheld and Ubiquitous Computing, pp. 12–27 (2000)
Cheok, A.D., et al.: Human Pacman: A mobile, Wide-area Entertainment System based on Physical, Social and Ubiquitous Computing. Personal And Ubiquitous Computing 8(2), 71–81 (2004)
Conrad, E.: Towards Embodied Spatial Interaction. In: CHI 2006, Workshop (2006), http://studio416.cfa.cmu.edu/CHI06workshop_AboutFace/Final_Papers/Conrad.pdf
Greenfield, A.: Everyware. New Riders, Indianapolis, Indiana (2006)
Ishii, H., et al.: AmbientROOM: Integrating Ambient Media with Architectural Space. In: Proceedings of Human Factors in Computing Systems (CHI 1998), pp. 173–174 (1998)
Jordan, P.W.: Designing Pleasurable Products. Taylor & Francis, London (2000)
Marks, L.E.: The Unity of the Senses. Academic Press, New York (1978)
Morville, P.: Ambient Findability. O'Reilly, Sebastopol (2005)
Norman, D.A.: The Design of Everyday Things. Basic Books, New York (1988)
Norman, D.A.: Emotional Design. Basic Books, New York (2004)
Norman, D.A.: The Design of Future Things. Basic Books, New York (2007)
O'Sullivan, D., Igoe, T.: Physical Computing. Thomson, Boston (2004)
Prante, T., et al.: Hello.Wall - Beyond Ambient Displays. In: Adjunct Proceedings of Ubi-com 2003, pp. 277–278 (2003)
Saffer, D.: Designing for Interaction. New Riders, Berkeley (2007)
Streitz, N.A., et al.: Designing Smart Artifacts for Smart Environments. IEEE Computer, 41–49 (2005)
Vasilakos, A., Pedrycz, W. (eds.): Ambient Intelligence, Wireless Networking, and Ubiquitous Computing. Artech House, Boston (2006)
Weiser, M.: The Computer for the Twenty-first Century, pp. 94–104. Scientific American (1991)
Weiser, M., Brown, J.S.: The Coming Age of Calm Technology. Xerox PARC (October 5, 1996), http://www.johnseelybrown.com/calmtech.pdf

8

Ubiquitous Gaming Interaction: Engaging Play Anywhere

Tiago Martins[1,2], Nuno Correia[1], Christa Sommerer[2], and Laurent Mignonneau[2]

[1] CITI-DI, Faculty of Sciences and Technology, New University of Lisbon, Quinta da Torre, 2829-516 Caparica, Portugal
{tms.martins,nmc}@di.fct.unl.pt
http://img.di.fct.unl.pt

[2] Interface Culture Lab, University of Art and Design Linz, Sonnensteinstraße 11-13, 4040 Linz, Austria
{Tiago.Martins,Christa.Sommerer,Laurent.Mignonneau}@ufg.ac.at
http://www.interface.ufg.ac.at

Abstract. Computer games have become a cultural phenomenon. Increasingly popular, they have made their way from PC's and consoles into the many small and sleek devices we carry with us daily, partaking of their mobility. When computers blend themselves into our daily environment, games will also move to colonize the new, ubiquitous paradigm of computation and may in this way manifest potentially through anything, anywhere. Ubiquitous games are already a popular topic of research, with focus on both technical and social aspects. Moreover they present a new and exciting challenge in interface design, as new forms of interaction are called for by a new paradigm. In this article we will see how current approaches to ubiquitous games and interaction break free of the traditional screens, mice and key pads and instead lead to reclaiming physical spaces, repurposing real objects and rediscovering social bonds – all of this while having fun.

8.1 Introduction

In their myriad manifestations, games have been part of human life ever since the dawn of times, having largely contributed to the establishment of social and cultural conventions [11]. They prompt us to do what we humans, by nature, excel at doing: analyzing a given scenario and acting upon it towards a desired outcome, through a process that involves much pattern analysis [15]. By cleverly disguising themselves as ritualistic pass-time activities, they turn what might otherwise be a tedious and repetitive process into a source of skill practice, entertainment and even frequent show-off. Over the past few decades a new kind of games has emerged. This new breed, powered by the very technologies that are shaping human culture in an unprecedented manner, reaches us through the very tools we use to work and communicate – computers.

Computers have come a very long way in a very short time. Once room-sized electromechanical machines used strictly for scientific purposes, they have rapidly evolved into palm-sized multimedia appliances with the processing, storage and communication capabilities orders of magnitude beyond their ancestors'. They are all around us. Flamboyantly displayed or hiding in plain sight, they connect us, inform us, guide us and even entertain us.

C. Sommerer et al. (Eds.): The Art & Sci. of Interface & Inter. Des., SCI 141, pp. 115–130, 2008.
springerlink.com © Springer-Verlag Berlin Heidelberg 2008

Computer games are not something that came along with faster processors and multimedia capabilities, however. They have been used both to entertain and to demonstrate technical capabilities ever since the very dawn of the computer age[1]. Or even before, if we consider the fact that in 1959 William Higinbotham had the unusual idea to set up a sort of virtual tennis game on an oscilloscope to illustrate its capabilities to the general public. Two years later Steve Russell, a young programmer from MIT, would code *Spacewar!* the first computer game ever known, on a punched-tape-driven mainframe computer. In 1967 videogames were already making their leap to black & white television sets, thanks to Ralph Bauer. And in 1971 Nolan Bushnell's arcade version of *Spacewar!* made video games leap even further onto public spaces [8].

Computer games have since then become a serious business. The still relatively young video game industry has grown to a point where ten million dollars may be considered a cheap development cost for a blockbuster game of the latest generation [14]. Some game genres are now taken seriously indeed not only by the industry and thousands (or even more) of affectionate players but also by educators and psychologists concerned with their level of violence and (perceived) lack of educational value.

Curiously enough, each other single device aside from PC's and game consoles that houses a small processor, display and a set of buttons is now also bound to have games played on it. Ever more commonly we may find games for graphing calculators, mobile phones, PDA's and MP3 players – not disregarding the existence of devices specifically designed as portable game consoles. We are to realize that wherever computational power becomes available, so do games seem compelled to follow [20].

Still, computer games made for either stationary or portable devices are very much constrained to take place in simulated, virtual worlds. On one hand they are securely contained inside a digital dimension that acts as magic circle of play, where they can trivially be identified as abstracted simulations and have little to no consequence on the physical world. On the other, they require that attention be diverted from reality, with considerable cognitive loads that can lead the player into forsaking significant social bonds – something that seems to have become commonplace [28].

Personal computers have long been bound to very much the same interfaces they had back in the 80's. Computer games were the driving force behind the first low-end sound and graphics cards and such input devices as joysticks and control pads, still popular nowadays with the addition of haptic feedback [8]. But they have equally struggled for long to make the best of these. As computer games break out of the screen and pervade reality, taking advantage of ubiquitous computing and context-awareness, completely new approaches to gaming interfaces are not only possible but also necessary. Consider the fact that personal devices already (and increasingly) sport built-in cameras, wireless network access, positioning systems and even motion sensors. As such devices become more and more common, with better data acquisition, processing and communication capabilities, they present themselves as one major channel for the next generation of computer games [23].

[1] It is interesting to point out that game playing is one of the preferred ways of testing AI algorithms.

8.2 The Ubiquitous Future of Games

The vision behind Ubiquitous Computation (Ubicomp, for short) goes far deeper than simply taking networked computers, shrinking them even more – something that is granted by the ongoing applicability of Moore's law[2] – and scattering them throughout our daily environment. Such a simplistic view could possibly mean a future made up of uncalled-for bursts of overwhelming information anywhere, at anytime, further disrupting users and adding up to their (already considerable) cognitive load. According to the serendipitous vision of Mark Weiser, Ubicomp must ensure a future of calm technology [29]. Ubiquity of computers both empowers and requires them to be context-aware, so that they may better blend into the environment and don't disturb us at every step with unnecessary information [9]. When capable to determine a meaningful context regarding users, the physical environment and even information from a previous context, computers should be able to better cater to our needs. Achieving this, however, is not only a matter of developing better databases, more tolerant networking structures and cleverer artificial intelligence, although these are indeed important steps. Nor is it only about what to display on a screen – as it ultimately was with PC's and many mobile devices – as screens may not even be necessary in many cases. In the vision of calm technology, computers truly blend in with the environment and provide information at the periphery of our attention, until called for in providing more detail. And so the conception of adequate human-computer interfaces is of the utmost importance [12].

8.2.1 Form, Nomenclature and Main Challenges

We have seen that ever since the very beginning have computers been increasingly tied to games, even before we would even dream of hearing music or watching videos on them. So it would only be natural that researchers and artists alike would eventually start investigating what benefits and issues arise from bringing games to the paradigm of Ubicomp.

Benford *et al.* point out [2] several forms in which these activities may occur: location-based reinterpretations of classic games; social interaction games; touring artistic games; educational games in physical spaces; and commercial pervasive games. Although such a discussion is not within our present scope, it is still noteworthy to point out that, in such a context, the terms "pervasive" and "ubiquitous" have largely been used interchangeably although, by definition, they do not share the same meaning [20]. Besides making this distinction, noted ubiquitous games researcher and designer Jane McGonigal alternatively suggests three categories, each with its own set of research aims, artistic intentions, and social impacts: ubicomp games, which focus on bringing computer games to this new paradigm; pervasive games, which focus on playful, thrilling disruption of reality, with maximum social weight; and ubiquitous games, which focus in rediscovering affordances of everyday objects and spaces through their use in games, made possible by Ubicomp technology. Throughout this text we will not make special distinction in nomenclature or categories however; our focus is on the interfaces and technologies used.

[2] In 1965 Gordon E. Moore, a co-founder of Intel, observed empirically that the number of transistors on a chip doubles about every two years.

The main challenges faced when developing ubiquitous games are enumerated by Benford *et al* [2]. The use of wireless networking and sensor technologies – whether they are location-sensing, computer vision or gesture recognition – introduces the challenge of dealing with uncertainty. The fact that ubiquitous activities involve hybrid architectures, with different devices and network configurations, is a second challenge. Ubiquitous games also span very different domains (including aspects of both traditional and computer games) and so require a careful approach in design, choosing what to represent as digital or physical and how to do it. Configuration is a further challenge, in the sense that ubiquitous games are many times inevitably tied to a physical location or otherwise constrained by complicated set up of technological resources. The last challenge is orchestrating the game in real-time, as it may not be possible to automate the whole process or to provide fail-safe gameplay, given the mingling of physical and digital. We shall now see how ubiquitous games have tackled these problems, and also what principles and technologies were used in their making.

8.2.2 Current State of the Art

Pirates!, by the Interactive Institute, was an early approach, based on a concern in regaining social aspects of traditional gaming, where players meet face-to-face [3]. In it each player takes on the role of commander of a pirate ship, sailing from island to island in search of treasure. Output is presented to the player by a handheld computer, which is also used for input when exploring an island. To sail between islands the players have to move in a real physical space, risking confrontation with other players in the form of open sea battles. To handle the question of positioning, radio-frequency beacons were used as proximity sensors, mounted on the mobile device and at the physical locations for the virtual islands. This enabled each device to know when it was close to an island or another player. Instead of relying on an absolute positioning system, the physical interaction functions through relative proximity detection. The fact that the stationary beacons don't need to be connected to the network is an advantage in ease of set-up.

The artist group Blast Theory together with the Mixed Reality Lab of the University of Nottingham devised *Can You See Me Now*, best described as a mix of pervasive game and performance [7]. In this game we may find online players pitted against performers on the streets. Online players access the game through a web-based interface where they control an avatar in a 3D model of a real location. Performers (known as "runners") roam the streets of the real location trying to catch the online players' avatars by referring to their position on a mobile computer. Runners can read the text messages exchanged between online players; these in turn are able to listen to the runners' radio communications – which is an interesting way to perceive their humanness. The Global Positioning System[3] (GPS) is used for tracking the runners. Its use proved unreliable, especially in narrow streets, due to lack of precision and occasional difficulty in obtaining fix data[4]. Runners were quick to deal with this

[3] A GPS receiver obtains its absolute geographical position through a triangulation method based on a mesh of 24 satellites orbiting the Earth. This method requires line of sight to at least three (four?) such satellites and thus cannot be used indoors.

[4] Nowadays, however, most commercially available GPS devices provide error-compensation for navigating in urban canyons.

Fig. 8.1. Can You See Me Now. Photo © Blast Theory.

as part of their tactics however, by realizing where on the city streets it was most likely to be experienced.

Uncle Roy All Around You is a later approach based on the same principle of combining street and on-line play, in a mix of ubiquitous game with live performance [1]. Street players (not performers) search for the character Uncle Roy by following clues on a handheld computer. Online players follow the progress of street players and can choose whether or not to help them. Ultimately a street player is led to Uncle Roy's office and afterwards invited to enter a white limousine. In the end both the online and the street player are asked if they are willing to make a commitment to support one another during the following year. This game performance explores the relationship of trust towards strangers and makes use of actors and real physical assets, such as a limousine, an office and a phone booth. Wireless communication is based on GPRS. Given the previous problems with GPS in *Can You See Me Now*, players must now explicitly indicate their geographical position through a map interface on the mobile device. An orchestrating room was set up to coordinate the game's physical resources and to identify players in need of assistance.

Positioning issues may be tackled with a mixed approach. *InStory*, by the Interactive Multimedia Group, is a location-based interactive narrative [5]. Players are offered an interactive experience which combines elements of narrative and gaming and takes strong advantage of the physical beauty and historical richness of a real location. As the player roams the physical space, interactive narrative segments and story-driven challenges are presented on a mobile device. By using the mobile device's image capture facility, she can also contribute with annotated images that may be repurposed in new narratives. Geographical position is obtained by three methods, chosen automatically. GPS positioning is used as a preferential method for outdoor navigation, as newer GPS antennas provide better results in areas with moderate foliage canopy. As a support for GPS and to provide indoors positioning a commercial

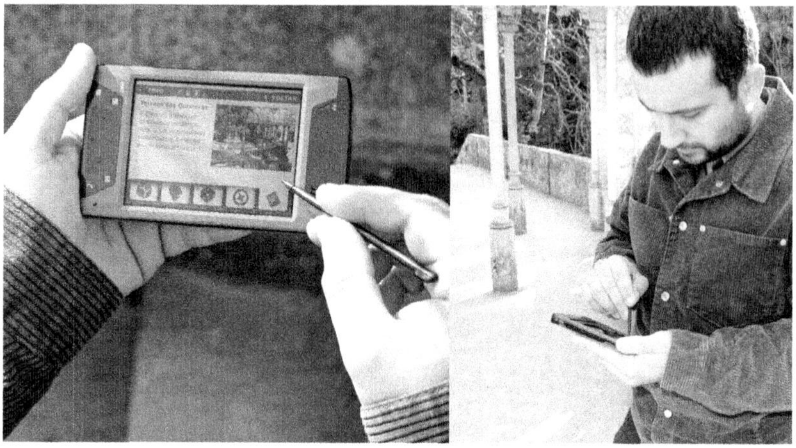

Fig. 8.2. InStory - Regaleira. Photos by Rute Frias, © Interactive Multimedia Group.

system based on Wi-Fi triangulation is used. As disadvantages, this technology requires a considerable number of access points to be deployed (at least 3 must be "visible" at any given location) and rigorous calibration. In the case that both automatic systems are unavailable, the user can explicitly indicate her position by tapping a map interface on the mobile device.

Not all approaches require setting up a specific location as a game board, though. *Road Rager* is designed for brief encounters on the road, making use of a gesture-driven tangible interface [4]. Players in separate vehicles become duelling wizards casting spells, shocking or throwing magical sludge at each other. In this approach by the Interactive Institute, the tangible interface is not only used as an intuitive metaphor for activating the different attacks but also allows players to better divide their attention between interface and real world while playing the game. This interface is attached to a PDA with wireless LAN capabilities, which uses this technology to detect and communicate with other players in the vicinity. The game is intended for passengers who are not engaged in the maneuvering of the car and designed not to disturb those who are. *Road Rager* needs no hardware to be deployed in any given space (since it uses *ad-hoc* networking between players' devices) and turns a potentially boring everyday situation into an exciting gaming experience.

RFID (Radio Frequency Identification) is an increasingly popular technology that allows automatic identification of objects, places and even animals marked with RFID transponders (commonly known as "tags"). These come in diverse varieties, diverging in size, form, read distance and the ability to store additional information. Lancaster University's Infolab21 research centre devised the first game in the world to use mobile phones equipped with RFID readers, dubbed *Pac-Lan* [23]. This game is a remapping of the classic *Pac-Man* game onto the pedestrian area of the Lancaster University campus. One player takes on the role of the Pac-Lan character and is chased by other four players taking on the role of ghosts. Placed throughout the play area are big yellow pills marked with RFID tags. By using a mobile phone equipped with an RFID reader to read these tags, players explicitly indicate their position, communicated to the game server via GPRS by the mobile phone. To effectively "kill" the main character, a ghost must approach her and use the mobile to read a tag placed on the back of her costume. The reverse applies whenever the main character collects a "power-pill". The Pac-Lan character is lead to indicate his position by his objective of "collecting" as many pills as possible; the ghost characters must regularly indicate their position or become unable to "kill" the Pac-Lan character. It is interesting to note how RFID may be used as a positioning method. It does not provide a completely ubiquitous positioning mechanism, but when placed on stationary objects (such as a lamp post) it can prove very accurate. Passive RFID tags do not need a power source[5], are quite unobtrusive and very easy to set-up.

The usage of multiple interfaces is something that can be explored in order to confer role-play aspects to a pervasive game and promote team work. *Epidemic Menace* presents such an approach [17]. Players form teams of "medical experts" in order to catch a menacing virus and devise a cure for it. Part of the team is stationed in a game room, where an overview of the game status is available, as well as a communications station to guide team members and a virus analysis station. Guided by the crew at the control room, mobile players can pinpoint a virus' location and behaviour by listening to a sound interface, or don a mobile Augmented Reality (AR) interface that allows them to actually see the viruses and attack them with a spray tool. Communication with team members is achieved through a mobile assistant, which can also be used to capture viruses. Positioning is achieved through GPS. The game makes brief use of live acting and an AIBO robot that aids the players at some point in the story. Also, a virus is something that can't be seen with the naked eye and so it is an "enemy" that dispenses physical representation.

One of the most extreme approaches to pervasive gaming was staged in Stockholm. The pervasive live action role-play game *Prosopopeia Bardo 2: Momentum* [27] spanned five weeks, merging with the real lives of thirty players. In this follow-up to a previous instalment [13], the players role-played themselves as possessed by ghosts of revolutionaries, being allowed to switch from the ghost personality to their own and *vice-versa*. As game events could take place anyplace at anytime, it is convenient that the player can "be herself" to deal with real-life situations that might arise. The game made extensive use of indexical propping – meaning that real places and objects represent themselves – leading players to establish a game contextualization for locations and objects that would otherwise be ignored by outsiders. Also several interesting interfaces were developed as "techno-magical" artefacts for use in the game. Stationed on an abandoned nuclear reactor core (which served as the players' headquarters) were several installations that allowed the players to receive messages from spirits, among them an EVP[6] device that required its user to be strapped down on a steel bed. Mobile interfaces included an RFID-reading glove, GPS receiver and mobile phones, which were used in searching and reclaiming "magical nodes" to the world beyond. This project has achieved meaningful insight on the choice between seamless and "seamful" interfaces, as well as on the complex effort of orchestration and ethical aspects of merging a game and real life.

---

[5] A passive RFID tag is powered by induction from the reader's antenna.

[6] EVP stands for Electronic Voice Phenomenon, a term referring seemingly perceptible voices (said by some to be of paranormal origin) captured in audio recordings.

Fig. 8.3. Epidemic Menace. Photo by Nina Dautzenberg, © Sony Netservices.

As we may see, the challenges of ubiquitous gaming are being overcome by inventive use of technology in providing gameplay and their potential as social activities being brought to light. In what interfaces are concerned, all these current examples of success make use of diverse positioning techniques, mobile devices, RFID, tangibles and even live actors. Other popular techniques also include use of fiducial markers (such as *Semacode* [25]), gesture input and AI-driven virtual characters.

8.3 Reinventing Game Interfaces

In the previous section we have seen how the notion of ubiquitous gaming is already being approached and how pioneering approaches provide hands-on experience to identify common issues and the insight to overcome them. We may argue nonetheless that the vast majority of interaction in such gaming activities is either based on the player's location, mediated by a mobile device or both. But there are reasons for this. Mobile devices already possess the computational power to comfortably provide network communication, audiovisual output and handle input both from the user and from peripheral devices such as GPS antennas, RFID readers and cameras. In some case these peripheral devices aren't exactly peripheral anymore, as they are embedded into the mobile computer. Another advantage of using mobile devices is that they have become accessible to a large audience. Still, the Ubicomp vision of coupling bits and atoms (as put by Ishii and Ullmer [12]) pushes us onwards, to explore a new and vast territory, extending well beyond small screens, head-mounted AR displays, physical displacement and tapping with a stylus. Whether they are research projects, interactive art installations, design concepts or commercial products, we may find promising interfaces for tomorrow's games, already today.

8.3.1 Unconventional Interfaces

Location-based games take advantage, to lesser or greater extent, of a real physical space. But to what extent could that be possible? If the concern is sheer size, than look no further. An entire building facade was used by the Chaos Computer Club as a monochrome pixel display in *Blinkenlights* [22]. Computer-controlled lamps were set up behind each of the one hundred and forty four windows of the Haus des Lehrers building in Berlin to form a gigantic 18x8 display. Passers-by could interact via their mobile phones to display text messages, custom animations and play the arcade classic *Pong*. The setup effort is tremendous, though. Similar approaches are possible, requiring less setup, such as Douglas Stanley's *Invaders!* [26]. In the original version, projected onto the facade of the Marseille World Trade Centre, passers-by were able to control a rehash of the classic *Space Invaders* by means of any light sources they might have, such as torches or car headlights[7]. Aside from providing an easy to use interface based on common devices, *Invaders!* was also a highly mobile installation, powered by laptop batteries and an adapter for a car's cigarette lighter plug.

Fig. 8.4. Blinkenlights. Photo by Thomas Fiedler - www.blinkenlights.de

Trees and plants, which are very common in public spaces, can serve as both input and output of information. In *Interactive Plant Growing*, Christa Sommerer and Laurent Mignonneau present us with the opportunity to create a virtual garden sporting over twenty five program-based plants by simply approaching and touching real ones [21]. The installation consists of a screen where the virtual garden is grown and five real plants each on a pedestal. One of the plants – the cactus – resets the virtual scene. Interaction is achieved by measuring, on the very plants themselves, the difference in electrical potential between them and the visitor. Conversely, *PlantDisplay* is meant to present information (although coarsely) by controlling water and lighting

[7] In later versions of the same installation, gesture input is used instead.

Fig. 8.5. Interactive Plant Growing. © 1992/93 Christa Sommerer and Laurent Mignonneau.

provided to living plants [16]. This approach to information visualization by the Faculty of Environmental Information of Keio University is illustrated with two interesting examples: visualizing activity in human relationships based on the amount of communication (phone calls, e-mails) between two friends; and visualizing the general happiness of a community based on content from posts on a weblog.

Games of all kinds presuppose physical and/or mental activity and are undeniably a source of excitement. But the Interactive Institute's Smart Studio[8] subverts such notions with the game *Brainball* [10]. One of the novelties of this game lies in the fact that it uses the players' brain waves (and nothing more) as input. The electrical activity of each player's brain is captured by a headband with electrodes. The other novelty (and main focus of the concept) is that players do not compete through physical or mental activity, nor are they even supposed to be excited. A ball on a table between the players is moved by brainwaves that occur when the brain is in a relaxed state; and so the game turns out to be all about relaxation. Besides the peculiarity and non-invasiveness of the interface itself, we find a mapping that is reversed in relation to other computer games – where excitement and rapid button press are a must. Such an approach, besides the inherent novelty, suggests that the notion of calm technology may analogously be applied in game interfaces, contrarily to what Weiser may have thought [29].

Other kinds of bio-signals can also be used as input, such as heart rate. *Dragons* is a collaborative project between Active Ingredient, the Lansdown Centre for Electronic Arts and the London Sport Institute, mingling health concerns and ubiquitous gaming [6]. The player isn't required to go anywhere in specific but as she moves a virtual landscape (reminiscent of old isometric games) is created based on her

[8] Smart Studio was transformed in 2006 into the Art & Technology Program.

Fig. 8.6. Brainball. Photo by Tobias Sjödin, © Interactive Instititute.

displacement and optimality of the heart rate. The landscape blossoms and becomes lush if the heart rate is at optimal levels; otherwise it becomes impoverished or darkly wooded. Position is obtained through GPS and heart rate by means of an inexpensive heart rate sensor. The fact that the game is location-independent and that it can easily be integrated into daily activity are also interesting concepts to explore in ubiquitous gaming.

Oddly enough, even the act of micturition has been used as a game interface for a game in a public space. MIT Media Lab's Dan Maynes-Aminzade and Hayes Raffle take on social aspects of urinating in *You're In Control*[9] [19]. The installation consists of a urinal whose surface detects the position of impact of a stream of liquid and a screen mounted at eye-level. Users are welcome to test their accuracy at a variant of the typical *Whack-a-Mole* game. Aside from the novel combination of play with the act of micturition (and other benefits pointed out its authors), *You're In Control* includes a custom controller for everyone to use – which is rather convenient, to say the least. This strong candidate for "strangest controller ever used in a game" is a simple, non-electronic device consisting of a rubber nozzle connected to two water reservoirs.

Ultimately, interfacing with a computer may be as simple as touching, stepping or sitting, without resort to wireless networking, computer vision or carefully placed sensors. *RedTacton* is a technology that uses weak electrical currents on the surface of the human body as safe, high speed transmission path [24]. This means that a person can literally act as a two-way data channel between devices. For instance, a user can have a personal device in permanent contact with her skin and simultaneously connect

[9] Sounding similar to "Urine Control".

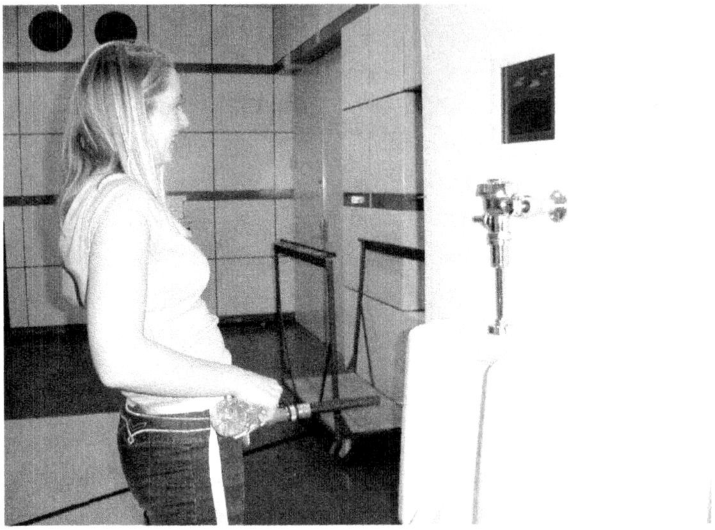

Fig. 8.7. You're In Control. Photo ©2003 Hayes Solos Raffle and Dan Maynes-Aminzade.

it to other computers simply by touching real objects or surfaces[10]. Such an innovative technology may easily be adopted in ubiquitous gaming not only as a network channel but also as the means to provide touch-based interaction, allowing many-to-many relationships between players, objects and surfaces.

We have seen that most approaches in bringing play to physical spaces and objects require a lot of setup. However, it is possible to turn everyday objects into game interfaces without great intervention. Haiyan Zhang's concept *Control Freaks* aims at opportunistic play by easily turning everyday objects into gaming interfaces [30]. Basically acting as a motion- and sound-sensitive "parasite", a *Control Freak* can be clamped to an object – anything from chairs to people – thus augmenting it into a game controller.

8.3.2 A Wearable Approach

Sharing a similar vision, the *Gauntlet* is a concept for wearable game controller, being developed at the Interface Culture Lab [18]. The idea behind this concept is allowing ubiquitous, gesture-driven interaction with real objects and spaces without the need for complicated setup. Instead of deploying sensors in real spaces and objects, these are placed on the user, embedded in an arm piece. Wearability is thus a key issue, as the arm piece must be comfortable enough to wear for long periods of time and as unobtrusive as possible to movement of the forearm and hand. A three-axis accelerometer acquires data for gesture recognition. A digital compass providing data to

[10] Support for simultaneous connections is achieved by a Carrier Sense Multiple Access With Collision Detection (CSMA/CD) algorithm. Similarly to what happens in Wi-Fi networks, this algorithm ensures that only one device is using the communications channel at any given time.

Fig. 8.8. Noon – A Secret Told By Objects. Photo by Thomas Wagner. © Interface Culture.

determine where the user is pointing (provided her exact location is known). An RFID reader allows automatic recognition of tagged objects and can also be used as a positioning method by reading a tag placed in a stationary object. Data from these sensors is read by a microcontroller and sent via Bluetooth to a mobile device, responsible for data processing, network communication and connecting to other devices, such as a GPS antenna. By donning the arm piece, players are free to interact through gesture, proximity, touch and manipulation of real objects. Furthermore, an object can be added to the game in real time by placing an RFID tag on it.

To demonstrate and conduct user testing of interactions using the Gauntlet, a single player game has been devised. "Noon – A Secret Told by Objects" is a game that uses objects and physical interaction to convey an interactive narrative. The player's objective is to unveil the truth behind a tragic fire by retrieving memories from salvaged objects, using the Gauntlet and a magical Tome – a PDA encased in a book, responsible for managing game logic and providing audiovisual output. To trigger memories the player must touch, hold or – in some cases – manipulate the objects. Different levels of interaction may yield different information. As not all memories are happy ones, some manifest primarily as aggressive poltergeist. In the presence of these harmful spirits the Tome produces audio noise and the displayed media appears burnt, becoming progressively garbled and unreliable. By using the Gauntlet to scan her periphery the player can determine the source of the attacks (the direction in which the noise is more intense) and deflect them. One of the objects, the clock, is very special for it allows shifting the time of all object's memories. By touching the object and then pointing at one of several candles surrounding the table the player can access key moments in the hours preceding the tragedy. As the player navigates memories in any order she may desire, a narrative is pieced together leading to the immediate cause of the fire and the hidden reasons behind it.

8.4 Conclusion

Computers will fade more and more out of our sight, and the less we see them the more we will experience their embodiment as pure interaction. Later generations will thus come to perceive computers in a very different way than we do now. It is ours the task to pioneer intuitive, engaging and aesthetically pleasing manifestations for this new form of computation, all the while overcoming issues of personal agency, privacy and sanity [9]. While based on the principles and technology of Ubicomp, ubiquitous games share these requirements, and provide a fertile ground in which artists and researchers can explore new forms of interaction and obtain important insights that may be used in other applications. Ubiquitous games may yet become an early paragon of the vision behind Ubicomp.

We have presented state-of-the art ubiquitous, pervasive and mixed-reality games exploring hands-on these new forms of play and highlighted concepts that suggest the world as a place filled with opportunities for human-computer interaction. It should be clearer now that the next breed of computer-mediated games hold much more than the promise of breaking out of the screen. In blending themselves with the environment they have the power of leading players to: rediscover physical spaces and take part in the continuity of their history; perceive new affordances (or rediscover old ones) in objects by using them as game elements; engage in play in anyplace, at anytime, through potentially anything; keep a desired amount of focus on the real world while playing; form new social relationships or strengthen existing ones; and exercise their body and mind.

All in all, such approaches to ubiquitous forms of play and to the way we experience them empower us to conceive a new breed of games, manifesting closer to us and savoured in ways never thought of before. Play will continue to be part of human culture, yes; but in an age of calm technology it can be a veritably engaging, entertaining and meaningful cultural experience.

References

1. Benford, S., et al.: Uncle Roy All Around You: Implicating the City in a Location-Based Performance. In: Proceedings of the International Conference on Advances in Computer Entertainment Technology 2004 (ACE 2004), Singapore (2004)
2. Benford, S., Magerkurth, C., Ljungstrand, P.: Bridging the physical and digital in pervasive gaming. Communications of the ACM 48(3), 54–57 (2005)
3. Björk, S., Falk, J., Hansson, R., Ljungstrand, P.: Pirates! Using the Physical World as a Game Board. In: Proc. of Interact 2001, Tokyo, Japan (2001)
4. Brunnberg, L., Juhlin, O.: Keep your eyes on the road and your finger on the trigger - Designing for mixed focus of attention in a mobile game for brief encounters. In: Proc. of the 4th International Conference on Pervasive Computing. Dublin, Ireland (2006)

5. Correia, N., et al.: InStory: A System for Mobile Information Access, Storytelling and Gaming Activities in Physical Spaces. In: Proceedings of the ACM SIGCHI International Conference on Advances in Computer Entertainment Technology 2005 (ACE 2005), Valencia, Spain (2005)
6. Davis, S.B., et al.: Ere be Dragons: Heart and Health. In: Proceedings of The Third International Workshop on Pervasive Gaming Applications (PerGames 2006), Brussels, Belgium (2006)
7. Flintham, M., et al.: Where on-line meets on the streets: experiences with mobile mixed reality games. In: Proceedings of the SIGCHI conference on Human factors in computing systems (CHI 2003), Florida, USA (2003)
8. Forster, W.: The Encyclopedia of Game Machines. GAMEplan (2005)
9. Greenfield, A.: Everyware: The Dawning Age of Ubiquitous Computing. New Riders Press, Berkeley (2006)
10. Hjelm, S.I.: Research + design: the making of Brainball. Interactions 10(1), 26–34 (2003)
11. Huizinga, J.: Homo Ludens. Beacon Press, Boston (1971)
12. Ishii, H., Ullmer, B.: Tangible bits: towards seamless interfaces between people, bits and atoms. In: Proceedings of the SIGCHI conference on Human factors in computing systems (CHI 1997), Atlanta, USA (1997)
13. Jonsson, S., Montola, M., Waern, A., Ericsson, M.: Prosopopeia: Experiences from a Pervasive Larp. In: Proceedings of ACM SIGCHI International Conference on Advances in Computer Entertainment Technology 2006 (ACE 2006), Hollywood (2006)
14. Kohler, C.: Interview: Epic's Mark Rein (2007), `http://blog.wired.com/games/2007/02/interview_epics.html` (retrieved March 12, 2007)
15. Koster, R.: A Theory of Fun for Game Design. Paraglyph Press, Scottsdale (2006)
16. Kuribayashi, S., Wakita, A.: PlantDisplay: Turning Houseplants into Ambient Display. In: Proceedings of the ACM SIGCHI International Conference on Advances in Computer Entertainment Technology 2006 (ACE 2006), Hollywood (2006)
17. Lindt, I., et al.: Combining Multiple Gaming Interfaces in Epidemic Menace. Experience Report. In: Proceedings of Conferences on Human Factors in Computing Systems 2006 (CHI 2006), Montréal, Québec, Canada (2006)
18. Martins, T., Heidecker, C., Sommerer, C., Correia, N.: Designing a Wearable Interface for Pervasive Games. In: Proceedings of the Fourth International Symposium on Pervasive Gaming Applications (PerGames 2007), Salzburg, Austria (2007)
19. Maynes-Aminzade, D., Raffle, H.S.: You're In Control: A Urinary User Interface. In: CHI 2003 Extended Abstracts on Human Factors in Computing Systems (2003)
20. McGonigal, J.: This Might Be a Game: Ubiquitous Play and Performance at the Turn of the 21st Century. PhD Thesis, University of California, Berkeley (2007), `http://www avantgame.com/dissertation.htm` (retrieved May 1, 2007)
21. Mignonneau, C., Sommerer, C.: Designing Interfaces for Interactive Artworks. In: Proceedings of the Fourth International Conference on Knowledge-Based Intelligent Engineering Systems and Allied Technologies (2000)
22. Project Blinkenlights Website, `http://www.blinkenlights.de/`
23. Rashid, O., et al.: PAC-LAN: mixed-reality gaming with RFID-enabled mobile phones. In: Computers in Entertainment (CIE), vol. 4(4), ACM Press, New York (2006)
24. RedTacton Website, `http://www.redtacton.com/`
25. Semacode Website, http://semacode.org/
26. Stanley, D.: Invaders! (2001), `http://www.abstractmachine.net/blog/invaders/` (retrieved April 10, 2007)

27. Stenros, J., Montola, M., Waern, A., Jonsson, S.: Momentum Evaluation Report. IPerG Deliverable D11.8 (2007)
28. Turkle, S.: Life on the Screen: Identity in the Age of the Internet. Simon & Schuster, New York (1995)
29. Weiser, M., Brown, J.S.: The Coming Age of Calm Technology. Xerox PARC (1996)
30. Zhang, H.: Control Freaks – Transforming everyday objects into opportunities for play (2006), `http://failedrobot.com/thesis/` (retrieved January 18, 2007)

9

Fashionable Technology – The Next Generation of Wearables

Sabine Seymour and Laura Beloff

Moondial, Fashionable Technology, Schottenfeldgasse 50/10, 1070 Wien, Austria
sabine.seymour@moondial.com, off@saunalahti.fi
www.moondial.com, www.realitydisfunction.org

Abstract. Wearable technologies in a mobile, networked environment will take the interface into the real world both literally and metaphorically, as our bodies become the interface, mediated through handheld, wearable, or embedded devices. Mobile, wearable media can be traced back to history f.e. with hidden cameras and wristwatches. By augmenting the physical through digital, wearable computers are constructing a new entity with its own specificities. A research in electronic textiles integrates technology into textile and thus allows it to become a dynamic interface. All wearables, objects or garments, must become more than mere mediators of perception. The future lies in a human-centered integration of man and machine.

Keywords: wearable technologies, wearable media, wearable computers, fashionable technologies, electronic textiles.

9.1 Introduction

Wearable technologies in a mobile, networked environment will take the interface into the real world both literally and metaphorically, as our bodies become the interface, mediated through handheld, wearable, or embedded devices.

We have invented glasses to augment our vision and wristwatches to better manage our time. Recently we have developed mobile phones to better manage our lives and our social networks. These devices became fashion items in themselves. The integration of these functionalities in garments and textiles is a current trend and requires a great understanding of fashion and its many facets. The technological character of such integration is a further element that requires attention to the body and the materials used. A new generation of wearables is developing. It shows that technology and fashion are not as distant as it might first seem. Even the thread-up and thread-down of the weaving process corresponds to the binary logic of 0 and 1 of a computer circuitry.

9.2 Wearables

The first known battery-run, mobile and wearable computer was built for predicting gambling results. In his article "The Invention of the First Wearable Computer" Edward O. Thorp writes about his collaboration with Claude Shannon in designing and constructing a roulette-predicting device, which is said to be the first wearable

C. Sommerer et al. (Eds.): The Art & Sci. of Interface & Inter. Des., SCI 141, pp. 131–140, 2008.
springerlink.com © Springer-Verlag Berlin Heidelberg 2008

computer. The wearable version of the computer was complete and operational in June of 1961. It used twelve transistors and was about a size of a cigarette pack. The device was hidden in a shoe. In 1961 it was shortly tested in a casino in Las Vegas by its creators and approved successful. The endeavor was kept secret until 1966, when Thorp finally announced the roulette system publicly. Based on results from this experiment a small group of scientists built in the early 1970s an operational wearable computer for the same purpose –predicting roulette- using the next generation of hardware and technology. [15] This version is known as The Eudaemon Shoe. By 1978 the computer was working and tested in Las Vegas with an average profit of 44% for every dollar.

9.2.1 Steve Mann's Definition of Wearables

Cyborgs are what is normally thought of when mentioning wearables suggesting to the terminology defined by Steve Mann: "A wearable computer is a computer that is subsumed into the personal space of the user, controlled by the user, and has both operational and interactional constancy, i.e. is always on and always accessible" [6].

Steve Mann is undoubtedly one of the main figures in the field of wearable computers. Starting in turn of the 1970s and 1980s by building a wearable "photographer's assistant", he has developed a series of wearable systems to the present day with features like body mounted cameras, lighting equipment and head mounted displays. One of his recent projects is EyeTap. He writes about the development: "Our wearable computer reality mediators have evolved from headsets of the 1970s, to EyeTaps with optics outside the glasses in the 1980s, to EyeTaps with the optics built inside the glasses in the 1990s to EyeTaps with mediation zones built into the frames, lens edges, or the cut lines of bifocal lenses in the year 2000 (e.g. exit pupil and associated optics concealed by the transition regions)." [7] This small size device affects the eye to become both a camera and a display. The authors write that: "These digital eye-glasses can help us remember better, through what is called a lifeglog (lifelong cyborglog) or 'glog, for short. A 'glog uses lifelong video capture to record what our eyes see over our entire lifetime." [7]

With a long history in wearable computers, Steve Mann has written extensively on the developments of his experiments and on wearable computers in general. He has defined three basic modes of operation as characteristic of wearable computers: constancy, augmentation and mediation. Constancy means that the device does not need to be turned on or opened up prior to use. Augmentation refers to the idea that computing is not the primary task, but the user can be doing something else simultaneously. This expected feature of wearable computers is often referred as "hands-free". Mediation means that the device allows the user to control inbound informational flow for solitude and outbound informational flow for privacy. [8]

The appearance of early wearable computers (or cyborgs using them) was clunky, had enormous power consumption, and used awkward interfaces in comparison to current design development. Human-centered factors like ergonomics, physiognomic, and feedback mechanisms had not yet been considered. This development is obvious, for example, when looking at the development of Steve Mann's experiments from the early 80s up to this day. Steve Mann is convinced that reality mediators, which are aimed at personal everyday use, such as hearing aids and personal eyeglasses must have an unobtrusive or hidden appearance, or be designed to be sleek and fashionable. [9]

These goals of making wearable technologies to become fashionable have influenced another strand of development within the field of wearable computers; the aim of making the computers disappear into our clothing and creating hardware which can be flexible, washable, small in size and light in weight. This area of research has grown hugely during the last years especially amongst practitioners of textile-, fashion-, and industrial design. They have been developing many new solutions for making electronic components from conductive yarns and textiles and embedding small components straight into the textiles.

9.2.2 Wearable Computers and Augementation

Ana Viseu makes a clear distinction between simulation and augmentation. She considers wearable computers to be augmentation. She writes: "... the digital is incorporated in the physical without necessarily showing its presence: Here, the human is no longer the measure of all things, the entity that machines are designed to imitate. The human body is viewed as being deficient, in need of improvement, of being enhanced with computing capabilities." She continues: "Rather than building self-contained machines, machines and humans are coupled together into a new hybrid actor." [17] She continues to argue that the discourse around the production of wearable technologies focuses mainly on "quantifiable, causal relationships, thus overlooking the fact that the augmentation of the physical through the digital does not result in physical plus digital, but in a new entity with its own specificities. An augmented human being has a distinct reality, and this raises new issues regarding the place of the human body and self in its relation to technological artefacts." [17] According to Viseu wearable computers are largely the product of ubiquitous computing and embedded computing, which offer a new way to interact with the environment, which is also expected to be responsive and communicative. [17] [18] Rather than considering wearable computers as tools, they have become more a technological companion.

9.2.3 Mobile Versus Wearable Media

Media archeologist Erkki Huhtamo has investigated the predecessors of mobile media devices. He writes: "Mobile media devices can be treated as "apparata", that are partly technological, partly psychological, partly cultural. The devices themselves incorporate certain "built-in" modes of usage that are then negotiated, perhaps embraced, perhaps contested, by the users themselves." [5] Huhtamo divides mobile media devices into three categories based on their usage; portable media, wearable media and vehicle-mounted media. The categories shift depending on how and where the device is used, for example if the iPod is listened to while running or while driving via a car-stereo.

Wearable media according to Huhtamo refers to things that become attached to the user's body in a more rigorous sense than portable things. Also a mobile phone in a pocket or attached to a belt and used with a hands-free headset would qualify as wearable according to Huhtamo. When one considers conventions of use and symbolic meanings as being equally important as functions, then the most common form of a wearable would be the wristwatch, and its predecessor the pocket-watch. The author's (Sabine Seymour) theory of the degree of body integration defines three stages of integration. The least integrated are handheld devices like cell phones with no actual body integration. Wearables have some type of integration and are mostly garments. Embedded devices are the most integrated and become more and more common in medical procedures.

An interesting historical example of wearable media - as well as it is an early form of surveillance or spy-technology, is the commercially successful device C.P. Strin's "Concealed Vest Camera" from the 1880s. It was a camera designed to be worn under the clothing that shot images through a buttonhole[1]. A recent examples using the same system is Burton's Espionage Jacket for snowboarding that was first released in 2005. Anna Novakov writes that during the 1870's there was "an enormous amount of interest in the detective camera, which were miniature cameras meant to be used by the flaneur to enter the city space and secretly take pictures. They were kept hidden in places such as the inside of a gentleman's hat. While walking down the street, he could be taking photographs, gathering evidence to be scientifically evaluated later. These cameras became extremely popular, and took many forms, often hidden in clothing." [11] Novakov sees a continuum from the hidden cameras of the 1870s to the contemporary surveillance technologies that have developed with the ability and preoccupation to watch and observe others all the time. The continuation of this desire can be seen in the recurrence of small size -hidden- cameras in mobile media devices like sunglasses with recording possibilities[2] or in mobile phones. Now all of us are equipped within our everyday life with the possibility to be detectives - whether we want it or not.

9.3 Fashionable Technology

Fashionable Technology is dealing with the next generation of wearables. The worlds of fashion, ubiquitous computing, design, science, and wearable technologies are rapidly converging. A new field is emerging that the author (Sabine Seymour) terms Fashionable Technology. With the appearance of transitive materials like electronic textiles, shape-memory alloys, and technologies like Skinplex novel interfaces for the body are created. Fashion as the art of creating for the body is thus an extremely important craft that needs further exploration in the field of Fashionable Technology.

Fashionable Technology looks at end users as fashionable beings, attentive to style, aesthetics, branding, and the expressive potential of wearable technologies. It refers to all design objects that are associated with aesthetics, wearability, or some degree of mobility. Thus including clothing and all types of accessories, jewellery, and gadgets from hearing aids to bags. In a stationary context Fashionable Technology represents objects that deal with dynamic interfaces, in particular referring to interior design objects. It includes experiments with enhanced materials, technologies, or electronic textiles and testing the effects of the interaction with the objects.

---

[1] http://www.boxcameras.com/stirnvestad.html (accessed December 17, 2006).
[2] http://www.spygadgets.com/undercover-cameras/sunglasses-camera.html (accessed December 31, 2007).

Fashionable Technology can easily be misinterpreted as dealing only with fashion and technology. The term focuses on 'able' and inspires to make technology fashionable and aesthetically pleasant. The extension of the human body through wearables confirms the need for many other disciplines as discussed in the conclusion.

9.3.1 Fashion and Technology

An example from haute couture, of using technology on the body today as a means to communicate an experience with technological enhancement is the Cypress fashion designer Hussein Chalayan. The airplane dress from the spring/summer collection 2000 is made from the same material used in aircraft construction and changes shape by remote control. It was one of the first examples to merge technology and fashion in a visible well-crafted way. Chalayan is of the few fashion designers who are experimenting with technology. Even more dynamic and expressive is Chalayan's spring/summer collection 2007 entitled Boing Boing with clothes beginning to twitch, move, change shape – reconfigure on their own. The garments morph through decades of fashion. Zippers close, hemlines rise. The perfect symbiosis of fashion and technology emerged on the runway. The dresses were driven electronically by controlled, geared motors concealed in hard cased containers on the buttocks of the models that also sheltered the microprocessors and battery pack.

Due to the current state of the technology bags or other larger accessories, are frequently used to hide the technical components needed for the envisioned functionality. Fashion Victims, a project by former students Agnelli Davide, Buzzini Dario, and Drori Tal of the Ivrea Interaction Design Institute proves just that. The idea of creating a collection of garments that react according to the surrounding mobile phone calls had to be modified into a bag to accommodate the physical functions derived from the technical components needed to express the concept, the idea. Still being a part of clothing the use of a bag was still accommodating the initial idea the project website notes: "The choice of this medium has to do with the already existing language and codes of apparel, that we use in order to communicate, self-express and position ourselves." The wearer can choose to switch off the bag's functionality, pull out the mechanism, and simple use it as a bag. Such empowerment is in particularly important for garments that a wearer cannot simply take off in public. The varying degrees of operational constancy and interactivity in particular refer to the fact that wearables or technologically enhanced garments do not necessarily have a computational component. These kinds of materials called 'smart textiles' feature scientific advances in materials research and include things such as better insulators or fabrics that resist stains. [2] Dynamic textiles surfaces for instance can be created without computational power by an intelligent use of ambient factors or materials. Thermochromic inks used in silk screening do not need any computational power to dynamically change its properties. A rise in temperature changes the color. Such change can be made by using conductive fibers that heat up or simply through the rise of body temperature depending on the temperature properties of the ink.

'Electronic textiles' refer to a material that incorporates capabilities for sensing (biometric or external), communication (usually wireless), power transmission, and

Fig. 9.1. Exchange-Dress. © Ammann+Siebrecht Fotografen AG.

interconnection technology to allow sensors other computational devices to be networked together within a fabric. [2] The various wearable technologies – sensors, micro-controllers, or embedded systems in general – are often the major component for Fashionable Technology projects to allow for functionalities that do require a computational calculation. The three garments of the thesis project Fashionation by Celine Studer[3] advised by the author (Sabine Seymour) rely on wearable technologies to recreate the stories of the garments physically. Humans communicate constantly with someone else and leave a mental mark. These impressions always fade away and are sometimes mixed with the marks other conversationalists leave.

The garment shown in Fig. 9.1 titled Exchange-Dress uses color as the metaphor to tell that story, to convey the concept. The visitors of the performance receive color-coded bracelets and depending in the proximity to the Exchange-Dress the dress reacts. The originally white Luminex fabric changes to the color communicated by the bracelet of the visitor. The color fades when the visitor parts again from the wearer of the Exchange-Dress. This interaction is only possible through the use of an IR transmitter and receiver, batteries, a micro-controller. Thus, simple computational power and the creation of a networked communication are necessary to allow for the story to unfold. The Exchange-Dress was designed by Marscha Jäggi und Françoise Adler.

Today wireless technologies are an integral part of space-human, human-human, and Body Area Network communication methods. 'Clothing That Arranges The Body' developed by Hannah Perner-Wilson in the class Fashionable Technology projects[4] conducted by the author (Sabine Seymour) deals with a different kind of body network. Hannah describes: "[Clothing That Arranges The Body] amounts to a

[3] Celine Studer who conducted the thesis is a student at Hyperwerk in Basel, Switzerland.

[4] The class runs in the department Interface Culture at the University of Applied Arts and Design in Linz, Austria.

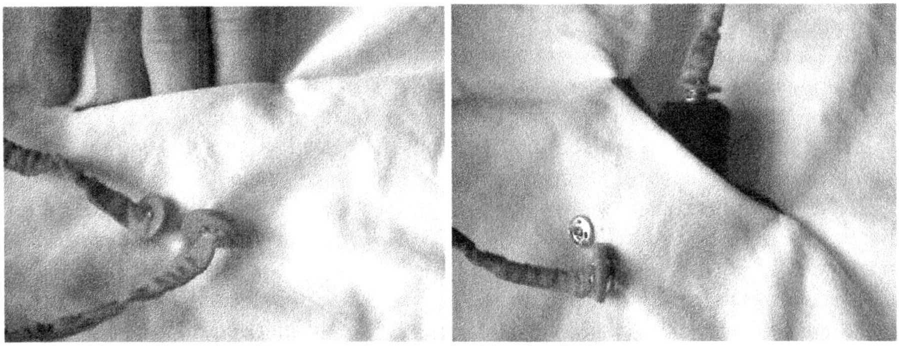

Fig. 9.2. Clothing That Arranges The Body. © Hannah Perner-Wilson.

way of connecting the flow of electricity between devices scattered in pockets around our body." Fig. 9.2 shows the cabling system that runs electricity from one pocket to another combining function and aesthetics.

The miniaturization of the components needed for cellular technologies, WIFI, RFIDs or Bluetooth allow for the integration in wearables, in objects worn on the body. If necessary, the computation can be transferred to stationary processors gathered in a physical space or portable devices like cell phones or PDAs. Such devices already have the ability of computation through programming languages and eliminate the necessity to pack all technical functionality into a wearable object or garment that thus might loose its practicality and beauty. Current technology and research have come a long way in making wearable computing a reality, with smaller, and even flexible batteries, sophisticated sensing mechanisms, and powerful, compact embedded systems.

9.3.2 A Reflection on Perception and Function in Fashion

In the chapter Lumbar Thoughts in Travels in Hyperreality, Umberto Eco describes his heightened awareness of his weight gain when wearing a pair of jeans that was just a bit too constraining. The slight discomfort changed the perception of himself but also his etiquette towards the opposite sex. "Not only did the garment impose a demeanor on me, by focusing my attention on demeanor, it obliged me to live towards the exterior world" [3]. In our contemporary world fashion is used as a synonym for style, dress, adornment, decoration, and clothing. "It could also be said that, while all clothing is an adornment, not all clothing is fashion, and that while all fashion is dress, not all dress I fashion, for the same reason" [12]. The research conducted by the author (Sabine Seymour) is based on the notion that garments are an immediate interface to the environment and thus a constant transmitter and receiver of messages, emotions, and experiences. The meaning of the surface of a garment "the cult of the body as an object of public display" [9] is once again apparent in our culture. T-shirts are a public display and allow making a statement to the exterior. The wide popularity seems to prove the human need for expression and communication. Besides LED signs or plaques, the artist Jenny Holzer also uses t-shirts as a medium for her statements like 'Abuse of Power Comes as no Surprise'. The use of wearable technology allows a dynamic reaction through the garment. Scent, sound, visuals, touch, and even taste are translated into immediate conversations with the environment or the body area network of the wearer. These real-time reactions are based on various inputs. They can be triggered through measuring vital signs of the wearer (passive input), interactions with physical switches on the garment (active input) or the existence of a specific substance in the environment.

Referring to his experience Eco notes: "Today they [jeans] are worn also for looks, but primarily they are very utilitarian" [3]. Fashionable Technology examines the intertwining of aesthetics and function in particular in regards to the use of technology. Marshall McLuhan describes clothing as our extended skin. "Clothing, as an extension of the skin, can be seen both as a heat-control mechanism and as a means of defining the self socially" [10]. Banard describes the functions of clothing or fashion as material or cultural. Material functions are protection, concealment, and attraction. Cultural functions are communication, individualistic expression, social or economic status, political symbol, or religious condition [1].

Through technology the functions of clothing can be enhanced and new ones are defined. Communication in Fashionable Technology refers not to a position in a cultural or social order but to the actual networked communication technologies. A contextual analysis reveals and defines such aspects of communication, and of aesthetics and functionality. Regardless of the tremendous communication aspects of fashion, it has rarely enjoyed a very good reputation in the past. "Despite its undeniable success as a social and commercial phenomenon, it remains the very exemplum of superficiality, frivolity and vanity" [16].

A view shared by the authors about the aspects of functionality within design and technology, J. Redström writes that: "Though phenomenological, sociological and other studies have challenged and expanded our understanding of technology, practice still seems to be dominated by an instrumental perspective. Central to our understanding of technology still lies notions of use, the idea that technology is the means for achieving certain ends, often by amplifying the power of our actions." [13] He thinks that it is problematic for the design processes that we have a strong tendency to describe technology in terms of its functionality. This kind of emphasis on functions can entail that design becomes a matter of fixating predefined ways of interpreting and using an object. Instead he proposes that thinking of design as crafting objects with forms, which require to be interpreted, positions designers (as well as users) quite differently in comparison to thinking of design as creating objects with functions, which need to be understood.

9.3.3 The Epidermis as Interface

"Fashionable Technologies enhance the cognitive characteristics of our epidermis – the surface of our body and the largest human organ. The epidermis, or the skin, is our principle communicator of emotional and physical states. The human skin has obvious communication abilities. It communicates through blushing, sweating, and variations in tension and temperature. These localized variants can be extended through the use of sensor and actuator technologies. The sensors are able to detect signals from the skin and the actuators can in turn produce certain types of sonic or haptic output. Reciprocally, this output can appeal to our physiological senses" [14].

"Helmut Lang's fashion often showed skin at the expense of figure. It played upon the theme of fabric as second skin" [19]. In Fashionable Technology the term second skin refers to the use of high-tech materials or 'enhanced' textiles. The skin, however, can also be a hinderer of functionality. When moist it conducts electricity and can conflict with the conductive properties of a conductive yarn. Such yarn is often best affixed on the surface of the textile to avoid the risk of touching the skin.

Electronic textiles or materials result from the integration of technology and thus enhancing its properties to become an interface. All wearable technologies are closely related to electronic textiles – both having technology as its core. "Many labels [fashion] are returning to traditional processes, couture principles, and craft techniques to distinguish their garments from mass-produced fashion brands, drawing on both new and traditional textile methodology to create an aesthetic for the future" [12]. This movement enables the transfer of knowledge in garment construction and the understanding of the body. The interaction with conductive metal-based buttons, zippers, or hooks is intuitive and can simply be transferred to the closing and opening of an electronic circuit. Such know-how can only be acquired through the collaboration of fashion and textiles designers and technologists.

Gemperle created a guideline for wearability noting placement, form language, human movement, proximity, sizing, attachments, containment, weight, accessibility, sensory interaction, thermal, aesthetics, and long-term use. The list is still a useful start for dealing with wearable technologies on our body to create an interface. Gemperle notes: "A product that is wearable should have wearability. Wearability is defined as the interaction between the human body and the wearable object. Dynamic wearability extends that definition to include the human body in motion" [4].

9.4 Conclusion

The author's (Sabine Seymour) focus on garments that are seen, felt, heard, and touched demand greater collaboration within scientists, designers, and artists to transform textile processes and products. The emergence of conductive and electronic materials enables the creation of fashionable electronically controlled garments, while ensuring their wearability. Such collaborations require the involvement of fields like biology as demonstrated by the projects Victimless Leather by Ionat Zurr and Oron Catts or BioCouture by Suzanne Lee. The need for inspiration through fashion design was perfectly demonstrated by Hussein Chalayan in his collaboration with the London-based studio that built the animatronics. Fabrican by Manuel Torres is based in science as well as in fashion. But also the involvement of philosophers, anthropologists, and psychologists needs to surface in Fashionable Technology.

The research and projects by the author (Sabine Seymour) and her students at the department Interface Culture at the University of Applied Arts and Design in Linz, Austria focus on the garment as interface. Transitive materials, ubiquitous computing, sensor technologies, and 'digital' craft are the main components of the works. The significance of Fashionable Technology as an emerging field is apparent. Fashionable wearables must engage the wearer through the mechanisms of fashion. Wearables, objects or garments, must become more than mere mediators of perception. They must become communicators of style.

References

1. Barnard, M.: Fashion as Communication, Routledge, London, vol. 10, pp. 49–71 (2002)
2. Berzowska, J.: Electronic Textiles: Wearables Computers, Reactive Fashion, and Soft Computation. In: Jeffries, J. (ed.) Textile, The Journal of Cloth & Culture, Digital Dialogues 2: Textiles and Technology, vol. 3(1), Berg Publishers, Biggleswade (2005)
3. Eco, U.: Travels in Hyperreality, pp. 191–195. Harvest Books, Fort Washington (1990)
4. Gemperle, F., Kasabach, C., Stivoric, J., Bauer, M., Martin, R.: Design for Wearability. Carnegie Mellon University, Pittsburg, PA (1998), http://www.ices.cmu.edu/design/wearability (retrieved April 22, 2002)
5. Huhtamo, E.: Pockets of Plenty: An Archaeology of Mobile Media. In: Rieser, M. (ed.) The Mobile Audience, Art and New Located Technologies of the Screen (forthcoming)
6. Mann, S.: Definition of Wearable Computer. University of Toronto, Toronto, Canada (1998), http://wearcomp.org/wearcompdef.html (retrieved October 18, 2006)
7. Mann, S., Sehgal, A.: Eyetap Designs for Continuous Lifelong Capture of Personal Experience: Capturing Reality and Merging virtuality. In: CARPE 2004. ACM, New York (2004)
8. Mann, S.: Fundamental Issues in Mediated Reality, WearComp, and Camera-Based Augmented Reality. In: Woodrow, Barfield, Caudell, T. (eds.) Fundamentals of Wearable Computers and Augmented Reality. Lawrence Erlbaum Associates, Inc., Mahwah (2001)
9. Mann, S.: Mediated Reality with implementations for everyday life at presenceconnect.com. Presence: Teleoperators and Virtual Environments (2002)
10. McLuhan, M.: Understanding Media – The Extension of Man, pp. 119–122. MIT Press, Cambridge (1995)
11. Novakov, A.: Private Dicks: The Detective in Art and Popular Culture (1998)
12. Quinn, B.: Techno Fashion, p. 163. Berg, Oxford (2002)
13. Redström, J.: On Technology as Material in Design. In: Redström, M., Redström, J., Maze, R. (eds.) IT + Textiles, Edita Publishing Oy (2005)
14. Seymour, S.: Fashionable Technology – The essential balance between aesthetics and function. In: Kuzmanovic, M., Maes, A., Bernard, Y. (eds.) x-med-a, pp. 38–41. FoAM, Brussels, Belgium (2006), http://xmeda.be (retrieved August 22, 2006)
15. Thorpe, E.O.: The Invention of the First Wearable Computer Second International Symposium on Wearable Computers (1998)
16. Vinken, B.: Fashion Zeitgeist – Trends and Cycles in the Fashion System, p. 137. Berg, Oxford (2005)
17. Viseu, A.: Augmented Bodies and Behaviour Bias Interfaces (2002)
18. Viseu, A.: Simulation and Augmentation: Issues of Wearable Computeres Ethics and Information Technology (2001)
19. Warwick, A., Cavallaro, D.: Fashioning the Frame. Boundries, Dress and the Body, p. 136. Berg, Oxford (2001)

10

The Hybrid City: Augmented Reality for Interactive Artworks in the Public Space

Clara Boj and Diego Díaz

www.lalalab.org

Abstract. Based on the idea of hybrid space (understood as the result of the transformation of the actual models of perception of reality through the effect of the development of technological systems), we propose to use mixed and augmented reality systems to create interactive artwork that facilitates the real and effective integration of information and communication systems into the physical public space. This will propitiate new ways of understanding and living in the city.

10.1 Introduction

Cities are elements in continuous transformation. The increasingly fast assimilation of technological devices in urban space, the advances in ubiquitous computing and embedded technologies, raise the question of how city space will respond to such transformations. What are the material and symbolic effects of the integration between the real and the synthetic on urban space? How can we effectively combine urbanism and architecture with information and communication networks?

New developments in mixed-reality technologies make a new configuration of hybrid space, between physical and digital, possible. The first part of this chapter analyzes the fundamentals of augmented reality, the main systems and applications. It also introduces some of the agents in this research area with special attention to outdoor augmented reality systems. Then we present the application of some of these systems and tools in the creation of an interactive artwork in the public space to stimulate public activity in the city.

10.2 Digital Perception of Reality: From Virtual Reality to Mixed Reality

Although the first investigations in virtual environments [1] were initiated in the 60's, it wasn't until the late 80's and early 90's when, due to the greater advances in graphic image processing in real time and sensor technologies, a more intense research in Virtual Reality [2] was undertaken.

At that time, Virtual Reality was understood as a connetion door to a new computer-generated world which was possible to transit (or surf) and experience through different, more or less immersive, visualization devices. The first experiences in virtual reality made us forget the perception of our own physical body and put us in a synthetic space, modeled according to a designer's criteria, which could introduce

C. Sommerer et al. (Eds.): The Art & Sci. of Interface & Inter. Des., SCI 141, pp. 141–161, 2008.
springerlink.com © Springer-Verlag Berlin Heidelberg 2008

some random parameters to modify or generate the environment while we were walking through it. The invention of virtual reality became a big revolution on a technical and symbolic level: it promised a new reality far away from bodily constrictions and related to a new model of sensorial perception [3].

Nevertheless, leaving the body behind, which was really appreciated in the beginning, rapidly became the main inconvenience of this new paradigm. Virtual environment perception was, in a way, mutilated by the lost of spatial references offered to the body by the senses, by the balance, and the feeling of one's own presence [4] in relation to the other elements in the environment.

Research in virtual reality has improved enormously since the earliest stages. But at the same time, great interest has arisen in exploring the intermediate territory between both paradigms, between real and virtual, between physical environments and digital environments, promising to solve some of the perceptive problems of the virtual reality [5].

Mixed Reality refers to the technology that combines, in a given space and time, physical elements from the real world and computer generated objects. The first approaches in that direction appeared accidentally during the development of the first Head Mounted Display by Sutherland [6], who was using a see-through device that allowed one to see the real surroundings and virtual objects simultaneously. As Sutherland main goals were others, this initial approach was forgotten and was only used much later when interest in Mixed Reality aroused.

At present, Mixed Reality is understood as the incorporation of virtual objects into a three-dimensional real space or, alternatively, the incorporation of real elements in a synthetic digitally-generated environment [7].

Some research approaches understand mixed reality as subcategories of virtual reality. However the classification most extended among researchers is based on the reality-virtuality continuum of Milgram and Kishino [8] that establishes a progression from the completely real to the completely virtual as opposite borders of a continuous line. On one extreme of this continuum we find virtual reality, where all the elements are digitally generated, and on the same line's other extreme we find reality as we perceive it through our senses, without using any technological devices.

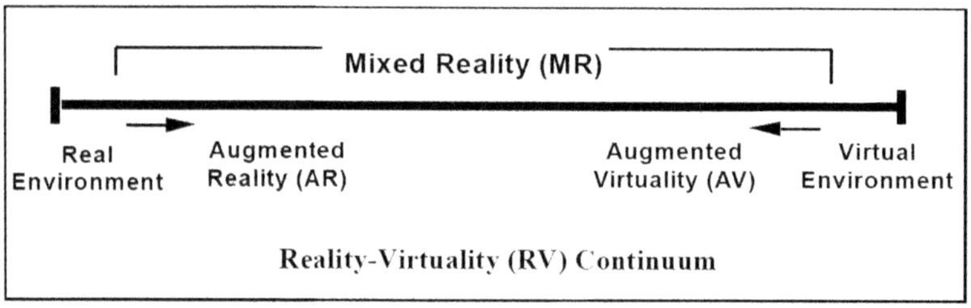

Fig. 10.1. Milgram's Reality-Virtuality Continuum

Mixed Reality is anywhere between the extremes of the continuum, formed by all those situations that combine real and virtual elements. Ranging in between, we find augmented reality and virtual reality. The position on the continuum offers us a clear reference as to the amount of virtuality or reality shared with the extreme. Augmented reality refers to the incorporation of digital elements in a three-dimensional real space. Augmented virtuality refers to the incorporation of real elements in a three-dimensional computer generated environment.

10.3 Defining Augmented Reality

Augmented Reality combines real and virtual elements which can be interacted with. In contrast with Virtual Reality, where the user is introduced inside a fully immersive digital environment, which takes away his real surroundings, AR allows us to see the real world at anytime, but augmented with virtual objects that, in an ideal situation, seem to be physically present in the space. As opposed to RV, which replaces reality, AR complements it.

An augmented reality experience must include 3 main characteristics:

1- Combines real and virtual elements.
2- Is interactive in real time.
3- Is registered in three dimensions [9].

Some authors add a fourth characteristic, suggesting that to have a satisfactory perception of the augmented environment, real and virtual elements must be aligned, that is, their positions must correspond with the coordinates of the real world.

Augmented Reality makes it possible to add virtual objects to the real world but it can also potentially make real elements disappear from the scene. Some researchers call this experience mediated or diminished reality, although it can be considered a subset of AR. Research has been developed more intensively in the former way (adding). It can be also applied to the other senses, not only the sense of sight. For example, by adding directional 3D sound to the environment or even by filtering some sounds [10]. Another direction of Augmented Reality research is haptics, using devices with tactile responses, in accordance with the position of virtual elements on the real space, augmenting their virtual presence [11].

Augmented reality technology applications combine different software and hardware techniques to make possible fast, stable, robust and efficient experiences of interactive real-time, three-dimensional environments where real and virtual objects are mixed together [12]. Several aspects must be taken into account but the key point of any AR application is to calculate with accuracy the position and orientation that virtual objects must have in the real space in relation to the observer's position. This must be visualized in real time. To address these issues several techniques have been developed: from using magnetic sensors and geo-spatial location systems to computer vision techniques for identifying visual tags. We will explain these later in this chapter.

In addition, at the present, it is also possible to integrate and use almost every interactive and graphical feature of 3D video games in an augmented reality system. Light and ambience effects, physics programming, complex systems interactions, graphic rendering and processing, artificial intelligence programming, etc. can be integrated in an augmented reality environment to make the experience more realistic.

10.3.1 Augmented Reality Interaction

Augmented reality creates a new interactive relationship between humans and computers. This new paradigm was expected to revolutionize the Graphical User Interface developments, taking away the traditional desk metaphor in the operative systems and replacing it with a user's interface, physically distributed on the space and controlled by tangible elements. [13] [14] [15].

Later, the goals of this initial research have expanded to other areas. Specific applications have been developed to integrate digital information on the real image to help workers do a complicated task, such as medical operations, mechanic reparations, architectural visualizations, etc...Those systems frequently incorporate physical elements like interactive devices which require the user's manipulation.

Augmented Reality presents several factors of interest to Human Computer Interaction research:

- Spatial interaction. virtual interactive space is no longer limited to the bi-dimensional resolution of the screen monitor. It is expanded to 360 degrees of physical three-dimensional space. Real physical space becomes an interactive environment and users can walk through it, interacting with the system. This makes it possible to potentially extend the system to any location.
- Tangible interaction. Using augmented reality we are able to see digital information directly annotated over real objects. Almost any real element can act as a navigation tool for manipulating data [16]. This creates a more symbolic relation between users and interface objects [17].

These capabilities can be extended by combining AR technologies with multi-user systems and remote interaction; connecting users in different locations, expanding the complexity and dynamism of AR applications.

10.3.2 Augmented Reality Processing

Augmented Reality experiences are created through different processes that take place independently of the technologies used.

- Identification. The system must know the position of orientation of both the user and objects surrounding him. Accuracy and rapidity are required, even if the user is in movement.
- Registration or calculation of the position and orientation of virtual objects in the space. We need to know the relationship between users and virtual elements and compare the data with that which has already been predefined in the system. Real and virtual objects must be correctly aligned or the illusion of the presence of virtual objects in the space will be compromised. To get a precise registration is one of the big challenges of AR, to allow an optimal stability of the system. Some filters can be used to eliminate unnecessary information.
- Rendering. Once the system knows the position and orientation data of every element, the rendering process is started in real time. Due to AR relays in the combination of graphical elements in real space, realistic and effective rendering methods are very important. In some AR applications we may need to combine virtual and real elements in a way that users can not distinguish. Photo-realistic

rendering techniques are therefore required. Elements must be integrated in a consistent way, resolving occlusion problems [18] [19], shadow casting and inter-reflection issues [20]. Realistic lighting behaviors, like ray-tracing and radiosity, are beneficial [21] [22]. At present, the integration of these rendering techniques can be used in an application that doesn't need a real time frame rate. For interactive applications in real time we will need to depreciate graphical process that can slow the system. The actual increase of real time processing in graphic hardware and the new possibilities of the Real Time Shading Languages open possibilities for more photo-realistic and complex rendering.

- Visual feedback. Once objects are rendered, they must be shown to the user in the position and orientation registered. There are several technologies to display this information. Head Mounted Display is maybe the most common but they still have some optical limitations, such as a limited vision field and fix focus, limited resolution and instability and weight and size, referring to human factors.
- Interaction. Users of an augmented reality system can interact with the environment by moving, walking through the space or touching real and virtual objects. This action will cause a continuous repetition of the process described above to actualize the visual information in real time [23]. It is possible to add other tools such as sensors to add more complex interaction capabilities to the system.

10.3.3 Common Standards in Augmented Reality

Augmented Reality has opened a vast field of applications. In 1997 Azuma [24] identified six classes of potential AR applications: medical, manufacturing and repair, annotation and visualization, robot path planning, entertainment, military aircraft navigation and targeting.

Present research has widely surpassed these expectations. Now different research is under development in investigation areas such as education, psychology, publicity, sports training, urban planning, etc. These applications combine tracking techniques and hardware setups and a significantly high number of technical variations have been developed.

We can classify these technologies depending on the different processes in which they act: tracking and sensing to identify and register the real environment and visual displays to represent the final information for the users.

10.3.3.1 Augmented Reality Tracking Technologies

Many existing technologies are used, from computer vision base techniques to a combination of different sensors for spatial location. Using one or another will depend on the application we want to develop and its special condition: placement, number of users, accuracy required, etc. We will use different tracking systems, for example, for outdoor or indoor applications. We will describe here the more widely used.

10.3.3.1.1 Marker Tracking. Computer Vision Based Pattern Recognition. This technique is based on the analysis of real time video images. The system analyzes the stream of images captured by the camera, looking for specific patterns, generally fiducials, and calculating their position and orientation in the geometrical three-dimensional space. A number of researchers are investigating fiducial vision based

Fig. 10.2. Example ARToolKit Fiducial

tracking [25]. ARToolKit developed by Kato and Billinghurst [26] is the first set of computer vision tracking libraries that can be used to calculate camera position and orientation in relationship to to physical markers in real time.

ARToolKit features include the use of a single camera for position/orientation tracking, fiducial tracking from simple black squares, pattern matching software that allows any marker patterns to be used, calibration code for video and optical see-through applications, and a sufficiently fast performance for real-time augmented reality applications. The fiducial markers are known-sized squares with high contrast patterns in their centers. The ARToolKit determines the relative distances and orientation of the marker from the camera. In addition, the ArToolKit incorporates a calibration application to determine the placement of the camera relative to the user's line of vision, thus the ARToolKit can determine proper placement of graphical objects for AR applications.

ARToolKit was released in the late 90's and has given rise to other tools such as ArtoolKitPlus [27], Artag [28], Dart [29], etc. Those tools have given way to a great number of applications in the field of education [30] and architectonic research [31]. Magic Book [32] is a very simple application in which several markers are integrated between the book pages. Several digital elements are superimposed to this marker and the reader accesses the augmented content of the book by using an HMD or through the computer screen. Fiducial marker tracking is also very useful in office applications such as the augmented reality desk [33] and in sound interfaces such as Reactable [34], an application that acts like a music synthesizer using specially designed markers that, once detected, configures the visual and musical composition.

These tools make possible a rapid development of low cost prototypes because it requires a basic system with a web cam and some printed markers and it offers a high degree of stability and robustness. They are also easily integrable with graphic programming libraries for video game development. Marker tracking systems are really extended between researchers but they present some constrictions due to real environment lighting limitations. Bad scene lighting will make the system work incorrectly. This technique is especially suitable for indoor situations with controlled lighting. Also, this system requires the addition of artificial elements to the environment, which is not always possible.

10.3.3.1.2 Markerless Tracking/Natural Feature Tracking. New advances in vision based tracking systems don't require fiducial markers to be placed on the scene [35] [36]. A new approach has been developed to detect and track the natural features of unprepared scenes from sequential images, such as textures or corners. This method uses different algorithms to find the most reliable feature points in the image. Then, those detected points are matched in different images. Furthermore, motion parameters are occasionally calculated and motion prediction is used to minimize the searching scope in the next step.

By using a database with the chosen environment's information, it is possible to search for predefined elements comparing real time camera images with existing data [37]. It is also possible to look for unknown elements on the image but these techniques are still under development [38]. This was usually used for visual effects during the post-production process, but now, due to advances in computer processing, it can be done in real time [39] [40].

The advantage of this method is that it allows a transparent and immediate tracking of the scene, which can theoretically be implemented in any place at a low cost, propitiating a great number of new applications. As other systems of vision based tracking, it depends on light conditions and dynamic environments can not be easily registered. Identification of mobile visual tags is unstable and it requires a great amount of processing power, making it useless for portable applications.

10.3.3.1.3 Sensorial Tracking. As mentioned before, in some situations it can be useful to use techniques that don't require modifying the real environment. The use of position and orientation sensors can be an alternative. For example the use of a compass/gyroscope tracker gives us motion-stabilized orientation measurements combined with Global Position System (GPS) which calculates the position of the user on the space. Those systems are commercially available and their cost, although not very expensive, limits their use to research context.

Their operation is based on physical principles [41] such as Time of Flight TOF, which consists in measuring the distance between different features attached on one side to a reference and on the other side to a moving target. Distances are determined by measuring the temporal difference in the signal reception from one emitter to several receivers. The implementation of these calculations can be done with ultrasonic sensors, pulsed infrared laser-diode, GPS, optical gyroscopes, etc.

They are frequently used in Virtual Reality applications such as CAVE [42] in which people wear an attached magnetic sensor to know their position and head orientation. We can see other examples in the first investigations of Livingston and Andrei [43].

The advantage of that system is its easy adaptation to any kind of application. However, some of these devices accumulate errors with time, which happens with magnetic sensors, or have a limited accuracy in a given context, which occurs with GPS. Their cost is relatively high and their use is not generalized. It is frequent to use hybrid tracking systems where data are registered through two or more devices. A traditional hybrid set-up combines a number of systems such as inertial, optical, electro-magnetic and GPS [44] [45].

10.3.3.2 Augmented Reality Display Technologies

Obtaining satisfactory results when creating AR depends on the visualization device used. There is still no such thing as a perfect display. 40 years after the development of Shuterland's first HMD [46] technical limitations still make this issue a challenge. We can differentiate between common approaches: body worn displays, hand-held devices and projection base displays.

10.3.3.2.1 Body Worn Displays. These systems have been specially designed for Mobile AR. Traditional backpack systems include a notebook computer, an HMD, cameras and additional support hardware. Mobility is imitated by energy consummation and the weight of the equipment; one approach is to move some of the computation load to remote servers, reducing the equipment the user must wear.

As for HMD, we can differentiate between 2 kinds: optical and video HMD. Optical or see-through HMD is composed of a transparent pane of glass on which digital information is projected. In this way the user can see virtual elements superimposed to the real image of the environment. Video HMD is composed of small LCD screens that prevent the user from seeing directly around him. Those devices use 1 or 2 cameras (for stereo vision) to show the image through the LCD [47]. Another method is the "virtual retinal display" [48] which creates images directly on the retina with low power lasers whose modulated beams are scanned by microelectro-mechanical mirror assemblies that sweep the beam horizontally and vertically. Its potential advantages are high brightness and contract, low power consumption and large depth of field in contrast with video or optical HMD, in which low resolution, brightness and contrast don't propitiate a good mixture of virtual and real images [49].

10.3.3.2.2 Hand Held Displays. Handheld computers, mobile phones and personal digital assistants have the potential to introduce Augmented Reality to large audiences. Thanks to the increase in the processing capacities of mobile devices, combined with utilities such as digital cameras and GPS, it has recently been possible to develop augmented reality applications which are much lighter and more portable, making them especially suitable for outdoor situations. This kind of interface is commonly known as the "magic lens metaphor" because users look through the small screen of the mobile device as if it were a lens, and they can see the surroundings augmented. Its interesting applications for gamming, education and entertainment are under development, including multi-user interaction. However, there are still some limitations due to the limited computation capabilities, the batteries' lifespan or the limited brightness and size of the screens.

Invisible Train [50] is the first real multi-user Augmented Reality application for handheld devices (PDAs), in which users control virtual trains on a real wooden miniature railroad track. This system was developed using Studierstube [51], a software framework designed to accelerate the task of developing and deploying collaborative Augmented Reality applications on handheld devices.

10.3.3.2.3 Projection Base Displays –Spatial Augmented Reality. Some AR applications, such as museum displays, don't always require mobile systems because the whole experience is developed under controlled indoor parameters. In these cases, projection base display systems can be a good solution.

Those systems offer several advantages in relation to image quality, resolution, field of view, focus, etc. and also in relation to tracking techniques and human factors, because conditions are stable and users are not forced to wear heavy displays, allowing greater freedom of movement. Those systems are nevertheless limited to static applications but some mobile projectors are under development. Researchers in this area define this technique as spatial augmented reality [52], consisting in several methods to project directly over the real objects or to create reflections using translucent glass or mirrors.

10.4 Integrating Augmented Reality Technologies to the Urban Space

Urban planning is one of the main application areas of augmented reality technologies. Investigation in that direction is usually limited to the visualization of scaled models of planned construction projects or archeological reconstruction of historical buildings.

Although these projects are very innovative, they don't take into account the possibilities for perception, interaction and participation that augmented reality can offer the user, which may be directly applied to the urban space. Not just as representation of ideas, but as direct actions within the city context.

Outdoor augmented reality research is a field in continuous expansion. At first, navigation maps and past and future events visualization were thought of as the main applications areas, but interest has risen in contexts such as video games in the public space, tourist attractions and even military projects for territory exploration.

Nevertheless, current developments have encountered difficulties, which have been especially detected when working with open air space, where environmental conditions create problems not detected in indoor systems: lighting, energy consummation, maintenance, etc. In an outdoor AR experience, the user is mobile and able to walk through the information space and usually needs to interact with the system with a simple hand interface, allowing him freedom of movement and stability.

In ideal conditions, outdoor AR application users should be able to wear an AR display and walk around with no restrictions. The scene generator, display and tracking system must be self-contained and capable of withstanding exposure to the environment.

Some challenges identified for outdoor AR [53]:

- Size, weight and power issues, and all issues related with the ergonomic aspects of a wearable PC system. When a user walks around the outdoors on his own and must carry batteries, the computer and all other resources, then size, weight and power issues become very significant concerns. Developments in laptops, wearable PCs, PDAs and other upcoming portable devices may provide improvements that outdoor AR systems can use to their advantage.
- Displays. Due to light conditions, displays are required which have sufficient contrast for working in outdoor settings. Attention must be paid to the choice of

screen colors for purely virtual images, which the application must display, because of the lighting conditions and background colors of the outdoors.

Also, the dynamic range of the real world must be compressed into the range of the output monitor, which is generally much smaller. The resolution of the display monitor is much lower than the resolution of the human eye. These cause a loss of detail that may not be acceptable in some applications. The display's field of vision must be calibrated to the physical world.

- Tracking. The user's position and orientation must be determined with sufficient accuracy to avoid significant registration errors. In open air situations we have much less control over the environment and we can not usually modify it to fit our needs. The tracking technologies available for the outdoors include GPS, Inertial Sensors, Active sources, such as transmitters and receivers, electronic compasses and optical sensors, such as web cams or surveillance cameras. None of these system provide a complete solution for the outdoors: GPS covers the user's 3D position but can not measure the orientation; Inertial sensors accumulate errors with time and other approaches, such as computer vision techniques, are not currently robust enough and are dependent on light conditions.

Present developments in outdoor AR work are surpassing these limitations by combining several techniques on wearable computers, displaying information and tracking. Hybrid tracking, using 2 or more technologies, has resulted in an overall system that behaves more robustly than with each sensor applied individually [54].

ARQuake is an extension of the desktop game Quake, and has been converted into an outdoor-indoor mobile augmented reality application [55]. Mobile Users wear a backpack system with Head Mounted Display and can walk through a limited open air area fighting with virtual monsters that appear in the real space. It combines GPS, magnetic compass and vision-base optical tracking. When GPS is unavailable, the system switches to visual tracking derived from the ARToolKit. The rendering of virtual elements has been adapted to daylight conditions, using colors easily seen over the real background.

LifeClipper [56] is an audiovisual walking experience in the public space. When walking around in a chosen culturally interesting area, the visitor's position and viewing direction is measured by means of a GPS and magnetic compass. The found situation is augmented with images displayed on HMD and sound through the headphones. Registration accuracy is not essential in this kind of application so minor misalignments between the physical world and the virtual elements don't drastically affect the user's final perception.

Vidente [57] is a project meant to support the field work process. Field workers equipped with a mobile device, such as a Mini Tablet PC, can point towards a particular surface of interest and are provided with real time visualization of the subsurface network of cables and pipes for the chosen spot. This system combines inertial sensors with GPS information.

A number of outdoor AR [58] [59] [60] [61] researchers agree that a new attitude is needed concerning AR technologies in the public space. Future applications must be built in parallel to the urban space, surpassing techniques limitations. We must consider the possibility of incorporating AR practice into urban space design, not only as a pre-visualization tool during design stages, but as an constructive element with its own identity, able to articulate the connection between the physical urban space of our cities' streets and buildings with virtual public spaces which have arisen in the context of Internet and the new mobile communication tools. In this way, AR would configure new communication and information capabilities in connection with the city's context.

Based on those considerations, we have created an interactive artwork, which shall be described here in more detail.

10.5 Free Network Visible Network: Data Visualization and Location Base Information in the City Using Augmented Reality

Considering hybrid space as the new paradigm for people's experiences, we have developed an interactive artwork whose main goal is to bring people closer to the knowledge of this new environment. It also encourages people to actively participate in city design through physical and digital approaches. Free Network Visible Network proposes the visualization with augmented reality of the flows of data between wireless networks. At the same time, it is presented as a new tool for digital communication on the public space; users are able to dynamically superimpose text messages, video, static images, 3D models and sound to the real space.

Fig. 10.3. A person using the FNVN project

In last 20 years, digital information has floated into our lives. No matter where we are, even if we are not able to see it, we know that we are completely surrounded by data. This data contains information, ideas, feelings and emotions. The space of digital wireless networks is also the space of invisible meanings that connect people.

The metaphoric representation of these invisible meanings within augmented reality intends to act on the urban landscape as a medium to create new strategies in the public domain and to re-think the concept of public space. By representing the information that is continuously circulating between the nodes of a wireless network, we will obtain a visual map of the hybrid city, of the layers and compositions that information creates while flowing and mixing with the physical environment.

In another approach, placing digital information in specific locations in urban space expands the narrative possibilities and communicative expressions of the traditional city space and improves the city-citizens interaction. Our cities' physical space and architecture have different levels of meanings, visible and invisible, depending on our approach to them, from a personal or historical point of view, their practical value, aesthetics, the tactile and visual emotions provided by the materials, etc. The digital information added to these sites is nourished with all this content. It is complemented by the symbolic value of the specific location. These strategies are also very interesting on a practical level, because the information comes to us as related to the place, which can be useful or symbolically important for us [62] [63].

10.5.1 Free Network Visible Network System Overview

Free Network Visible Network is composed of different kinds of tools and actions that, as a whole, create an experience for urban interaction and communication. It has been installed in different contexts with small variations, to better adapt to the specific configuration of the place. It is possible to see these versions in http://www.lalalab.org/redvisible.htm

Basically, this is a project acting in areas where a free-access urban wireless network exists. By demarcating the different nodes of the network with different visual tags, our system gives users the opportunity to send information directly to specific locations while other users make a tour through the city, accessing this information in situ.

Free Network Visible Network consists of different elements:

- A signalization system for urban wireless networks
- A Visible network server, allowing communication and the exchange of information between all the connected visible network clients.
- A Visible network client, for interaction with the network and the public space.
- A wearable computing system to walk through the city and access the information

10.5.1.1 Urban Demarcation of Wireless Networks

Since the beginning of the expansion of the wireless network for digital data interchange, several social initiatives have raised to encourage the collective use of the wireless networks. These proposals demand the freedom of citizens to administrate the digital space in a free manner. The property of wave space and the necessity of universal access to the resources and information circulating on the networks is the leitmotiv of projects like warchalking [64], a term that refers to the action of walking through the streets looking for WIFI nodes, drawing a small chalk icon to help other people identify them without difficulties. Warchalking movement uses some standards symbols to identify different kind of nodes. In our project we use different designs and messages inside a predefined frame. This frame corresponds with the fiducial marker design of MXRToolKit, printed on large format papers. Some authors [65] consider that in the outdoors and in unprepared environments using mobile AR applications, it generally isn't practical to cover the environment with markers. However we choose this method because markers on the street will have a multiple

Fig. 10.4. FNVN demarcation of wireless network

symbolic and practical function: first, they are a direct message depending on their design, also they indicate a free node of a network and finally, when using the free network software, they can be used as activators of the whole augmented reality system.

10.5.1.2 Visible Network Server

With this technology we can establish communication dynamics between all the users of the system connected to the network. Ideally, a server program must be run by an independent computer connected to the network. This software will be turned on, listening for any client demand. Each time a software client asks to be connected to the server, the last updated configuration of the whole system will be sent out. Each time a client/user modifies the information superimposed on any of the markers, this information is sent to the server, which automatically resends it to all the clients connected. Each information package contains, depending on the situation, an image, a 3D model, sound, video or a text message. The server program will send only the last modification sent by a user. For each new user connected to the network, the server will send the whole data package in use at that time, that is, one for every visual marker with associated information.

10.5.1.3 Visible Network Client

This software establishes direct relationships between physical markers placed on the street and the virtual digital information that is floating in the air or superimposed on the marker. It is mainly based on computer vision using pattern recognition; we mainly use the MXRToolKit, an open source library, adapted to our project's needs and combined with other functions to extend the interaction between system and user.

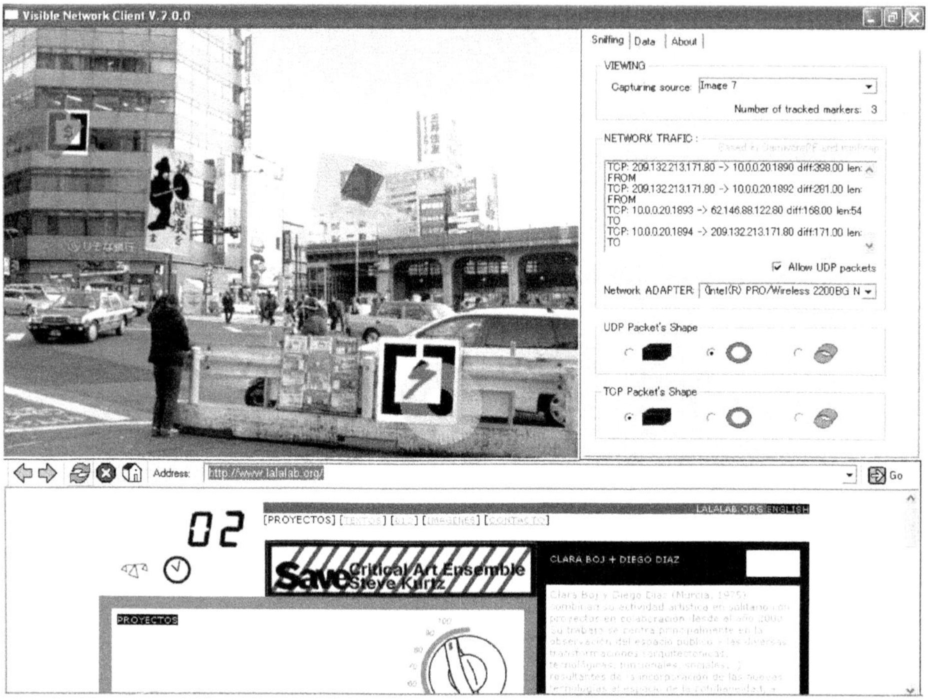

Fig. 10.5. FNVN Client

MxRtoolkit [66] is used for the identification and registration process of the real time video image, looking for visual markers in public space. Parallel to these processes, other tools will give us the information that is flowing on the network and using OPENGL [67]. It will be represented and rendered over the video image.

The main elements of this software are:

- The Network sniffer, which gives us information about the data circulating on the wireless card. Initially this tool was based on the Carnivore PE [68] server, a project by Radical Software Group and in the WINPCAP [69]. In later versions we have improved and adapted this element to our specific needs.

  Once we get the network data, this information is processed and transformed in shapes of various sizes and bright colors. In the configuration menu we have the possibility of choosing between different kinds of shapes: cubes, cylinders, cones or rings. We can distinguish what kind of information is circulating on the network depending on those different shapes.

  If there is some data flow, that information will be stored in a data structure, such as the sender and receiver's IP address, port number, data type, the length of each data package, etc. We use port number information to decide whether the virtual object is a cube, a cone, a cylinder or a ring.

  The last three entries in the IP information are used to determine the scale of the virtual object in x, y and z direction. Every data package transmission detected by the sniffer will be used to create a virtual object. The color will remain the same if the data transmission is done through the same port number, otherwise the system will randomly select a new color for the virtual object. If the sender IP address on the data package is the same as the user's local IP, the corresponding virtual object will be flying from the camera position to the marker position, indicating the direction of data flow. If the receiver's IP address is the same as the user's local IP, the virtual objects will be flying from the marker towards the camera.

- The configuration menu and input processing, on the right side of the GUI. Using this tool we can choose a specific marker, from a thumbnail list. Using a small text editor, we can write messages and send them directly to this marker's specific location. In this menu we can also browse any image file, 3D model, sound or video from our hard disk and send it to the marker.
- The graphic display window, on the upper center side of the GUI. This window displays the visual experience of augmented reality. We can visualize the real time images of the web cam with elements superimposed on the visual markers. It also shows the representation of the data packages registered by the sniffer, flying from one marker to another.
- A web browser, at the bottom of the graphical interface. We can surf the www and create data traffic that will be processed, transformed in colorful shapes and represented floating in the air in the graphic display window.

10.5.1.4 Wearable Computing System

Different kinds of users can experience Free Network Visible Network: those who use a laptop equipped with a web cam and connected to the wireless network and those who wear a specially designed wearable computer system and experience the augmented reality through a head mounted display while walking through the streets.

As described previously, in the visible network client software there is a graphic display window to visualize the augmented environment. Displaying information in this manner is easily accessible and economic for every user, however, it does not have the same immersion possibilities and it is limited by the mobility of the laptop and the screen visibility in the open air. We make a difference between static users who interact through a conventional laptop and dynamic users who utilize the mobile wearable computer system.

This system consists in a laptop inserted in a specially adapted, rigid backpack which provides stability for the functioning of the system. This computer is equipped with a wireless card and connected to a head mounted display with a web cam on the top. Due to the resolution limitations of these kinds of devices, a special version of the visible network client has been designed to adapt the screen size. In this version, users can only visualize the graphic display window in which virtual elements are superimposed on to the video in real time. The configuration window and web browser have been removed.

Dynamic users can walk through the city using the backpack. When observing the areas signaled with visual markers, the software connects automatically to the server and asks for the information available for this marker. At the same time, all the communication between client and server is represented as geometric shapes floating in the air. Dynamic users access the information, updated by other users from their static positions, which can then be connected to any part of the network.

Fig. 10.6. Dynamic user

10.6 Conclusions and Outlook

We have described here an augmented reality interactive system for a public space which enables visualizations of the digital data flows between wireless networks. At the same time this system becomes a communication media that combines digital elements with specific locations on the public space. While other researchers in outdoor augmented reality focus their attention on surpassing the technical limitations of open air systems, our project is especially committed to configuring a new understanding of public spaces, such as the constant transformation of elements and paradigms specific to the tangible city augmented with digital information dynamically updated through the networks.

Ubiquitous and mobile computing developments bring citizens the possibility of accessing digital information anytime, anyplace. However, locating this information in specific places on the urban landscape enriches the perception of both environments, virtual and real, and generates new approaches for the perception of reality. Our system is designed as a democratic media for the collective use of the hybrid public space. Any user is able to add information to a specific site and this information will be accessible in situ to other users. Research in this area is fundamentally designed for architectural or entertainment purposes. Free Network Visible Network establishes a direct relationship with citizens, offering people a tool of free expression for the new public spaces. Free Network Visible Network is distributed freely through the project website.

Technically, our project is quite similar to previous applications for outdoor Augmented Reality. Several similarities can be found with projects analyzed in this chapter, such as ARQuake, LifeClipper, etc. Conceptually it is more related to the current state of locative media investigations, and is directly related to other geo-terrestrial annotation projects, some of them commercially available, such as the popular Google Earth [70], wikimaps [71], Stickymap [72] or OpenStreetMap [73]. All these systems are available through Internet and some of them allow direct access through mobile phones and PDAs [74]. Information can be downloaded even on the street, such as google map data, but some of these systems don't give users the ability to add information to the system. These projects use 2D or 3D maps visualization or satellite images where information is annotated. Considering all these pre-existing applications, we think our project adds an interesting approach to location base information technologies, combining the same strategies with augmented reality technologies which allows users to directly access information in situ and superimposed on the real surroundings.

Future research will expand the system to new interaction forms, offering users the possibility of adding or modifying information directly on site. A new version of the client-server software will be developed in conjunction with wikimap, to create AR-wikimap, in which users can see the information and geo-annotate data directly on the public space but also through the Internet. In this way, it will be possible to access to the project website and add information that will be geo-annotated to the real space and to the digital map. The system will also be expanded to other devices such PDAs and mobile phones.

Acknowledgments

Free Network Visible Network has been developed in part with the collaboration of Interaction and Entertainment Research Center at Nanyang Technological University. We would like to especially thank the researcher Duy Nguyen for his intense participation and valuable knowledge.

References

1. Rheingold, H.: Virtual Reality. Simon & Schuster, New York (1992)
2. Brooks Jr., F.P.: What's Real About Virtual Reality? IEEE Computer Graphics And Applications 19(6), 16–27 (1999)
3. Goslin, M., Morie, J.F.: Virtopia: Emotional experiences in Virtual Environments. Leonardo 29(2), 95–100 (1996)
4. Witmer, B., Singer, M.: Measuring presence in virtual environments: A Presence Questionnaire. Presence: Teleoperators and Virtual Environments 7(3), 225–240 (1998)
5. Drascic, D., Milgram, P.: Perceptual Issues in Augmented Reality. In: Proc. SPIE, Stereoscopic Displays VII and Virtual Systems III, vol. 2653, pp. 123–134. SPIE Press, Bellingham, Wash. (1996)
6. Sutherland, I.E.: The Ultimate Display. In: Proc. IFIP 1965, vol. 2, pp. 506–508 (1965)
7. Milgram, P., Takemura, H., Utsumi, A., Kishino, F.: Augmented Reality: A class of displays on the reality-virtuality continuum. In: Proc. Telemanipulator and Telepresence Technologies, vol. 2351, pp. 42–48 (1994)
8. Milgram, P., Kishino, F.: A Taxonomy of Mixed Reality Visual Displays. IEICE Transactions on Information Systems E77-D(12), 1321–1329 (1994)
9. Azuma, R.T.: A Survey of Augmented Reality. Presence: Teleoperators and Virtual Environments 6(4), 355–385 (1997)
10. Durlach, N.I., Anne, S.M. (eds.): Virtual Reality: Scientific and Technological Challenges. National Academy Press (1995)

11. Hayward, V., Astley, O., Cruz-Hernandez, M., Grant, D., Robles-De-La-Torre, G.: Haptic interfaces and devices. Sensor Review 24(1), 16–29 (2004)
12. Bimber, O., Raskar, R.: Modern Approaches to Augmented Reality. In: 25th Annual Conference of the European Association for Computer Graphics. Interacting with Virtual Worlds. Tu., vol. 8, pp. 1–86 (2004)
13. Billinghurst, M., Kato, H., Poupyrev, I.: Collaboration with Tangible Augmented Reality Interfaces. In: HCI International 2001, New Orleans, LA (2001)
14. Lee, G.A., Nelles, C., Billinghurst, M., Kim, G.J.: Immersive Authoring of Tangible Augmented Reality Applications. In: Proc. 3rd IEEE and ACM International Symposium on Mixed and Augmented Reality 2004 (ISMAR 2004), Arlington, VA, pp. 172–181 (2004)
15. Raskar, R., Welch, G., Cutts, M., et al.: The Office of the Future: A Unified Approach to Image-Based Modeling and Spatially Immersive Displays. In: ACM SIGGRAPH, Orlando, FL (1998)
16. Kenneth, P.F.: A taxonomy for and analysis of tangible interfaces. Personal Ubiquitous Computing 8(5), 347–358 (2004)
17. Ishii, H., Ullmer, B.: Tangible Bits: Towards Seamless Interfaces between People, Bits and Atoms. In: Proc. CHI 1997, Atlanta, Georgia, pp. 234–241. ACM Press, New York (1997)
18. Lepetit, V., Berger, M.-O.: Handling Occlusions in Augmented Reality Systems: A Semi-Automatic Method. In: Proc. Int'l Symp. Augmented Reality 2000 (ISAR 2000), pp. 137–146. IEEE CS Press, Los Alamitos (2000)
19. Fuhrmann, A., et al.: Occlusion in Collaborative Augmented Environments. Computers and Graphics 23(6), 809–819 (1999)
20. Mukaigawa, Y., Mihashi, S., Shakunaga, T.: Photometric Image-Based Rendering for Virtual Lighting Image Synthesis. In: Proc. 2nd Int'l Workshop Augmented Reality (IWAR 1999), pp. 115–124. IEEE CS Press, Los Alamitos (1999)
21. Stauder, J.: Augmented Reality with Automatic Illumination Control Incorporating Ellipsoidal Models. IEEE Trans. Multimedia 1(2), 136–143 (1999)
22. Debevec, P.: Rendering Synthetic Objects into Real Scenes: Bridging Traditional and Image-Based Graphics with Global Illumination and High Dynamic Range Photography. In: Proc. Siggraph 1998, pp. 189–198. ACM Press, New York (1998)
23. Ellis, S.R., et al.: Factors Influencing Operator Interaction with Virtual Objects Viewed via Head-Mounted Seethrough Displays: Viewing Conditions and Rendering Latency. In: Proc. Virtual Reality Ann. Int'l Symp. (VRAIS 1997), pp. 138–145. IEEE CS Press, Los Alamitos (1997)
24. Azuma, R.T.: A Survey on Augmented Reality. Presence 6(4) (1997)
25. Owen, C.B., Xiao, F., Middlin, P.: What is the best fiducial? In: First IEEE International Augmented Reality Toolkit Workshop, pp. 98–105 (2002)
26. Kato, H., Billinghurst, M.: Marker Tracking and HMD Calibration for a Video-based Augmented Reality Conferencing System. In: Proc. 2nd IEEE and ACM International Workshop on Augmented Reality (Iwar 1999), San Francisco, pp. 85–94 (1999)
27. Daniel, W., Dieter, S.: ARToolKitPlus for Pose Tracking on Mobile Devices. In: Proc. 12th Computer Vision Winter Workshop (CVWW 2007), St. Lambrecht, Austria (February 2007)
28. Fiala, M.: Artag, an improved marker system based on Artoolkit. NRC/ERB-1111 NRC 47166. National Research Council Canada (2004)
29. MacIntyre, B., Gandy, M., Dow, S., David, B.J.: DART: a toolkit for rapid design exploration of augmented reality experiences. In: Proc. 17th annual ACM symposium on User interface software and technology, Santa Fe, NM (2004)

30. Kaufmann, H., Schmalstieg, D.: Mathematics and Geometry Education With Collaborative Augmented Reality. Computers&Graphics 27(3), 339–345 (2003)
31. Broll, W., Lindt, I., Ohlenburg, J., Wittkämper, M., Yuan, C., et al.: Arthur: a collaborative augmented environment for architectural design and urban planning. Journal of Virtual Reality and Broadcasting (2004)
32. Billinghurst, M., Kato, H., Poupyrev, I.: The MagicBook - Moving Seamlessly between Reality and Virtuality. Computer Graphics and Applications 21(3), 2–4 (2001)
33. Kato, H., et al.: Virtual Object Manipulation of a Table-Top AR Environment. In: Proc. Int'l Symp. Augmented Reality 2000 (ISAR 2000), pp. 111–119. IEEE CS Press, Los Alamitos (2000)
34. Jordà, S.M., Geiger, G., Alonso, M.: The reacTable*: A Collaborative Musical Instrument, Kaltenbrunner. In: Proc. 15th International IEEE Workshops on Enabling Technologies (WETICE 2006), Manchester, U.K. (2006)
35. Azuma, R., et al.: Tracking in Unprepared Environments for Augmented Reality Systems. Computers and Graphics 23(6), 787–793 (1999)
36. Neumann, U., You, S.: Natural Feature Tracking for Augmented Reality. IEEE Trans. Multimedia 1(1), 53–64 (1999)
37. Coors, V., Huch, T., Kretschmer, U.: Matching Buildings: Pose Estimation in an Urban Environment. In: Proc. Int'l Symp. Augmented Reality 2000 (ISAR 2000), pp. 89–92. IEEE CS Press, Los Alamitos (2000)
38. Simon, G., Fitzgibbon, A.W., Zisserman, A.: Markerless Tracking Using Planar Structures in the Scene. In: Proc. Int'l Symp. Augmented Reality 2000 (ISAR 2000), pp. 120–128. IEEE CS Press, Los Alamitos (2000)
39. Reitmayr, G., Drummond, T.: Going Out: Robust Model-based Tracking for Outdoor Augmented Reality. In: Proc. IEEE ISMAR 2006, Santa Barbara, California (2006)
40. Lepetit, V., Lagger, P., Fua, P.: Randomized Trees for Real-Time Keypoint Recognition. In: Proc. Conference on Computer Vision and Pattern Recognition, San Diego, CA (2005)
41. Rolland, J.P., Baillot, Y., Goon, A.: A Survey of Tracking Technology for Virtual Environments. In: Fundamentals of Wearable Computers and Augmented Reality, ch. 3, pp. 67–112. Lawrence Erlbaum Assoc. Inc., Mahwah (2001)
42. Cruz-Neira, C., Sandin, D.J., DeFanti, T.A., Kenyon, R.V., Hart, J.C.: The CAVE: Audio Visual Experience Automatic Virtual Environment. Communications of the ACM 35(6), 65–72 (1992)
43. Livingston, M.A., Andrei, S.: Magnetic Tracker Calibration for Improved Augmented Reality Registration. Presence: Teleoperators and Virtual Environments 6(5), 532–546 (1997)
44. You, U.N., Azuma, R.: Hybrid Inertial and Vision Tracking for Augmented Reality Registration. In: Proc. IEEE Virtual Reality, pp. 260–267. IEEE CS Press, Los Alamitos (1999)
45. Behringer, R.: Registration for Outdoor Augmented Reality Applications Using Computer Vision Techniques and Hybrid Sensors. In: Proc. IEEE Virtual Reality, pp. 244–251. IEEE CS Press, Los Alamitos (1999)
46. Ibid. 6
47. Rolland, J.P., Holloway, R.L., Fuchs, H.: Comparison of optical and video see-through headmounted displays. In: Proc. SPIE Telemanipulator and Telepresence Technologies, vol. 2351-35 (1994)
48. Pryor, H.L., Furness, T.A., Viirre, E.: The Virtual Retinal Display: A New Display Technology Using Scanned Laser Light. In: Proc. 42nd Human Factors Ergonomics Soc., Santa Monica, Calif., pp. 1570–1574 (1998)

49. Satoh, K., et al.: Case Studies of See-Through Augmentation in Mixed Reality Projects. In: Proc. Int'l Workshop Augmented Reality (IWAR 1998), pp. 3–18. A.K. Peters, Natick (1998)
50. Wagner, D., Pintaric, T., Ledermann, F., Schmalstieg, D.: Towards Massively Multi-User Augmented Reality on Handheld Devices. In: Proc. Third International Conference on Pervasive Computing, Munich, Germany (2005)
51. Schmalstieg, D., Fuhrmann, A., Szalavari, Z., Hesina, G., et al.: The Studierstube Augmented Reality Project. Presence 11(1), 33–54 (2002)
52. Bimber, O., Raskar, R.: Spatial Augmented Reality: Merging Real and Virtual Worlds, 360 pages. A K Peters (July 2005)
53. Azuma, R.T.: The Challenge of Making Augmented Reality Work Outdoors. In: Ohta, Y., Tamura, H. (eds.) Mixed Reality: Merging Real and Virtual Worlds, ch. 21, pp. 379–390. Springer, Heidelberg (1999)
54. You, U.N., Azuma, R.: Hybrid Inertial and Vision Tracking for Augmented Reality Registration. In: Proc. IEEE Virtual Reality, pp. 260–267. IEEE CS Press, Los Alamitos (1999)
55. Thomas, B., et al.: ARQuake: An Outdoor/Indoor Augmented Reality First Person Application. In: Proc. 4th Int'l Symp. Wearable Computers (ISWC 2000), pp. 139–146 (2000)
56. http://www.lifeclipper.net/
57. Junghanns, S., Mendez, E., Schmalstieg, D.: Vidente - Ein Augmented-Reality-System zur Echtzeitvisualisierung unterirdischer Ver- und Entsorgungsinfrastruktur. In: Strobl, J., Blaschke, T., Griesebener, G. (eds.) Angewandte Geoinformatik 2006 - Beiträge zum 18. AGIT-Symposium Salzburg. Wichmann Verlag, Heidelberg (2006)
58. Mann, S.: Wearable Computing: A First Step Toward Personal Imaging. Computer 30(2), 25–32 (1997)
59. Piekarski, W., Gunther, B., Thomas, B.: Integrating Virtual and Augmented Realities in an Outdoor Application. In: Proc. 2nd Int'l Workshop Augmented Reality (IWAR 1999), pp. 45–54. IEEE CS Press, Los Alamitos (1999)
60. Starner, T., et al.: Augmented Reality through Wearable Computing. Presence: Teleoperators and Virtual Environments 6(4), 386–398 (1997)
61. Satoh, K., et al.: Townwear: An Outdoor Wearable MR System with High-Precision Registration. In: Proc. 2nd Int'l Symp. Mixed Reality (ISMR 2001), pp. 210–211. MR Systems Lab, Yokohama, Japan (2001)
62. Saskia, S.: Public interventions – The shifting meanings of the urban condition. In: Open 11 Hybrid Space – How wireless media mobilize pubic space, pp. 18–26. Nai Publishers Skor (2006)
63. Assia, Kraan: To act in public through Geo-Annotation- Social encounters through locative media art. In: Open 11 Hybrid Space – How wireless media mobilize pubic space, pp. 38–47. Nai Publishers Skor (2006)
64. http://en.wikipedia.org/wiki/Warchalking
65. Ibid. 35
66. http://mxrtoolkit.sourceforge.net/
67. http://www.opengl.org/
68. http://r-s-g.org/carnivore/
69. http://www.winpcap.org/
70. http://earth.google.com/
71. http://wikimap.sourceforge.net/Main_Page
72. http://www.stickymap.com/
73. http://www.openstreetmap.org/
74. http://www.google.com/gmm

Appendix 10.1: Toolkits and Authoring for Augmented Reality

AMIRE: http://www.amire.net/
ARTAG: http://www.artag.net/
ARTOOLKIT: http://artoolkit.sourceforge.net/
BAZAR: http://cvlab.epfl.ch/software/bazar/
DART: http://www.gvu.gatech.edu/dart/
JARTOOLKIT: http://jerry.c-lab.de/jartoolkit/
MARS: http://www1.cs.columbia.edu/graphics/projects/mars/
MXRTOOLKIT: http://mxrtoolkit.sourceforge.net/
MIXED REALITY TOOLKIT: http://www.cs.ucl.ac.uk/staff/r.freeman/
MRPLANET: http://sourceforge.net/projects/mrplanet/
OPENILLUSIONIST: http://www.openillusionist.org.uk
OSGART: http://www.artoolworks.com/community/osgart/
STUDIERSTUBE with ARTOOLKITPLUS: http://studierstube.icg.tu-graz.ac.at/

Appendix 10.2: Selection of Research Groups

- Augmented Reality - research projects, Bauhaus University. http://www.uni-weimar.de/~bimber/research.php
- Augmented Environments Lab, Georgia Institute of Technology: http://www-static.cc.gatech.edu/gvu/ael/
- Carnegie-Mellon Entertainment Technology lab: http://www.etc.cmu.edu/about/index.html
- Computer Graphics and User Interfaces Lab, Columbia University. http://www1.cs.columbia.edu/graphics/top.html
- Fraunhofer Institute for Applied Information Technology FIT, Collaborative Virtual and Augmented Environments: http://www.fit.fraunhofer.de/services/cvae_en.html
- HIT Lab, Human Interface Technology Lab Washington: http://www.hitl.washington.edu/
- HIT Lab NZ Human Interface Technology Lab, New Zealand: http://www.hitlabnz.org/route.php?r=home
- Mixed Reality Laboratory, University of Nottingham: http://www.mrl.nott.ac.uk/
- Mixed Reality Laboratory, Singapore http://www.mixedrealitylab.org
- STUDIERSTUBE, Wien and Graz University of Technology: http://studierstube.icg.tu-graz.ac.at/
- Computer Aided Medical Procedures & Augmented Reality, Technischen Univesity. Manchen, Munich: http://campar.in.tum.de/Chair/ResearchAr
- Wearable Computer Lab. University, South Australia. http://www.tinmith.net/wearable.htm

11

Digital Art/Public Art: Governance and Agency in the Networked Commons

Christiane Paul

Adjunct Curator of New Media Arts
Whitney Museum of American Art
945 Madison Ave. New York NY 10021

Abstract. Digital art has expanded, challenged, and even redefined notions of public art and supported the concept of a networked commons. The nature of agency within online, networked "systems" and "communities" is crucial to these developments. Electronic networks enable exchange and collectivist strategies that can question existing structures of power and governance. Networks are public spaces that offer enhanced possibilities of interventions into the social world and of archiving and filtering these interventions over time in an ongoing process. Networked activism and tactical response as well as artistic practice that merges physical and virtual space and augments physical sites and existing architectures are among the practices that are important to the impact of digital public art on governance.

11.1 Introduction

Digital technologies and new media art have expanded, challenged, or even redefined concepts of what constitutes public space, the public domain, and public art. Today's culture is to a large extent revolving around flows — of data (texts, images, and sounds), technologies, communication, and interaction — and supports the concept of a networked commons, which raises questions about agency, control, and governance. As David Garcia has pointed out, "these flows are not just one element in the social organization, they are an expression of processes *dominating* our economic, political, and social life." [1]

This essay will examine how digital art has used electronic networks to redefine the notion of public space by enhancing possibilities of various kinds of interventions. These interventions can take the form of an archiving and filtering of public contributions; a merging of physical and virtual space; an augmentation of physical sites and architectures; social softwares, or collectivist and activist strategies and tactical response.

In this context, it is necessary to consider artistic approaches to the mass media in general, as well as possibilities of understanding the networked commons in relationship to concepts such as authority, control, and governance.

Electronic networks have brought about formal redefinitions of what we understand as "public" and opened new spaces for artistic intervention. So-called "public art" has a long history, and the term has traditionally been used for art that is displayed in public spaces existing outside of a designated art context (in this sense, the museum and

C. Sommerer et al. (Eds.): The Art & Sci. of Interface & Inter. Des., SCI 141, pp. 163–185, 2008.
springerlink.com © Springer-Verlag Berlin Heidelberg 2008

gallery are not a public space); or for performative events in public space (for example works created by art movements such as Fluxus or the Situationists).

An important element in all public art is the varying degree of audience participation and agency. Agency manifests itself in the possibilities for influencing, changing, or creating institutions and events, or acting as a proxy. Degrees of agency are measured by the ability to have a meaningful effect in the world and in a social context, which naturally entails responsibilities.

In "The Artist as Ethnographer," [2] Hal Foster has outlined one of the inherent dangers of "public art" practice: that an artist engaging communities or sites outside of an art context might simply appropriate a community in the creation of a personal or autobiographical narrative of the artist's identity. The worst-case scenario being that a colonizing and romanticized appropriation of a community ultimately becomes a representation that the public identifies with the community itself.

The fact that digital art is inherently interactive, participatory, or even collaborative and — in its networked manifestation — potentially open to exchanges with trans-local communities, makes questions surrounding agency and the authority of authorship a central element of new media art practice. In media art, any form of agency is necessarily mediated. The degree of agency is therefore partly determined by the levels of mediation unfolding within an artwork. The agency of the creator / user / public / audience is also highly dependent on the extent of control over production and distribution of a work, which has been a central issue of the discourse on mass media.

11.2 "Technologies for the People": The Democratization of Mass Media and Its Discontents

Affordable software and hardware, the internet, and mobile devices such as PDAs (Personal Digital Assistants such as Palm Pilots) have brought about a new era for the creation and distribution of media content. The utopian promise of this era is "technologies for the people" and a many-to-many (as opposed to one-to-many) broadcasting system that returns the power over distribution to the individual and has a democratizing effect. The internet promised immediate access to and transparency of data and, in its early days, was dominated by research and educational institutions and a playground for artistic experimentation. The dream of a "network for the people" did not last long and from the very beginning, it obscured the more complex issues of power and control over media. While the internet is hailed as a "global" network, only a portion of the world is connected to it. At a time when the traffic on the information superhighway was consistently increasing in the US, many other countries didn't yet come along for the ride, largely due to the lack of local access and the fees charged by telecommunications companies; wide areas of the world do not have access to the internet and some countries have been subject to government-imposed access restrictions. The internet itself quickly became a mirror of the actual world, with corporations and e-commerce colonializing the landscape. While the burst of the "dot com" bubble ended a lot of the hype surrounding the internet economy and led to reconsiderations of e-commerce, the industry of digital technologies is very much alive.

The potential of a shift to many-to-many distribution networks was recognized much earlier and artists had started to expand the possibilities of the one-to-many broadcasting media at a time when the concept of many-to-many distribution systems was hardly recognized by the public in general. In the 1960s, Max Neuhaus defined new arenas for music performance by staging sound works in public arenas and experimenting with networked sound as a form of "virtual architecture." In the first installment of his project *Public Supply* (1966), he established a connection between the WBAI radio station in New York and the telephone network, implementing a 20-mile aural space around New York City, where participants could intervene in the performance by making a phone call.

Many-to-many distribution also was one of the dreams of video art, the "new" media of the late 20th and 21st century. When Sony portapaks became available in the late 1960s, artists and activists used this portable recording power for establishing alternative media networks, addressing issues of documentation and representation in the context of control over media distribution. Cooperatives and collectives such as Paper Tiger Television, Downtown Community Access Center, Video In (Vancouver), Amelia Productions, Electronic Café International, and the Western Front established and used public video-production facilities and telephone networking, media training initiatives, and cable access for the creation of alternative media networks, linking artists and communities. However, the attempt to establish distribution systems for the public at a larger scale ultimately failed. Apart from the fact that media systems can only be reconfigured with the combined creative endeavors of many individuals, earlier technologies such as video also still required far more complicated processing and distribution facilities than today's new media do.

Using "new technology" such as video and satellites, artists in the 1970s also began to experiment with live, networked performances that anticipated the interactions now taking place on the internet and through the use of streaming media. The focus of these projects ranged from the application of satellite technology for extending the mass dissemination of a television broadcast to the aesthetic potential of video teleconferencing and the exploration of real-time virtual space that collapsed geographic boundaries. At Documenta VI in Kassel, Germany, in 1977, Douglas Davis organized a satellite telecast to more than twenty-five countries, which included performances by Davis himself, Nam June Paik, Fluxus artist and musician Charlotte Moorman, and Joseph Beuys. In the same year, a collaboration between artists in New York (Liza Bear and Willoughby Sharp) and San Francisco (Sharon Grace and Carl Loeffler) resulted in *Send/Receive*, a fifteen-hour, two-way, interactive satellite transmission between the two cities. Also in 1977, what became known as "the world's first interactive satellite dance performance" — a three-location, live-feed composite performance involving performers on the Atlantic and Pacific coasts of the United States — was organized by Kit Galloway and Sherrie Rabinowitz, in conjunction with NASA and the Educational Television Center in Menlo Park, California. The project established what the creators called an "image as place," a composite reality that immersed performers in remote places into a new form of 'virtual' space. In 1982, the Canadian artist Robert Adrian, who had begun working with communication technology in 1979 and created various projects involving fax, slow-scan TV, and radio, organized the event The World in 24 Hours, during which artists in sixteen cities on three continents were connected for twenty-four hours by fax, computers, and videophone and exchanged and created 'multimedia' artworks. All of these performative events were first explorations of the connectivity that is an inherent characteristic of networked digital art.

Digital networks finally allowed a fairly fluent and broader implementation of this many-to-many model. While it would be problematic to forget about the limitations of access to the internet or digital technologies that still exist in large parts of the world, today's networking capabilities by far extend the reach that any of the previously mentioned artistic projects achieved. However, digital technologies are deeply embedded in various layers of commercial systems, and media control does by no means fully lie in the hands of the individual. However, it is the nature of digital technology itself that makes the boundaries of industrial, governmental, and legal control more porous and has redefined traditional systems of media control.

In order to trace the promise and reality of digital technology's potential to remediate these systems, it seems opportune to take a look back at past evaluations of this type of technological promise, most notably the "exchange" that took place in the 1970s between Hans Magnus Enzensberger and Jean Baudrillard on the then "new" media.

Hans Magnus Enzensberger's landmark essay "Constituents of a Theory of New Media" [3] — originally published in the *New Left Review* in 1970 — offered a perspective on the new electronic media of the time that in retrospect seems remarkably visionary and dated at the same time. Informed by an essentially Marxist perspective, Enzensberger saw the media of the 70s as a major reconfiguration of the production process: "For the first time in history, the media are making possible mass participation in a social and socialized productive process, the practical means of which are in the hands of the masses themselves." [4]

Enzensberger sees television or film as media that prevent rather than enable communication since they allow no reciprocal action between transmitter and receiver but reduce feedback to a lowest common denominator. As he points out, this limitation of the communication process mostly is not inherent to the technology itself, which would allow for the reconfiguration of the transistor radio from a receiver into a potential transmitter by circuit reversal. Media equipment is therefore both a means of consumption and production, and the boundary between the distribution and communications medium is a fluid one. The division between receiver and transmitter, as Enzensberger makes clear, reflects the one between producer and consumer. [5]

Revisiting Enzenberger's essay today, it often is easy to forget that he was not writing about the inherent potential of digital networks or the World Wide Web as communications medium. Other conclusions he draws, however, come as a surprise — among them the assumption that the great advantage of a switchable network is that it can no longer be centrally controlled [6] and thus undermines authoritarian, top-down systems. In the age of Echelon and packet-sniffing — the monitoring of network traffic and 'eavesdropping' on the information exchanged — by federal agencies, it is hard to imagine that Enzensberger could not see that control itself can rely on decentralized systems (as Baudrillard would point out in his reply).

The most debatable assumption Enzensberger makes may very well be that "The new media are egalitarian in structure." [7] As Jean Baudrillard points out in "Requiem for the Media," his reply to Enzensberger's essay, "the media are not even, somewhere else or potentially, neutral or non-ideological." [8] Particularly in the context of today's new media, it is crucial to be aware of the encoded agenda — political, commercial etc. — of any hardware or software, which has become a prominent topic in software art.

While Baudrillard appreciates Enzensberger's attempt to go beyond a "dialectic" of transmitter and receiver, he is fundamentally critical of the concept that the media allow mass participation in a productive process: "The mass media are anti-mediatory and intransitive. They fabricate non-communication — this is what characterizes them, if one agrees to define communication as an exchange, as a reciprocal space of a speech and response." [9]

11.3 Governance, Protocol and the "Terrorism of the Code"

Baudrillard's main criticism concerns the very structure of media itself, the transmission-reception process, which — in his opinion — does not allow for response or an exchange of speech. The problem according to Baudrillard lies in the ideological matrix embraced by communication theory (and formalized most notably by Roman Jacobson), which is based on the following sequence:

TRANSMITTER - MESSAGE - RECEIVER
(ENCODER - MESSAGE - DECODER) [10]

Baudrillard calls the above "matrix" a *simulation model* of communication since it supposedly excludes reciprocity of interlocutors and makes a message impossible since it would only exist within the categories of "emitted" and "received." "Terrorism of the code" is how Baudrillard describes this condition since the code — at least in his model — becomes the only agency that speaks. [11]

In the context of today's networked exchanges (be they e-mail, real-time chat, or any other from of communication), Baudrillard's argument at times becomes difficult to follow. Apart from the fact that these exchanges allow for an immediate, real-time response, it is debatable whether the process of "encoding" applies only to technology. One could argue that verbal human exchanges are highly reliant on codes (be they linguistic or social) and therefore are encoded and decoded on the speaker's and listeners end. The ambiguity of the "pure," verbal message is not erased through technological transmission but the latter adds further layers of mediation, which very often increase ambiguity. The seeming need for the so-called emoticons in e-mail messages is one indicator of the insecurities surrounding the proper reception of a message.

What Baudrillard proposes as a solution is a "symbolic exchange relation," in which there is a simultaneity of response: "The symbolic consists precisely [...] in restoring he ambivalence of meaning and in demolishing in the same stroke the agency of the code." [12] Interestingly, Baudrillard seems to see graffiti (a fairly static "response") as such a form of symbolic exchange.

In today's digital societies, it is hard to imagine how to escape the trap of controlled communication, be it on a technological or symbolic level. In contemporary theory, both humans and machines are frequently understood as "coded devices." According to theorists such as Katherine Hayles, we have already become "post-human" — technologically or biologically extended humans. In the context of digital technologies, agency has become distributed in terms of location, and as interconnected with code as with natural language. There seems to be no escape from code (in the broadest sense) and while its agency is much discussed today, it is neither perceived nor examined as a form of terrorism.

In the networked digital world, one layer of control and authority consists of the multiple protocols that enable and determine exchanges. In his book *Protocol - How control exists after decentralization*, Alex Galloway describes protocols — the sets of rules that govern networked relations — as based on two opposing technologies: one distributing control into autonomous locales, the other centralizing it in defined hierarchies, with the tension between the two creating the conditions for protocological control. [13]

Among the many protocols that control network relations are those enabling data transmission over the internet, such as TCP / IP (Transmission Control Protocol / Internet Protocol) and UDP (User Datagram Protocol); the Domain Name System (DNS), which handles internet addresses; and the Hypertext Transfer Protocol (http), which enables the retrieval of documents over the World Wide Web. The tension between autonomy and hierarchy on the internet becomes obvious in the difference between client-server relationships (allowing a client to retrieve information from a server via a personal computer) and peer-to-peer ones (providing a direct link for exchange between computing devices). Peer-to-peer as opposed to client-server becomes a philosophical as well as political issue: peer-to-peer is the promise of the liberation from the server as a hierarchical structure.

It would be a misconstruction to understand digital networks as either democratizing and empowering the consumer or completely determined by control mechanisms and the technological industry. The reality is closer to a "both / and." The existence of networks has opened up new spaces for autonomous producers and DIY culture, as well as the industry of market-driven media. The same technologies can often be applied to very different ends and effects, as the project *Carnivore* by Alex Galloway and the RSG (Radical Software Group) perfectly illustrates. Inspired by the packet-sniffing software DCS1000 (once nicknamed "Carnivore") that is used by the FBI to perform electronic wiretaps and search for certain "suspicious" keywords, *Carnivore* consists of an application that performs packet-sniffing on a specific local area network and serves the resulting data stream, as well as the "client" applications created by numerous artists, which interpret the data in visual ways. The project makes the source code of the software available to anyone interested in using it — as opposed to limiting its use for the purpose of surveillance — and defies an easy categorization of surveillance as either positive or negative.

In *Empire*, Hardt and Negri argue that the new paradigm of the global world order is configured as a dynamic and flexible systemic structure that is constructed horizontally — a "governance without government" that subsumes any "actor" (and, one would assume, agency) under the totality of the order of the whole. The supreme authority of the ordering effectively integrates everything and at the same time calls for more central authority. Hardt and Negri think of this "governance without government" as a machine that predetermines the exercise of authority and action across the entire social space where every movement can find its designated place only within the hierarchical relationship imposed on it by the system itself. [14]

While one can understand Hardt's and Negri's imperial world order as a unique mode of economic, political and cultural organization in general, it seems harder to apply it to the technological network of the internet in specific. The previously mentioned "both / and" structure of the internet certainly involves numerous protocols but at the same time, every module or protocol in this structure inherently encapsulates the possibility of both command and control and is configurable: what constructs control and authority also encapsulates the possibility of undermining and dismantling it. In the context of the networked commons, the concept of "governance without government" could also be revisited in terms of the interplay between openness to public participation vs. rules and mechanisms of access.

11.4 The Networked Commons

Main Entry: common
Function: noun
Date: 14th century

1 plural : the common people
2 plural but singular in construction: a dining hall
3 plural but singular or plural in construction, often capitalized
  a: the political group or estate comprising the commoners
  b: the parliamentary representatives of the commoners
  c: HOUSE OF COMMONS
4 the legal right of taking a profit in another's land in common with the owner or others
5 : a piece of land subject to common use: as
  a: undivided land used especially for pasture
  b: a public open area in a municipality
(WWWebster dictionary) [15]

In its original meaning, the term "commons" refers to land or a public area that is open to common use, the group of the commoners or their parliamentary representatives. In 2001, the founders of the Sarai New Media Initiative in Delhi published a reader on the public domain and introduced the term "Digital Commons." [16]

The idea of the digital or networked commons obviously requires a reconsideration of traditional definitions: the public space here is not a shared territory but a non-locality consisting of global communication systems that, while subject to protocols and regulations, largely exist outside of a single nation's or state's jurisdiction; the "commoners" also can not be defined strictly in terms of physical location but often are communities of interest that share ideas and knowledge and are dispersed around the world.

The concept of the (networked) commons is also inextricably interconnected with the notion of the public domain, which — as a social and cultural space — can be understood as a shared site of ideas in the broadest sense. In 1998, the Society for Old and New Media (De Waag) in Amsterdam started a research project titled "Public Domain 2.0," which was an attempt to reassert public agency in the information age and "address the conditions of the unfolding era of global information and communication systems." [17] The goal of the project is to design future public spaces in digital media environments that are monopolized by neither commercial interests nor a state and driven by active public participation.

The narrower, juridical and computing definitions of the public domain are rooted in notions of property right and copyright and point to the complex legal issues raised by digital technologies and networks and their inherent capabilities for appropriation and sharing.

Juridical Definition:
1: land owned directly by the government
2: the realm embracing property rights that belong to the community at large, are unprotected by copyright or patent, and are subject to appropriation by anyone
(Date: 1832)
(WWWebster dictionary) [18]

Computing definition:
(PD) The total absence of {copyright} protection. If something is "in the public domain" then anyone can copy it or use it in any way they wish. The author has none of the exclusive rights which apply to a copyright work.

The phrase "public domain" is often used incorrectly to refer to {freeware} or {shareware} (software which is copyrighted but is distributed without (advance) payment). Public domain means no copyright — no exclusive rights. In fact the phrase "public domain" has no legal status at all in the UK.
[The Free On-line Dictionary of Computing] [19]

In "Constructing the Digital Commons," [20] Eric Kluitenberg refers to writer and policy strategist David Bollier's argument that the concept of the public domain and the commons should be differentiated from each other. [21] Bollier distinguishes between the public domain as a passive open space that can be shared by anyone and everyone, implies no boundaries and ownership and therefore does not require responsibility for resources; and the commons as a space of shared resources (land, means of production, information) that is collectively owned by a more or less well-defined community and therefore implies boundaries: "There are rules and mechanism of access, and limitations on use that are defined by the shared values of the community sharing these resources." [22]

While Bollier's distinction is helpful and makes an important point, the boundaries between the public domain and digital commons can still be fluid. When it comes to art in the public space of networks, concepts such as passive vs. active space (agency), collective owner- and authorship, as well as rules and mechanisms of access are a complex interplay between technologies, software, authors, and users.

11.5 Art in the Networked Commons

Networked new media art that exists in the public space of networks — be it internet art or art involving mobile media such as cellphones and PDAs — can be understood as public art. Compared to more traditional forms of public art practice, internet art, which is accessible from the privacy of one's home, introduces a shift from the site-specific to the global, collapses boundaries between the private and public, and exists in a non-local space.

As other manifestations of new media art (such as networked installation or virtual / augmented reality works), net art can support varying degrees of interaction, ranging a from trigger-response interaction within a closed system or a relatively preconfigured database of elements; and participation within parameters set by artists; to the creation of these parameters and rules.

Fostering audience agency is an activist goal for many artists, and digital technologies have at least increased the technological possibilities for agency, even if these possibilities are not necessarily fulfilled. In the context of technological environments, one always needs to consider the respective degrees of agency of authors and participants, softwares, and systems.

Art in the networked space enables various kinds of interventions, which will be discussed in the following with regard to the idea of the shared production, and information resources of the digital commons.

11.5.1 Filtering and Archiving Public Contributions

Digital technologies offer enhanced possibilities of archiving and filtering public contributions over time in an ongoing process, which has become an underlying mechanism of many net art projects. In these cases, the creation of meaning is obviously dependent on both public participation and the respective filtering mechanisms.

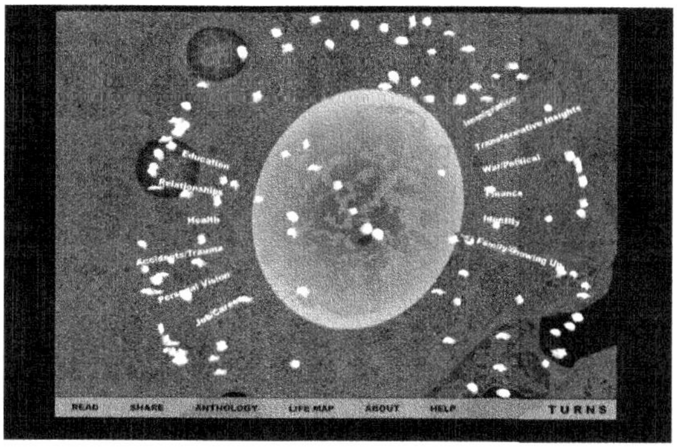

Fig. 11.1. Margot Lovejoy, *Turns*, screenshot

Margot Lovejoy's *Turns* [Fig. 11.1], for example, invites visitors to share personal stories of life turning points. On the website, stories are represented as pebble-like shapes that can be opened and returned to the narrative pool. The stories can be browsed according to 12 categories (such as education, relationships, health, trauma, family or war) and can be filtered according to gender, ethnicity, or the time at which the turning point was experienced. While the individual stories may be of varying quality or interest, they become a more complex social memory through the relational filters and lenses. The site becomes a reflection on the ways in which new media are influencing and changing notions of the individual in a social context.

A very different type of filtering unfolds in Warren Sack's *Agonistics - A Language Game* [23] [Fig. 11.2], which creates a space of interactive, graphical objects and dynamics inspired by the concept of "agonistics" (the science of athletic combats, or contests in public games). Theorists such as Chantal Mouffe have been interested in the democratic potential of agonistic contests, using metaphorical images and actions to describe verbal contests as a language game. Sack's project draws on these ideas and applies them to online discussion forums.

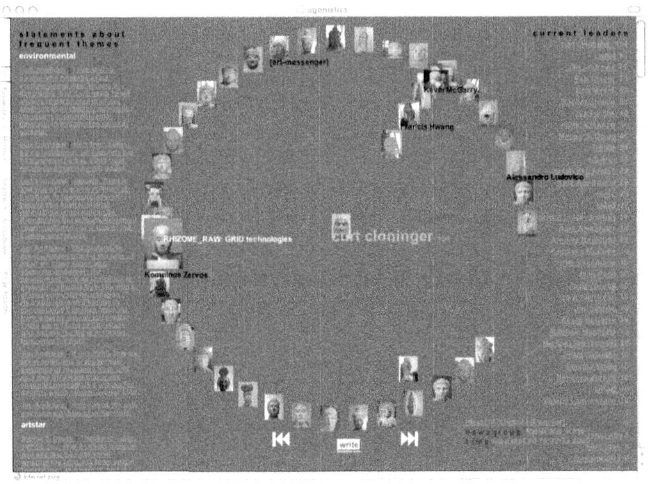

Fig. 11.2. Warren Sack, *Agonistics — A Language Game*, screenshot

Using any e-mail program, players can post to online, public discussions (for example, Usenet newsgroups or the Rhizome mailing list), and the project then translates players' messages into a graphical display where participants are represented by thumbnail images. Players are assigned a position on "fields" depending on the content of their message and are placed in relation to the other players who posted a message on the same theme. After each new message posted to the discussion, everyone's position is algorithmically recomputed. By posting a message to the discussion that voices a specific opinion about a theme, players can move themselves closer to or farther away from certain other players.

While both *Turns* and *Agonistics* enable participation and filtering on the basis of rules that are established by the artists (and the algorithms they use) and can be performed by participants, they create an enhanced awareness of an individual's "positioning," be it in a social context or in the ways they express their opinion. "Systems" and "communities" are traditionally understood in opposition to the privileging of the individual but systems can create narratives that highlight relationships between individuals.

As Sharon Daniel has argued, the increasing reliance of culture(s) and social systems on networks of exchange and economies of relation has induced a shift in art practice from individual authorship to models based on self-organizing systems. [24] However, the openness of so-called self-organizing system still varies considerably. Katherine Hayles has pointed out that such systems are still often "informationally closed" since they respond to stimuli based on their own, internal self-organization. [25] The transformation of a system through input from collaborating participants occurs in the acts of interpretation, translation, manipulation, contribution, and recombination of data.

11.5.2 Collaborative Creation

The shift from individual authorship to a collaborative creation process can manifest itself in various models, including public contributions to systems established by artists or the collaborative creation of the underlying system for the artwork itself. An example would be Natalie Bookchin's *agoraXchange* [26], an online community for designing a massive multi-player global politics game aimed at questioning violence and inequality of present political systems and exploring issues of government and governance. The project was commissioned by Tate Online and launched on March 15, 2004. The project explicitly invites participation by individuals, groups, classes, or organizations and the development takes place in a collaborative virtual space called the "Game Design Room." It is interesting to note that *agoraXchange* establishes certain governing rules by asking that proposals must be consistent with the four Decrees of the project (citizenship by choice, not birth; no inheritance; no rules for kinship relations established by a state; no private landrights).

Issues similar to the ones addressed in *agoraXchange* are explored in the online simulation game *NationStates* [27], which was not conceived as an art project but as a promotion for Max Barry's novel *Jennifer Government*, on which the game is based. While the book is set in an ultra-privatized world, the game allows players to create any type of nation. They are asked to choose a name, motto, national animal, and currency for their nation and then have to answer a short questionnaire about their politics. The questionnaire determines the type of nation, for example authoritarian or permissive; left-wing or right-wing; compassionate or psychotic. Once a day, players are faced with an issue — written by the author or a player and/or edited by a moderator — and need to make a decision about it, which in turn determines how the nation evolves.

NationStates functions on the basis of three main scales: personal, economic, and political, which each can be authoritarian (moral, or restrictive) or libertarian (liberal, or laissez-faire). On each of the three main scales, nations are ranked as having high, average, or low amounts of freedom (or permissiveness). On the basis of the rankings, nations are assigned one of 27 possible labels by the UN, the world's governing body that proposes and votes on resolutions, which are then binding on all member nations.

There is no way of "winning" the game per se although making it onto the top rungs of a United Nations report certainly is a measure of success. The reports, which rank nations on anything from economic strength to the most liberal public nudity laws, are compiled once per day, one for each Region and one for the entire world. While *NationStates* relies on ultimately very simple rules and governing systems, and gives players only limited control over the design of the system itself, it is an interesting take on the interplay of freedom and control (and governance without government).

While *NationStates* relies on contributors submitting to an established system (that gives limited influence over shaping its framework), *agoraXchange* increases the degree of agency by allowing participants to develop the system itself. This form of distributed creativity obviously raises questions regarding authorship and crediting, particularly in an art context, which traditionally favors the "single creator" model. A

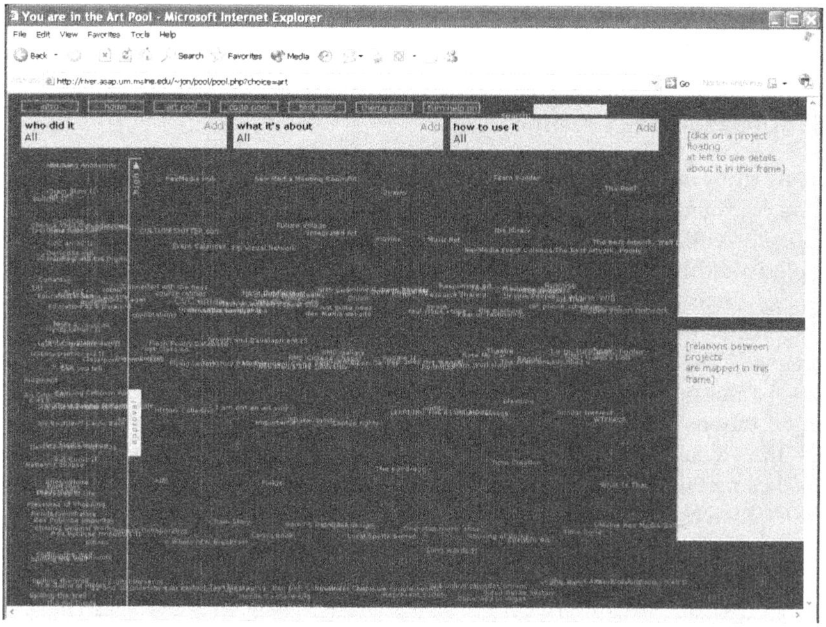

Fig. 11.3. U Maine Still Water Lab, *The Pool*, interface screenshot

project addressing these issues is *The Pool* [Fig. 11.3], developed by Jon Ippolito, Joline Blais and collaborators at the University of Maine's Still Water Lab. [28]

The Pool was specifically designed as an architecture for asynchronous and distributed creativity and documents the creative process in different stages: the "Intent," a description of what the artwork might be, an "Approach" to how it could be implemented and a "Release" of the artwork online. The architecture also includes a scaling system that allows visitors to the site to rate any given project. *The Pool* supplies descriptions of projects' versions, reviews of the projects, as well as relationships to other works in the database. Tags to contributors make it possible to credit all the artists who have worked on a project at any given stage. *The Pool* illustrates the shifts in the paradigm of culture production induced by the digital commons where a whole culture can be built upon seed ideas and different iterations of a particular project.

11.5.3 Interventions in Virtual Public Spaces

Within the networked commons, public space can take several forms. On the one hand, one could consider the whole internet as a public space — governed by multiple layers of protocols and providing different levels of access. Within this macrocosm, individual projects and sites can again create public spaces, dependent on their openness to public contribution. As in physical public space, these environments allow for different kinds of interventions, ranging from activist (public protests and civil disobedience) to more aesthetically oriented ones (similar to interventions in public, performances or those by Fluxus or the Situationists). These interventions often take place in gaming environments, which often support a fairly high level of agency due to the openness of their architecture.

In 1997, for example, a naked riot was staged in the popular game *Ultima Online* in order to demand bug fixes and server upgrades. Another famous online protest was the "adbusting" campaign Big Mac Attacked, which took place in the online version of the game *The SIMS*. [29] The intervention was prompted by the fact that EA, the company who created *The SIMS*, incorporated McDonald's kiosks in their online game. This form of deeply-integrated marketing led to an organized protest that asked players to picket McDonald's kiosks and tell visitors what they thought of McDonald's food and business strategies; to order and consume virtual McDonald's food and then use *The SIMS Online*'s "expressive gestures" feature to emote vomiting, sickness, or fatigue; or to open an independent restaurant and ask other Simians to support small business people instead of McDonald's franchise-machine.

Fig. 11.4. Schleiner, Leandre, Condon, *Velvet-Strike*, screenshot

Another critical intervention in an existing game is the activist art project *Velvet-Strike* [Fig. 11.4] by Anne-Marie Schleiner, Joan Leandre and Brody Condon [30], which was conceptualized as a direct response to President Bush's War on Terrorism. *Velvet-Strike* is a collection of graffiti that can be "sprayed" on the walls and rooms of the shooter game *Counter-Strike*, a multi-user game that allows participants to play members of either a terrorist group or counter-terrorist commandos. Putting the "weapon" of public opinion back into the hands of the players, *Velvet-Strike* enables users to spray their anti-war graffitis (one of them reads "Hostages of Military Fantasy") onto the walls of the game environment. The project led to massive protests and hate mail campaigns by the devoted players of *Counter-Strike*.

An online art project less focused on protest but intended as a parody on different forms of popular entertainment and their respective "territories," was Joseph DeLappe's *Quake/Friends* [Fig. 11.5] performance series, in which he staged episodes from the TV sit-com *Friends* in the online shooter game *Quake III Arena*. DeLappe describes the work as "a temporal occurrence of clashing inanities." [31]

The first *Quake/Friends* performance took place on October 18th, 2002. Seven performers connected to the same *Quake III Arena* game server online and — instead of participating in the 3D game — recreated an episode from *Friends* by logging in as one of the characters from the show, and then using the game's online messaging system by typing, while at the same time reciting, the lines from the show. The performers were constantly killed and reincarnated to continue the performance. A few days before the second performance was to take place in front of a live audience

Fig. 11.5. Joseph DeLappe, *Quake/Friends*

on March 8th, 2003, at the Sheppard Fine Arts Gallery of the University of Nevada in Reno, DeLappe received notice from the Warner Brothers legal department that they were concerned over possible copyright infringement over the use of *Friends* material in his performance works. Warner Brothers requested that the artist would cease to use *Friends* material in his work and remove all copyrighted *Friends* material from his website. DeLappe considers his work parody and thus protected by "fair use" standards established by the United States Supreme Court but reached a compromise with WB: he agreed to not perform a new script from the *Friends* TV show but use the same episode that was previously performed.

Similar projects are Joseph DeLappe's readings of selected works of WWI poet Siegfried Sassoon in the online WWII war game *Medal of Honor* and Adriene Jenik's and Lisa Brenneis' *Desktop Theater* [32] where the artists "invade" *The Palace* environment and use their avatars — graphic representations of themselves — to stage performances, such as an adaptation of Beckett's *Waiting for Godot*.

While the above mentioned projects essentially use the same strategies and methods as protests and performances in public space, they sometimes directly disrupt or "rewrite" a commercial software environment.

11.5.4 Collaborative Mapping of Physical Space

Network technologies have become all-pervasive and it would be problematic to understand the internet and networks in general as a purely virtual territory that has no connection to our physical environment. Wireless networks and the use of "nomadic devices" such as cell phones and PDAs, in particular, have blurred the boundaries between the non-local and locative, and locative media has become one of the most active areas in new media art.

A number of projects have focused on mapping and enhancing existing physical spaces and architectures. *PDPal* [Fig. 11.6] by Scott Paterson, Marina Zurkow and Julian Bleecker, for example, is a mapping tool for recording personal experiences of public space, more specifically, the Times Square area in New York City and the Twin Cities, Minnesota. [33] The tool is available on the Web and can also be downloaded to one's PDA. Users create maps by marking locations with little graphic symbols and giving them attributes and ratings. While the categories for mapping are relatively preconfigured, the prescription of certain categories or meta-tags also allows a more

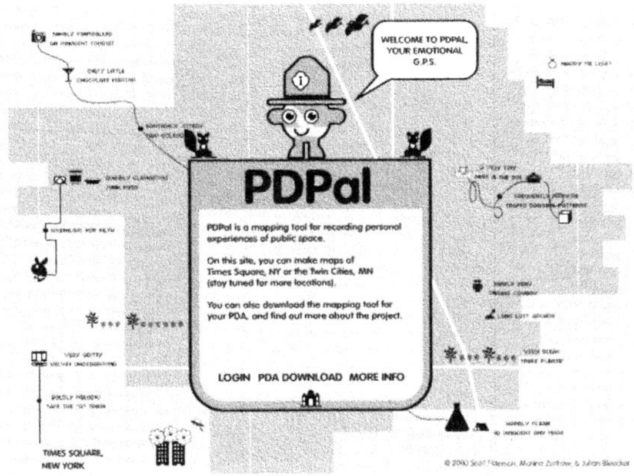

Fig. 11.6. Paterson, Zurkow, Bleecker, *PDPal*, screenshot

effective mapping of all the contributions. *PDPal* is inspired by the idea of emotional geographies — as opposed to a traditional cartography based on static sites — and the Situationists' concept of "psychogeography," which is frequently referenced in the realm of locative and mobile media.

Similar works are *MapHub*™ [34] — developed by members of the media arts and writing collective Carbon Defense League and supported by the Studio for Creative Inquiry at Carnegie Mellon University — and Q.S. Serafijn's *D-tower* [35], an art piece commissioned by the city of Doetinchem in the Netherlands and co-developed by V2_lab. Exploring the idea of shared urban storytelling, *MapHub*™, currently only available in Pittsburgh, allows users to place people, places, events, or notes on an interactive map and attach audio, video, or images. People can also use their mobile phones to call into the system and add a note. *D-tower* maps the emotions of the inhabitants of Doetinchem in a more specific way than *PDPal* by concentrating on happiness, love, fear, and hate. In *D-tower* human values and feelings become networked entities that also manifest themselves in physical space. The project consists of a physical tower, a questionnaire, and a website. Participants receive four questions every other day and the project translates their answers directly into a graph. The Web interface consists of "landcapes" that, respectively, represent an accumulation of all answers and the answers to every single question in the form of peaks and valleys. All results are connected to the map of the city of Doetinchem according to the zip codes of participants and show where in the city people are most scared or in love, and for what reason. The physical tower [Fig. 11.7], designed by NOX, is a 36 feet-high structure of geometries formed by polyester surface that has been computer-generated (through CNC milling). The four emotions charted in the projected are represented by four colors, green, red, blue and yellow, which determine the color of the lamps illuminating the building. Driving through the city, people can see which emotion is most deeply felt on that particular day.

In different ways, all of these mapping projects create a virtual, public repository for information that supplements physical sites. In terms of the definition of the commons, they consist of shared information resources that are collectively built by a

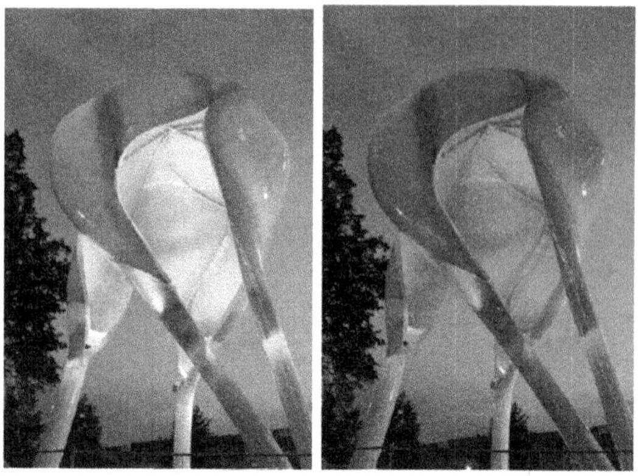

Fig. 11.7. Q.S. Serafijn, NOX, *D-tower*

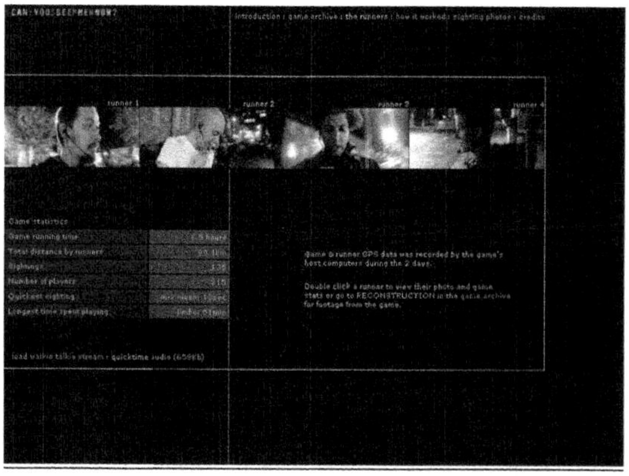

Fig. 11.8. Blast Theory, *Can you see me now?*, screenshot of Web interface

more or less well-defined community and involve boundaries established through rules and mechanisms of access. In the case of *PDPal*, the information can also be accessed at the physical site itself and in *D-tower*, a physical structure is transformed through the virtual repository. The enhancement of physical sites also unfolds in the form of locative media projects that provide public access to location-specific information via wireless networks.

At this point in time, there are relatively few artworks that have attempted to merge virtual and physical space, creating a one-to-one relationship between the spaces or a so-called mixed reality (which has been explored mostly within a gaming context). An example of this type of artwork would be Blast Theory's mobile game *Can you see me now?* [Fig. 11.8], which essentially takes the form of a chase where online players navigate their avatar through the streets of a city map in order to escape from "runners" in an actual city who are hunting them. The runners — equipped with a handheld computer cum GPS tracker that sends their position to online players via a wireless network — attempt to "catch" the online players whose position is in turn sent to the runners' computers. The virtual players can send text messages to each other and receive a live audio stream from the runners' walkie talkies. The game is over when runners "sight" their virtual opponents and shoot a photo of them (which obviously just captures empty space). The game achieves a noteworthy level of merging and collapsing physical and virtual space and raises profound questions about embodiment in these different sites. Blast Theory's project operates on the boundaries of telepresence and -absence: through networking, absence creates a presence in its own right that is absurdly documented in the sightings photos. Photography, an established mode of technological representation, becomes obsolete in the face of a presence — unfolding through virtual movements — that leaves no physical trace. As its title indicates, the project questions the very process of seeing itself, suggesting a form of perception independent of embodiment. At a time where GPS technology and "networked cells" are mostly associated with destructive or negative potential (surveillance, war machinery, and terrorism), Blast Theory emphasizes the creative possibilities of the human and technological network.

11.5.5 Remote Intervention in a Site-Specific Environment

A connection between the physical and virtual also occurs in the numerous telepresence or telerobotics projects that establish connections between remote locations or allow users to intervene in a site-specific installation via the internet. Examples would be the "Relational Architecture" projects by Raphael Lozano-Hemmer [36], among them *Vectorial Elevation*, which allowed the public to transform an urban landscape by means of more than a dozen robotically controlled gigantic searchlights that could be positioned via a website; and *Amodal Suspension* [Fig. 11.9], a large-scale interactive installation developed for the opening of the Yamaguchi Center for Arts and Media (YCAM) in Japan in 2003. Using a cell phone or web interface, people could send short text messages to each other, which were encoded as unique sequences of flashes by 20 robotically controlled searchlights, which created a giant communication switchboard in the sky around the YCAM Center and transformed the materiality of text messaging. Messages were removed from the sky if someone would "catch" them with a cell phone or the 3D Web interface.

Fig. 11.9. Raphael Lozano-Hemmer and team, *Amodal Suspension*

Fig. 11.10. Goldberg, Santarromana (and project team), *Telegarden*, networked telerobotic installation

A classic of these telerobotic projects is Ken Goldberg's and Joseph Santarromana's *Telegarden* [Fig. 11.10] installation, which was accessible online from 1995 until 2005. The work, which was originally located at the University of Berkeley and then permanently installed at the Ars Electronica Center in Linz, Austria, consists of a small garden with living plants and an industrial robot arm that could be controlled through the project website. Remote visitors, through moving the arm, could view and monitor the garden as well as water it and plant seedlings. *Telegarden* explicitly emphasizes the aspect of community by inviting people around the world to collectively cultivate a small ecosystem. Survival of the ecology is dependent on a remote social network. The *Telegarden* is particularly interesting to consider in the context of the commons *as* "a piece of land subject to common use" since it transcends the temporal and spatial continuity that is characteristic of agriculture.

11.5.6 Social Software — Tools for Representing Communities

The concept of networked commons also features prominently in so-called "artware," that is, alternative models for media systems and tools that are "not just art" but proposals for the restructuring or critique of existing media systems.

The inherent hope and promise here is that software production can be seen in the broader context of cultural production or, as Pit Schultz has put it, "that writing code has more meaning than making a program run or crash or sell." [37] Software always has to be seen as cultural construct, and the creation of artware addresses this construct from various angles, including the enhancement or re-engineering of existing software products; the creation of alternative, community-driven platforms of exchange; and the examination of agency, autonomy, or political agendas in software.

A wide area of artware consists of "social software" — tools that are aimed at providing platforms for community-based exchanges and publishing. An example of this type of project would be *Nine(9)* by the British collaborative Mongrel. *Nine(9)* is a continuation of Mongrel's project *Linker* [38] and was created by Mongrel member Harwood while he was artist-in-residence at the Waag Society Amsterdam. The project is an open-source software structure that allows individuals and communities to "map" their experiences and "social geographies." *Nine(9)* consists of a server-based application that can incorporate 9 groups x 9 archives x 9 maps = 729 collective knowledge maps. An important part of the project as "social software" is an ongoing dialogue between users and programmers in order to transcend standardized social relations. *Nine(9)* obviously plays with limitations — in structure or functionality, respectively — to test and explore possibilities of software.

Other projects, such as *Liken* by criticalartware (Ben Syverson, Jon Cates, Jon Satrom and Blithe Riley — core developers) investigate community-driven interfaces for social software. [39] *Liken* is a Web interface (with various different manifestations) to criticalartware's database of shared resources that present themselves as self-connecting nodes to which users can contribute. The pathways connecting the nodes change on the basis of usage, with more "traveled" paths growing stronger and paths attracting less interest fading away. Criticalartware's approach is that of hybridization, a self-reflexive crossbreeding of interfaces and connected threads that becomes a social document in itself.

An excellent portal for exploring free software tools for collaborative networking and media production is the *DIVE* CD-ROM, which was created by <KOP> (Kingdom of Piracy) and commissioned by the VirtualCentre-Media.Net and FACT, UK. The CD-ROM includes projects such as Mongrel's *Nine(9),* Radioqualia's *Frequency Clock*, and *LAST.FM*, a peer-to-peer network for streaming customized selections of music. [40]

11.5.7 Political Activism, Hacktivism and Tactical Media

Social softwares can be considered a subcategory of activism, which has a long history in art and has continuously addressed issues such as support of underrepresented communities, racism, gender-roles, and the control of media and information. The possibilities of exchange in the networked commons have revitalized collectivist strategies by providing new ways of interconnecting individuals and groups, as well as new means of challenging established structures of governance and power. The philosophy of free data and information exchange is a driving force behind the open source (and Copyleft) movement, which promotes unrestricted redistribution and modification of source code (provided that all copies and derivatives retain the same permissions).

The growing importance of control over information, privacy, and data protection and the public debate surrounding these issues have made activist art a new force that the art world cannot afford to ignore and several art exhibitions have been dedicated to activism, hacking, and open source in the information age. [41] The presentation of this type of work in an institutional context obviously raises its own set of issues since the work itself may run counter to what an institution represents or create legal conflicts that museums are not prepared to face.

Among the artists' groups that have critically examined authoritarian culture as it manifests itself through the use of media are the Surveillance Camera Players — who have staged performances in front of surveillance cameras — and the Critical Art Ensemble (CAE). The charges brought against CAE's Steve Kurtz and his continuing legal battle perfectly illustrate both the US authorities' sensitivity to critical investigations of policies and the restrictive authoritarian logic governing the current political climate.

Activist art collectives addressing issues of control systems and mechanisms obviously also have to pose questions regarding their own internal organization. As McKenzie Wark has pointed out, the internal structure of these groups can easily come to reflect the power relations of the outside world; a more "external" problem might be that a group attracts media attention and is pushed into merely reacting to their own media image. [42] As Wark rightly states, avant-garde groups of the past — Dada, Surrealism, the Situationists, the Living Theater, Art & Language — have often had a fairly troubled history when it came to their internal structure, relying on hierarchical or even dictatorial practices, or exhibiting a "Warhol syndrome of factory production with underpaid laborers." CAE, for example, prefer to work in cellular structures with a floating hierarchy rather than a "community," which still often has a hierarchy of representatives.

Activist projects in the realm of digital art are frequently using strategies such as appropriation, remixing, and the cloning of websites, or employ digital technologies as "tactical media" in order to reflect on the impact and control mechanisms of these technologies. A popular strategy is to turn the technology back on itself, as the Institute for Applied Autonomy does in its *iSee* project. [43] *iSee* is a Web-based application that creates maps of the positions of closed-circuit television surveillance cameras in urban environments. The maps allow users to find routes through the city that avoid these cameras and to choose "paths of least surveillance."

Activist interventions occasionally take the form of hacktivism, a method of engagement that uses hacking — the breaking, reformatting and re-engineering of data and systems — as creative rather than merely destructive strategy. The spectrum of hacktivism encompasses projects that are harmless pranks and interventions that operate on the border of legality. The Electronic Disturbance Theater, for example, frames its actions as "electronic civil disobedience" [44] and has staged a number of virtual "sit-ins" in support of the Zapatista rebels in Chiapas, Mexico, by using self-authored Web-based software called FloodNet for disrupting the service of targeted websites (such as the sites of the president of Mexico and the US Department of Defense).

Etoy [45] as well as the artists collective ®TMark [46] are examples of artists group that use corporate strategies and a corporate image to frame and construct their 'intellectual product.' ®TMark uses a mutual funds model to raise fund for and support project ideas suggested by internet users. The projects are usually aimed at undermining or providing a counterbalance to corporate interests. Corporate models in activist art point to the fact that the internet radically reconfigures context and the boundaries of the physical world: on the Web, every artist's project is always embedded in (only one click away from) the context of corporate sites and e-commerce.

The alternative space of the internet resists our traditional, physical model of ownership, copyright, and branding. As an open system and archive of reproducible data, the Web invites or allows for instant recontextualization of any information. The virtual real estate of a company or institution can easily be copied ("cloned") and reinserted into new contexts, a tactic that many artists, net activists or hacktivists have pursued. When Documenta X decided to "close down" its website after the end of the physical exhibition, the artist Vuk Cosic cloned the site, which remains available online until today. [47] Cosic also created 7-11.org, mimicking the website of the popular American convenience store to create a "convenient" platform for exchange among artists. The project *Uncomfortable Proximity* [48] by Mongrel's Harwood, the first piece of net art commissioned by the Tate Museum for its website, is a perfect example of shifting institutional contexts: reproducing the Tate website's layout, logos, and design, Harwood tells a history of the British art system that may be less than comfortable for an art institution.

Due to their focus on control and authority, activist art practice and tactical media in the public space of the networked commons make issues of governance surface most prominently. The networked commons has certainly redefined notions of what "public art" is and can be, particularly when it comes to the notion of space, which becomes a distributed non-locality. One can argue that a networked environment increases the public's agency in several respects — for example through enhanced distribution, filtering, and archiving mechanisms that give importance to an "individual's voice;" through the fact that intervention is not necessarily bound to a geographic space any more; and through a largely decentralized rather than hierarchical structure. This obviously does not mean that authority itself has been eliminated. As Charles Bernstein has put it, "Authority is never abolished but constantly reinscribes itself in new places. ... Decentralization allows for multiple, conflicting authorities, not the absence of authority." [49] In general, agency has become considerably more complex through the process of technological mediation.

One of the most fundamental differences between the degrees of control and agency in analog and digital media lies in the nature and specifics of the technology itself. Media such as radio, video or television mostly relied on a technological superstructure of production, transmission, and reception that was relatively defined. The modularity and variability of the digital medium however constitutes a far broader and more scattered landscape of production and distribution. Not only is there a plethora of technologies and softwares, each responsible for different tasks (such as image manipulation, 3D modeling, Web browsing etc.) but due to the modularity of the medium, these softwares can also potentially be manipulated or expanded. As a result, a certain singularization occurs in the numerous potential points of intervention for artistic practice. In this respect, the "new media" may not completely redefine connections between art and media but they certainly have opened the field for artistic engagement, agency, and conflicting authorities.

References

1. Garcia, D.: Some thoughts on the Public Domain (1998), http://www.debalie.nl/artikel.jsp?articleid=12829
2. Foster, H.: The Artist as Ethnographer. In: The Return of the Real, pp. 171–204. The MIT Press, Cambridge (1996)
3. Enzensberger, H.M.: Constituents of a Theory of New Media. In: Wardrip-Fruin, N., Montfort, N. (eds.) The New Media Reader, p. 261. The MIT Press, Cambridge (2003); Original Publication in: New Left Review 64, 13–36 (November/December 1970)
4. Ibid. 262
5. Ibid. 262-165
6. Ibid. 262
7. Ibid. 272

8. Baudrillard, J.: Requiem for the Media. In: Wardrip-Fruin, N., Montfort, N. (eds.) The New Media Reader, p. 280. MIT Press, Cambridge (2003); Original Publication in: For a Critique of the Political Economy of the Sign., pp. 164–168. Telos Press, St. Louis, MO (1981)
9. Ibid. 280
10. Ibid. 284
11. Ibid. 285
12. Ibid. 287
13. Galloway, A.: Protocol – How control exists after decentralization. The MIT Press, Cambridge (2004)
14. Hardt, M., Negri, A.: Empire. Harvard University Press, Cambridge (2000)
15. WWWebster Dictionary, http://www.m-w.com/dictionary
16. Sarai Reader 2001 – The Public Domain, http://www.sarai.net/journal/reader1.html
17. Frequently Asked Questions about the Public Domain (Version 6.0 – February 2004), This FAQ about the public domain has been re-edited for the fourth time after it first appeared in Dutch language. The original Dutch version was the result of an extensive Public Research called Public Domain 2.0, carried out by the Society for Old and New Media (De Waag) in Amsterdam in the beginning of 1998, http://www.debalie.nl/artikel.jsp?articleid=12829
18. WWWebster Dictionary, http://www.m-w.com/dictionary
19. Free On-line Dictionary of Computing, http://wombat.doc.ic.ac.uk/foldoc/
20. Kluitenberg, E.: Constructing the Digital Commons (2003), http://www.n5m4.org/index6f5c.html?118+575+3411
21. Bollier, D.: http://www.bollier.org
22. Kluitenberg, E.: Constructing the Digital Commons (2003), http://www.n5m4.org/index6f5c.html?118+575+3411
23. Sack, W.: Agonistics - A Language Game, http://hybrid.ucsc.edu/Agonistics/RHIZOME_RAW/Interface/agon1.html http://hybrid.ucsc.edu/Agonistics/
24. Daniel, S.: Systems and Subjects: Redefining Public Art. In: Lovejoy, M., Paul, C., Vesna, V. (eds.) Context Providers, University of Minnesota Press, Minneapolis (forthcoming)
25. Hayles, N.K.: Liberal Subjectivity Imperiled: Norbert Wiener and Cybernetic Anxiety. In: How We Became Posthuman: Virtual Bodies in Cybernetics, Literature, and Informatics. University of Chicago Press, Chicago (1999)
26. Bookchin, N.: agoraXchange, http://www.agoraXchange.net
27. Barry, M.: NationStates, http://www.nationstates.net/cgi-bin/index.cgi
28. The Pool, http://river.asap.um.maine.edu/~jon/pool/splash.html
29. Walsh, T.: Big Mac Attacked, http://www.alternet.org/story/14530
30. Schleiner, A.-M., Leandre, J., Condon, B.: Velvet-Strike, http://www.opensorcery.net/velvet-strike/
31. DeLappe, J.: Quake/Friends, http://www.unr.edu/art/DELAPPE/Recent%20Works%20In%20Progress/Quake%20Friends/QUAKE%20FRIENDS%20MAIN.html
32. Jenik, A., Brenneis, L.: Desktop Theater, http://desktoptheater.org/
33. Paterson, S., Zurkow, M., Bleecker, J.: PDPal, http://www.pdpal.com
34. Carbon Defense League: MapHub™, http://www.maphub.org
35. Serafijn, Q.S.: D-tower, http://lab.v2.nl/projects/dtower.html http://www.d-toren.nl/site/
36. Lozano-Hemmer, R.: http://www.lozano-hemmer.com
37. Runme: QuickView on Software Art, http://runme.org/project/+quickview
38. Mongrel: Nine, http://9.waag.org, Linker http://www.linker.org.uk
39. Criticalartware: Liken, http://www.criticalartware.net/lib/liken/

40. <KOP>: DIVE, http://kop.fact.co.uk/DIVE/cd/dive/index.html, Radioqualia: Frequency Clock, http://www.frequencyclock.net, Breidenbruecker, M., Miller, F., Stiksel, M., Willomitzer, T.: LAST.FM, http://last.fm
41. Marketou, J., Dietz, S. (curators): Open_Source_Art_Hack. New Museum of Contemporary Art, New York, http://netartcommons.walkerart.org/, Kingdom of Piracy. Ars Electronica, Linz, Austria (2002), FACT, UK (2003), http://kop.fact.co.uk/KOP/html/kop_fact.html, Nori, F.: I Love You - Computer_Viren_Hacker_Kultur. Museum for Applied Art, Frankfurt, Germany, http://www.digitalcraft.org/iloveyou/index.htm
42. Wark, M.: Digital Resistance: Explorations in Tactical Media. Autonomedia, New York (2002), http://rhizome.org/thread.rhiz?thread=4976&text=10078
43. Institute for Applied Autonomy: iSee, http://www.appliedautonomy.com/isee.html
44. Electronic Disturbance Theater, http://www.thing.net/~rdom/ecd/ecd.html
45. Etoy, http://www.etoy.com/
46. ®TMark, http://www.rtmark.com/
47. Cosic, V.: Documenta Done, http://www.ljudmila.org/~vuk/dx/
48. Harwood: Uncomfortable Proximity, http://www.tate.org.uk/netart/mongrel/home/default.htm
49. Bernstein, C.: Electronic Pies in the Poetry Skies. In: Bousquet, M., Wills, K.: The Politics of Information: The Electronic Mediation of Social Change, p. 7. Alt-X Press (2003), http://www.altx.com/ebooks/infopol.html

Author Index

Editors

Professor Christa Sommerer is currently the Head of the Interface Cultures program at the University of Art and Industrial Design in Linz Austria. She has worked as researcher and professor at ATR, IAMAS and Kyoto University in Japan and at the NCSA and MIT CAVS in the US and is considered one of the pioneers of interactive art. Her award winning interactive media installations are part of media museums and collections around the world. Together with Laurent Mignonneau she has created around 20 interactive art pieces and exhibited in around 200 exhibitions world wide.

Professor Lakhmi C. Jain is a Director/Founder of the Knowledge-Based Intelligent Engineering Systems (KES) Centre, located in the University of South Australia. He is a fellow of the Institution of Engineers Australia.

His interests focus on the artificial intelligence paradigms and their applications in complex systems, art-science fusion, e-education, e-healthcare, unmanned air vehicles and intelligent agents.

GPSR Compliance

The European Union's (EU) General Product Safety Regulation (GPSR) is a set of rules that requires consumer products to be safe and our obligations to ensure this.

If you have any concerns about our products, you can contact us on ProductSafety@springernature.com

In case Publisher is established outside the EU, the EU authorized representative is:

Springer Nature Customer Service Center GmbH
Europaplatz 3
69115 Heidelberg, Germany

Batch number: 10755246

Printed by Printforce, the Netherlands

Vorwort.

Nachdem die im Jahre 1888 „nach amtlichen Quellen bearbeitete" Ausgabe des Feldpolizeistrafgesetzes im Buchhandel bereits längere Zeit vergriffen war, trat an mich das ehrenvolle Anerbieten, eine neue Ausgabe des Gesetzes mit Anmerkungen zu besorgen. Die Aufgabe des neuen Kommentars mußte eine andere sein, als die der ersten Bearbeitung, welche in der Hauptsache nur einen Auszug aus den Gesetzesmaterialien brachte. Damals war es vor allem wichtig, bei den einzelnen Paragraphen das Verhältnis des neuen Rechts gegenüber dem früheren Zustand darzustellen und die Gründe anzugeben, welche den Gesetzgeber zur Beibehaltung oder Abänderung des früheren Gesetzesinhalts bestimmt hatten. Heute dagegen, wo bei den meisten das alte Recht in Vergessenheit geraten ist, mußte in erster Linie die Rechtsprechung, die sich an das Gesetz während der achtzehn Jahre seines Bestehens geknüpft hat, tunlichst erschöpfend berücksichtigt werden. Daneben war ich bestrebt, zum Verständnis des Gesetzes durch kurze Erläuterung der mit ihm im Zusammenhang stehenden Bestimmungen anderer Gesetze, insbesondere durch Hinweis auf die wichtigsten in Betracht kommenden Grundsätze des geltenden gemeinen Strafrechts beizutragen.

In der Anlage der Arbeit hat mir als Muster der bewährte Kommentar von Daude zum preußischen Feld- und Forstpolizeigesetz vom 1. April 1880 gedient. Letzteres ist ja auch für die Mehrzahl der Bestimmungen des Feldpolizeistrafgesetzes selbst vorbildlich gewesen.

Das Werkchen soll den an erster Stelle mit der Verfolgung von Feldfreveln betrauten Ortspolizeibeamten, Gendarmen, Bürgermeistern und Amtsanwälten die Handhabung des Gesetzes erleichtern. Es will aber nicht minder auch dem Richter eine brauchbare Hilfe bieten.

Etwaige Mängel bitte ich mit Nachsicht beurteilen zu wollen. Für die Mitteilung vorhandener Irrtümer oder Unvollständigkeiten werde ich dankbar sein.

Straßburg, im Dezember 1906.

Dr. Stieve.

Inhalt.

Abkürzungen.

ABl. = Zentral- und Bezirks-Amtsblatt für Elsaß-Lothringen.
AG. = Ausführungsgesetz.
Begr. = Begründung zum Entwurf eines Feldpolizeistrafgesetzes für Elsaß-Lothringen in den Verhandlungen des Landesausschusses XV. Session, 1. Band, Vorlage Nr. 8.
BGB. = Bürgerliches Gesetzbuch für das Deutsche Reich.
Bruck. = Die Gemeindeordnung für Elsaß-Lothringen, zweite auf Grund des Halleyschen Kommentars neu bearbeitete Auflage von Bruck.
CPO. = Civilprozeßordnung für das Deutsche Reich.
Dalloz Rép. = Dalloz, Répertoire méthodique et alphabétique de législation de doctrine et de jurisprudence.
Daude = Das (preußische) Feld- und Forstpolizeigesetz vom 1. April 1880 mit Erläuterungen von Daude, vierte Auflage.
E. = Entscheidungen des Reichsgerichts in Strafsachen.
EG. = Einführungsgesetz.
Forststg. = Elsaß-lothringisches Gesetz, betreffend das Forststrafrecht und das Forststrafverfahren vom 28. April 1880.
FPStG. = Feldpolizeistrafgesetz für Elsaß-Lothringen vom 9. Juli 1888.
GBl. = Gesetzblatt für Elsaß-Lothringen.
Gem.-O. = Gemeindeordnung für Elsaß-Lothringen vom 6. Juni 1895.
GVG. = Gerichtsverfassungsgesetz für das Deutsche Reich.
Huber = Huber, Die Wassergesetze Elsaß-Lothringens, zweite Auflage.
Jacob und Fecht = Das Gesetz für Elsaß-Lothringen, betr. Wasserbenutzung und Wasserschutz, erläutert von Jacob und Fecht.
Joh. = Jahrbuch der Entscheidungen des preußischen Kammergerichts in Sachen der nichtstreitigen Gerichtsbarkeit und in Strafsachen von Johow und Ring, jeweils Abteilung C (Strafsachen).

K. B. = Bericht der Spezialkommission zur Beratung des Entwurfes eines Feldpolizeistrafgesetzes in den Verhandlungen des Landesausschusses XV. Session, 2. Band, Seite 397 ff.
KG. = Kammergericht.
Kisch = Kisch, Elsaß-lothringisches Landesprivatrecht.
LA. = Verhandlungen des Landesausschusses XV. Session, 2. Band, Sitzungsberichte.
Leoni-Mandel = Das öffentliche Recht des Reichslandes Elsaß-Lothringen von Leoni und Mandel.
Molitor = Das Gesetz, betreffend die Ausführung des Bürgerlichen Gesetzbuches in Elsaß-Lothringen, erläutert von Molitor.
OLG. = Oberlandesgericht Colmar.
Olshausen = Kommentar zum Strafgesetzbuch von Olshausen.
Oppenhoff = Das Strafgesetzbuch für das Deutsche Reich, erläutert von Oppenhoff.
pr. F. u. FPG. = preußisches Feld- und Forstpolizeigesetz vom 1. April 1880.
R. = Rechtsprechung des deutschen Reichsgerichts in Strafsachen, herausgegeben von den Mitgliedern der Reichsanwaltschaft.
RG. = Reichsgericht.
RGBl. = Reichsgesetzblatt.
Slg. = Sammlung von Gesetzen, Verordnungen, Erlassen und Verfügungen, betreffend die Justizverwaltung in Elsaß-Lothringen.
StGB. = Strafgesetzbuch für das Deutsche Reich.
StPO. = Strafprozeßordnung für das Deutsche Reich.
Ztschr. = Juristische Zeitschrift für das Reichsland Elsaß-Lothringen.

Einleitende Vorbemerkung.

Die Gestaltung des Feldpolizeistrafrechts steht im wesentlichen der Landesgesetzgebung zu; dies bestimmt bezüglich des materiellen Rechts der Artikel II Absatz 2 des Einführungsgesetzes zum Strafgesetzbuch vom 30. August 1871, wonach die besonderen Vorschriften über die durch das Strafgesetzbuch nicht berührten Materien, insbesondere über strafbare Verletzungen der Feldpolizeigesetze in Kraft bleiben sollen und demzufolge auch von der Landesgesetzgebung durch neue Vorschriften ersetzt werden können, bezüglich des Verfahrens der § 3 Absatz 3 des Einführungsgesetzes zur Strafprozeßordnung, wonach die Landesgesetze anordnen können, daß Feldrügesachen durch die Amtsgerichte in einem besonderen Verfahren sowie ohne Zuziehung von Schöffen verhandelt und entschieden werden.

Vor der Einführung des Feldpolizeistrafgesetzes waren in Elsaß-Lothringen für das Verfahren lediglich die Vorschriften der Reichsstrafprozeßordnung maßgebend. Der Hauptsitz des materiellen Rechts dagegen war das Gesetz vom 6. Oktober 1791, betreffend die ländlichen Güter und Gebräuche und die Feldpolizei, der sogenannte code rural, insbesondere dessen zweiter Teil, der speziell von der Feldpolizei handelte; daneben kamen aber noch Spezialgesetze und einzelne Bestimmungen des code pénal von 1810 in Betracht. Schließlich hatte auch das Reichsstrafgesetzbuch in seinen §§ 368 Ziff. 1—9 und 370 Ziff. 1 und 2 die strafrechtlichen Bestimmungen hinsichtlich der Feldpolizei vermehrt. Das Verhältnis all dieser Gesetzesbestimmungen zu einander war ein überaus unklares. Die fortdauernde Geltung nicht weniger Bestimmungen des code rural gegenüber dem code pénal und dem Reichsstrafgesetzbuch war zweifelhaft, so daß es selbst dem Juristen kaum möglich war, über das bestehende Recht zuverlässige Auskunft zu geben. Zudem waren einzelne Bestimmungen des code rural veraltet und paßten nicht mehr in die veränderten wirtschaftlichen Verhältnisse der Neuzeit.

Dem in der Praxis sich am meisten fühlbar machenden Übelstande, daß nämlich nach dem code rural in Verbindung mit dem Gesetz vom 23. thermidor des Jahres IV die Geldstrafen nach dem Wert von Arbeitstagen festgesetzt werden mußten, und daß als Mindeststrafe der Wert von drei Arbeitstagen bestimmt war, hatte bereits das Gesetz, betreffend die Bestrafung von Zuwiderhandlungen gegen die Vorschriften der Feld- und Fährpolizei vom 30. März 1887 (GBl. S. 39), abgeholfen. Gleichzeitig mit diesem Gesetz aber hatte die Landesregierung von Elsaß-Lothringen in Würdigung der vorgedachten Mißstände die Vorbereitung eines das Feldpolizeistrafrecht durchgreifend regelnden Gesetzentwurfes in Angriff genommen.

Bei der Ausarbeitung des Entwurfs, mit welcher der damalige Oberlandesgerichtsrat, jetzige Senatspräsident am Reichsgericht Förtsch beauftragt war, wurde neben dem Forststrafgesetz vom 28. April 1880 (GBl. S. 75), das preußische Feld- und Forstpolizeigesetz vom 1. April 1880 (Preuß. Ges.-Sam. S. 230), welches bereits bei dem erst-genannten elsaß-lothringischen Gesetz als Vorbild gedient hatte, zum Muster genommen. Weiter ist auch den das Feldpolizeirecht betreffenden Artikeln 111—124 des bayerischen Polizeistrafgesetzbuchs vom 26. Dezember 1871, der Gesetzgebung des Nachbarlandes Baden, insbesondere der Feldpolizeiordnung für den Amtsbezirk Kehl und einem dem französischen Senate im Jahre 1886 vorgelegten Projet de loi sur le code rural (livre III de la police rural) gebührende Berücksichtigung geschenkt worden.

Der am 9. Februar 1888 vom Bundesrat genehmigte Gesetzentwurf nebst Begründung wurde bereits am 13. Februar 1888 dem Landesausschuß vorgelegt (Vorlage 8 der XV. Session des Landesausschusses). Die erste Lesung fand am 23. Februar 1888 statt und schloß mit der Überweisung des Entwurfes an eine Spezialkommission. Berichterstatter der Kommission war der Abgeordnete Dr. Gunzert. Der Entwurf wurde von der Kommission in neun Sitzungen durchberaten und mit einigen sachlichen Änderungen versehen dem Plenum des Landesausschusses von neuem überwiesen. Der Bericht der Kommission befindet sich S. 397 ff. der Verhandlungen des Landesausschusses (XV. Session, 2. Band, Sitzungsberichte). Der Landesausschuß erteilte in den Sitzungen vom 12., 13. und 19. April 1888 in zweiter und dritter Lesung dem Gesetzentwurf, nachdem ihm mehrere, darunter jedoch nur wenige wesentliche Änderungen eingefügt waren, seine Zustimmung (s. die angeführten Verhandlungen S. 374 ff., 435 ff., 543 ff.). Der Bundesrat stimmte den vom Landesausschuß beschlossenen Änderungen in seiner Sitzung vom 7. Juni 1888 zu. Die Sanktion des Gesetzentwurfes durch den Kaiser erfolgte am 9. Juli 1888.

Das FPStG. hat den code rural nicht vollständig ersetzt. Letzterer enthält außer feldpolizeilichen Strafbestimmungen eine Reihe von Vorschriften agrarrechtlichen Charakters. Diese gelten neben dem FPStG. fort.

Der Zweck des FPStG. ist in erster Linie der Schutz der Bodenerzeugnisse und der Landwirtschaft, sowie die Aufrechterhaltung der Ordnung auf dem Lande (vergl. § 50). Doch darf der Name des Gesetzes nicht zu der Annahme verleiten, daß seine Bestimmungen ausschließlich nur auf Felder im engeren Sinne des Wortes und nur auf ländliche Verhältnisse Anwendung fänden. U. a. enthält das FPStG. auch Bestimmungen über die Wasserpolizei, die in gleicher Weise den Interessen der Landwirtschaft wie der Industrie dienen (vergl. §§ 30, 31, 35), ferner Vorschriften, welche die allgemeine Sicherheits- und Gesundheitspolizei sowohl auf dem Lande als auch in den Städten betreffen (§§ 32, 34, 35, 36). Die besonders wichtigen Vorschriften über den Felddiebstahl (§§ 9 ff.) finden Anwendung, auch wenn der Begehungsort eine Stadt ist. Hinwiederum ist die Anwendbarkeit anderer Bestimmungen, so z. B. der §§ 26 und 27 auf Feldmarken beschränkt und in Städten sowie in zusammenhängend gebauten Ortschaften ausgeschlossen. Als Grundsatz ist aufzustellen, daß nach dem besonderen Inhalt jeder Vorschrift, insbesondere aber auch nach ihrer Entstehungsgeschichte, zu prüfen ist, ob sie sich nur auf ländliche Orte und Feldmarken oder auch auf städtische Grundstücke bezieht.

Durch spätere Landesgesetze ist auf das FPStG. in zweifacher Beziehung eingewirkt worden. Zunächst hat § 31 des Gesetzes eine strafrechtlich wichtige Ergänzung in § 46 des Gesetzes, betreffend die Wassernutzung und den Wasserschutz vom 2. Juli 1891 (GBl. S. 82), erfahren. Ferner ist § 35 Ziff. 5 durch § 54 des Gesetzes, betreffend die Fischerei vom 2. Juli 1891 (GBl. S. 69), aufgehoben worden. Des näheren vergleiche die Bemerkungen zu den bezeichneten Paragraphen.

Durch das Bürgerliche Gesetzbuch ist das FPStG. nur in soweit berührt worden, als die Vorschriften des ersteren auf die Bestimmungen des allgemeinen Teils des Reichsstrafgesetzbuches, welche gemäß § 1 FPStG. in beschränktem Maße auch auf die in diesem Gesetz mit Strafe bedrohten Handlungen Anwendung finden, eingewirkt haben. Soweit das FPStG. in das Gebiet des bürgerlichen Rechts übergreift, enthält das Einführungsgesetz zum Bürgerlichen Gesetzbuch entsprechende Vorbehalte. So sind die Bestimmungen des FPStG. über die Buße (§ 5) und die Haftbarkeit der Gewalthaber und Dienstherren für die durch ihre Untergebenen verwirkten Bußen (§§ 3, 4, 23) aufrecht erhalten worden durch Art. 107 EG., wonach unberührt bleiben die landesgesetzlichen Vorschriften über die Verpflichtung zum Ersatz des Schadens, der durch das Zuwiderhandeln gegen ein zum Schutz von Grundstücken erlassenes Strafgesetz verursacht wird (vergl. Kisch, S. 326 ff.). Zugunsten der Bestimmung des FPStG. über die Pfändung von Weidevieh (§ 42 ff.) ist ein Vorbehalt in Art. 89 EG. z. BGB. enthalten. Das Recht, Tauben zu töten, die während der Saat- oder Erntezeit einer ortspolizeilichen Verordnung über das Einschließen der Tauben zuwider auf einem fremden Grundstück angetroffen werden (§ 24, Abs. 2), ist durch Art. 130 EG. z. BGB. aufrecht erhalten worden. Die Befugnis der Eigentümer von Weinbergen zum beliebigen Abernten und Betreten ihrer Rebstücke kann auch nach Einführung des Bürgerlichen Gesetzbuchs im öffentlichen Interesse Beschränkungen unterworfen werden (vergl. EG. z. BGB. Art. 111 und §§ 38, 48 dieses Gesetzes). Schließlich ist hier noch hervorzuheben, daß die vom FPStG. nach der polizeilichen Seite geregelten, das Grundeigentum beschränkenden Rechte der Nachlese (§ 19) und die in § 25 erwähnten Weiderechte, soweit letztere privatrechtlicher Natur sind, gemäß Art. 184, 115 EG. z. BGB. bestehen geblieben sind.

Feldpolizeistrafgesetz

vom 9. Juli 1888

(Gesetzblatt für Elsaß-Lothringen 1888 S. 73 ff.)

Wir **Wilhelm**, von Gottes Gnaden Deutscher Kaiser, König von Preußen ꝛc.
verordnen im Namen des Reichs, für Elsaß-Lothringen, nach erfolgter Zustimmung des Bundesrats und des Landesausschusses, was folgt:

Erster Titel.

Strafbestimmungen.

Allgemeine Regeln des Strafgesetzbuchs.

§ 1.

Die in diesem Gesetz mit Strafe bedrohten Handlungen unterliegen den nachstehenden Bestimmungen und, insoweit in denselben Abweichungen nicht vorgesehen sind, den Vorschriften des Strafgesetzbuchs.

1. Durch § 1, welcher dem § 1 des Forststg. gleichlautet, werden, unter dem Vorbehalt von Besonderheiten, namentlich die allgemeinen Lehren des deutschen Strafrechts, die in den „Einleitenden Bestimmungen" des StGB. und in dessen „Erstem Teil" enthalten sind, auf die im FPStG. mit Strafe bedrohten Handlungen für anwendbar erklärt. Besondere Bedeutung für das FPStG. wohnt folgenden allgemeinen Grundsätzen inne:

a) Nur objektiv rechtswidrige, d. h. vom Gesetz verbotene Handlungen sind strafbar. Die Rechtswidrigkeit einer vom Gesetz für gewöhnlich mit Strafe bedrohten Handlung kann im einzelnen Falle aus besonderen Gründen ausgeschlossen sein. Die wichtigsten und im Ersten Teil des StGB. aufgeführten Gründe des Ausschlusses der Rechtswidrigkeit sind: Nötigung (§ 52 StGB.), Notwehr (§ 53 StGB.) und Notstand (§ 54 StGB.). Der Begriff des die Rechtswidrigkeit ausschließenden Notstandes war in § 54 StGB. eng umgrenzt. Nur diejenige Handlung war gestattet, die in einem unverschuldeten auf andere Weise nicht zu beseitigenden Notstande zur Rettung aus einer gegenwärtigen Gefahr für Leib oder Leben des Täters oder eines Angehörigen begangen worden ist. Durch zwei in das Strafrecht eingreifende Bestimmungen des BGB. hat dieser Begriff des Notstandes aber eine wesentliche Erweiterung erfahren. Zunächst gestattet § 228 BGB. allgemein die Beschädigung oder Zerstörung einer fremden gefahrdrohenden Sache, wenn die Handlung zur Abwendung der Gefahr erforderlich ist, und der Schaden nicht außer Verhältnis zu der Gefahr steht. Noch weiter geht der auch für das Feldpolizeistrafrecht besonders wichtige § 904 BGB. Hiernach ist die Einwirkung auf eine fremde Sache (z. B. das Betreten von Grundstücken, das Zerstören von Bodenerzeugnissen) auch dann gestattet, wenn von der Sache selbst keine Gefahr droht, die Einwirkung aber zur Abwendung irgend einer gegenwärtigen Gefahr notwendig ist, und der drohende Schaden gegenüber dem aus der Einwirkung entstehenden Schaden unverhältnismäßig groß ist.

Andere für das Feldpolizeistrafrecht wichtige Fälle ausgeschlossener Rechtswidrigkeit sind das Selbsthülferecht zum Zwecke der Verwirklichung eines Anspruchs, wenn obrigkeitliche Hülfe nicht rechtzeitig zu erlangen ist (§ 228 BGB.), das Recht des Eigentümers eines Bienenschwarms, bei dessen Verfolgung fremde Grundstücke zu betreten (§ 962 BGB.), schließlich das Recht des Militärs, Grundstücke zu Uebungszwecken zu benutzen (§§ 11 ff. des Ges. über die Naturalleistungen für die bewaffnete Macht im Frieden vom 13. Februar 1875 in der Fassung vom 24. Mai 1898 (RGBl. S. 361). Vgl. auch noch Bem. 2 u. 5 zu § 29.

Zu unterscheiden von der Rechtswidrigkeit und damit Strafbarkeit ist die bürgerlichrechtliche Schadensersatzpflicht. Letztere besteht nach ausdrücklicher Gesetzesvorschrift häufig, auch wenn keine Rechtswidrigkeit vorliegt, so insbesondere im Falle des § 904 BGB.

b) Neben der objektiven Rechtswidrigkeit der Handlung ist erforderlich, daß den Täter ein subjektives Verschulden trifft. Zu beachten ist aber, daß die Mehrzahl der in dem FPStG. mit Strafe bedrohten Handlungen unter den Begriff der sog. Polizeidelikte fällt, deren Wesen nicht sowohl in der Verletzung eines bestimmten Rechtsgutes als vielmehr in der Uebertretung der zur äußeren Aufrechterhaltung der öffentlichen Ordnung dienenden Vorschriften besteht. Die früher weit verbreitete Auffassung, wonach es bei Polizeidelikten nur auf die äußere Verwirklichung des gesetzlichen Tatbestandes, nicht aber darauf ankomme, ob sie schuldhafter Weise geschehen sei, ist heute allerdings in Wissenschaft und Rechtsprechung meist aufgegeben. Doch ist bei den Polizeidelikten nicht notwendig, daß sie immer vorsätzlich begangen sind. Ohne daß es einer besonderen Hervorhebung im Gesetze bedarf, ist vielmehr auch Fahrlässigkeit, sofern eine solche nach dem gesetzlichen Tatbestand überhaupt denkbar ist, strafbar. Soweit eine im FPStG. bedrohte Handlung nicht unter den Begriff des reinen Polizeideliktes fällt, was insbesondere von dem Felddiebstahl (§ 9) gilt, muß vorsätzliche Verwirklichung des Tatbestandes gefordert werden.

Nach dem StGB. ist ein Verschulden nicht vorhanden in den Fällen der sog. Zurechnungsunfähigkeit, d. h. bei Geisteskrankheit und anderen Bewußtseinsstörungen unter den Voraussetzungen des § 51 StGB., bei Jugendlichkeit des Täters nach Maßgabe der §§ 55 und 56 StGB., und wenn es sich um einen Taubstummen handelt, der die zur Erkenntnis der Strafbarkeit seines Handelns erforderliche Einsicht nicht besaß (§ 58 StGB.). Vgl. des näheren noch die Bemerkungen zu §§ 2, 3 und 4 dieses Ges.

Der Irrtum des Täters über das Vorhandensein von Tatumständen, welche zum gesetzlichen Tatbestand gehören oder die Strafbarkeit erhöhen, schließt die Strafbarkeit überhaupt oder die erhöhte Strafbarkeit nach der strengeren gesetzlichen Bestimmung aus (§ 59 StGB.). Soweit aber die im FPStG. bedrohten Handlungen auch im Falle der fahrlässigen Verursachung strafbar sind, gilt diese Bestimmung nur, wenn der Irrtum des Täters nicht selbst durch Fahrlässigkeit verschuldet war.

Nach einem allgemeinen Grundsatz des Strafrechts ist nicht erfordert, daß der Täter das Bewußtsein der Rechtswidrigkeit seiner Handlung hatte. Hiernach kann man sich insbesondere nicht darauf berufen, man habe nicht gewußt, daß das Gesetz die begangene Handlung mit Strafe bedrohe („Unkenntnis des Gesetzes schützt nicht vor Strafe"). Indessen hat das FPStG. mehrfach das Merkmal der Rechtswidrigkeit in einer nach dem Sinne der betreffenden gesetzlichen Vorschrift näher zu bestimmenden Richtung in den Tatbestand der strafbaren Handlung aufgenommen. Dies ist insbesondere in den §§ 17, 18, 20, 26, 27 Ziff. 1, 28, 29, 30 Ziff. 2, 35 durch die Beifügung des Wortes „unbefugt" oder „ohne Befugnis" geschehen. In solchen Fällen muß der Täter das Bewußtsein der mangelnden Befugnis gehabt haben. Sofern aber, was die Regel bildet, auch fahrlässige Begehung der Handlung strafbar macht, steht dem Bewußtsein der mangelnden Befugnis die durch Fahrlässigkeit verursachte Unkenntnis gleich.

c) Die weitaus überwiegende Mehrzahl der nach diesem Gesetz strafbaren Handlungen sind Uebertretungen, da sie regelmäßig nur mit Haft oder mit Geldstrafe bis höchstens einhundertfünfzig Mark bedroht sind (§ 1 Abs. 3 StGB.). Ausnahmen bilden nur die im § 12 und 13 bezeichneten Arten erschwerten Felddiebstahls, die, weil für sie Gefängnis angedroht ist, unter den Begriff der Vergehen fallen (§ 1 Abs. 2 StGB.). Der Mindestbetrag einer Geldstrafe bei Uebertretungen ist Eine Mark (§ 27 StGB.). Eine wegen einer Uebertretung

1*

erkannte nicht beizutreibende Geldstrafe ist in Haft umzuwandeln (§ 28 Abs. 1 StGB.). Dies geschieht, anders wie im Forststg. (§ 5), ausschließlich nach Maßgabe des § 29 StGB., wo es heißt:

„Bei Umwandlung einer wegen einer Uebertretung erkannten Geldstrafe ist der Betrag von einer bis zu fünfzehn Mark einer eintägigen Freiheitsstrafe gleich zu achten.

Der Mindestbetrag der an Stelle einer Geldstrafe tretenden Freiheitsstrafe ist ein Tag, ihr Höchstbetrag bei Haft sechs Wochen.... Wenn jedoch eine neben der Geldstrafe wahlweise angedrohte Freiheitsstrafe ihrer Dauer nach den vorgedachten Höchstbetrag nicht erreicht, so darf die an Stelle der Geldstrafe tretende Freiheitsstrafe den angedrohten Höchstbetrag jener Freiheitsstrafe nicht übersteigen."

Eine Befreiung von der an Stelle der Geldstrafe verwirkten Freiheitsstrafe durch Gemeindearbeiten, wie im Forststg. (§ 6) ist im FPStG. nicht zugelassen.

d) Bei einigen der im FPStG. mit Strafe bedrohten Handlungen tritt die Verfolgung nur auf Antrag ein. Es sind das die Fälle der §§ 15 Abs. 1 und 2, 17 Ziffer 2, 18 und 29. Ueberall ist die Zurücknahme des Antrags zulässig. Das Nähere über den Strafantrag ist in Bem. 3 zu § 15 enthalten.

e) Ueber die Verjährung der im FPStG. bedrohten Handlungen ist folgendes zu bemerken:

Die Strafverfolgung verjährt im allgemeinen, da es sich meist um Uebertretungen handelt, gemäß § 67 Abs. 3 StGB. in drei Monten. Im Falle des § 12 tritt aber die Verjährung gemäß § 67 Abs. 2 StGB. erst in drei Jahren, im Falle des § 13 erst in fünf Jahren ein. Die Verjährung beginnt mit dem Tage, an welchem die Handlung begangen ist, ohne Rücksicht auf den Zeitpunkt des eingetretenen Erfolges (§ 67 Abs. 4 StGB.). Sie wird unterbrochen durch jede richterliche Handlung, welche wegen der begangenen Tat gegen den Täter gerichtet ist; nach der Unterbrechung beginnt eine neue Verjährung (§ 68 StGB.). Unabhängig von der Verjährung der Strafverfolgung ist die Verjährung des Schadensersatzanspruchs, für welche die Vorschriften des bürgerlichen Rechts maßgebend sind (§ 852 BGB.). Auf eine Buße an Stelle des Schadensersatzes kann aber natürlich nach Verjährung der Strafverfolgung nicht mehr erkannt werden, da ja gemäß § 5 Abs. 1 die Verpflichtung zur Zahlung einer Buße nur vom Strafrichter „neben der Strafe" ausgesprochen werden kann.

Die Strafvollstreckung verjährt, wenn gemäß §§ 12 oder 13 auf Gefängnisstrafe erkannt worden ist, in fünf Jahren, in allen übrigen Fällen in zwei Jahren (§ 70 Ziff. 5 und 6 StGB.). Die Verjährung beginnt mit dem Tage, an welchem das Urteil rechtskräftig geworden ist (§ 70 Abs. 2 StGB.). Sie wird unterbrochen durch jede auf Vollstreckung der Strafe gerichtete Handlung derjenigen

Behörde, welcher die Vollstreckung obliegt, sowie durch die zum Zwecke der Vollstreckung erfolgende Festnahme des Verurteilten. Nach der Unterbrechung der Vollstreckung der Strafe beginnt eine neue Verjährung (§ 72 StGB.).

2. Vorschriften, welche von den Bestimmungen des StGB. abweichen, sind außer bei den einzelnen strafbaren Handlungen enthalten in § 2 bezüglich der Strafbarkeit der Jugendlichen, in § 7 bezüglich der Behandlung der Beihülfe und in § 8 bezüglich der Behandlung des Versuchs, der Begünstigung und der Hehlerei.

3. Im Ausland von einem Deutschen begangene Zuwiderhandlungen gegen das FPStG. können in den Fällen der §§ 12 und 13, da es sich hier um Vergehen handelt, entsprechend der Regel des § 4 Ziff. 3 StGB. nach dem deutschen Gesetz verfolgt werden, sofern sie durch die Gesetze des Ortes, an welchem sie begangen wurden, mit Strafe bedroht sind, und sofern keine der die Strafverfolgung ausschließenden Voraussetzungen des § 5 StGB. vorliegen. Soweit dagegen die im Ausland begangenen Zuwiderhandlungen gegen das FPStG. Uebertretungen im Sinne des StGB. sind, kann eine Verfolgung in Deutschland gemäß § 6 StGB. nur dann stattfinden, wenn dies durch ein besonderes Gesetz oder durch einen Vertrag angeordnet ist. Ein besonderes Gesetz in dieser Hinsicht gibt es nicht. Auch bestehen in Ansehung der unmittelbar an Elsaß-Lothringen grenzenden Nachbarstaaten, Luxemburg, Frankreich und der Schweiz, keine Verträge, welche die Verfolgung der auf dem fremden Gebiete von eigenen Staatsangehörigen begangenen Zuwiderhandlungen gegen die feldpolizeilichen Vorschriften seitens des Heimatsstaates anordnen. Dagegen ordnet ein zwischen dem Deutschen Reich und Belgien am 29. April 1885 abgeschlossener Vertrag (RGBl. S. 251) an, daß Deutsche, welche in Belgien und Belgier, welche in Deutschland sich eines Forst-, Feld-, Fischerei- und Jagdfrevels schuldig gemacht haben, in dem Gebiete desjenigen Teiles, welchem sie angehören, nach den dort geltenden Gesetzen und unter den darin bestimmten Voraussetzungen bestraft werden sollen.

4. Die Geldstrafen, welche in Feldpolizeistrafsachen von den Gerichten ausgesprochen worden sind, sind grundsätzlich der Gemeinde des Tatortes zuzuweisen. Dies bestimmen der heute noch geltende Art. 466 code pénal und Art. 4 der Ordonnanz vom 30. Dezember 1823. Eine Ausnahme gilt hinsichtlich derjenigen Geldstrafen, welche wegen Beschädigung und Verunreinigung der zum großen Straßenwesen gehörigen Staats- und Bezirksstraßen aus § 26 und 27 FPStG. verwirkt sind. Diese gehören zu einem Drittel der Gemeinde des Tatortes und zu zwei Dritteln der Staatskasse (Art. 115 des Dekrets vom 16. Dezember 1811 in Verbindung mit dem els.-lothr. Landesgesetz vom 18. Juni 1891 — GBl. S. 39 —, wodurch die früher gesetzlich fixierten Anzeigebelohnungen in Wegfall gekommen sind). Mehrfach haben die Gemeinden

auf ihren Anteil an den Strafgeldern zugunsten der Staatskasse als Entgelt für die Uebernahme gemeindlicher Lasten, z. B. der Unterhaltung der Gerichtsgebäude, verzichtet. Die Einziehung und Zuweisung der Geldstrafen erfolgt durch die Verkehrssteuerämter. (Vgl. auch Min. Verf. für die Verkehrssteuerbehörden, betr. das Gerichtskosten- und Strafwesen vom 10. Dezember 1896 ABl. 1897 S. 9 ff.)

Jugendliche Frevler.

§ 2.

Die in § 57 Nr. 3 des Strafgesetzbuchs bei der Verurteilung von Personen, welche zur Zeit der Begehung der Tat das zwölfte, aber nicht das achtzehnte Lebensjahr vollendet hatten, vorgesehene Strafermäßigung findet bei Zuwiderhandlungen gegen dieses Gesetz keine Anwendung.

1. Nach § 57 Nr. 3 StGB. ist bei der Verurteilung eines Angeklagten, der zur Zeit der Begehung der strafbaren Handlung das zwölfte aber nicht das achtzehnte Lebensjahr vollendet hatte, die Strafe zwischen dem gesetzlichen Mindestbetrag der angedrohten Strafart und der Hälfte des Höchstbetrages der angedrohten Strafe zu bestimmen. Diese Strafermäßigung findet bei Zuwiderhandlungen der bezeichneten Jugendlichen gegen die Bestimmungen des FPStG. keine Anwendung.

2. Anwendung dagegen findet § 57 Nr. 4 StGB., wonach gegen Personen, die sich im Alter zwischen zwölf und achtzehn Jahren einer strafbaren Handlung schuldig gemacht haben, in besonders leichten Fällen auf Verweis erkannt werden kann. Hierdurch unterscheidet sich das FPStG. von § 2 Forststg., wonach die Strafe des Verweises in Forststrafsachen ausgeschlossen ist. Auf Verweis kann aber nur durch Urteil, nicht durch Strafbefehl erkannt werden (vgl. § 447 Abs. 2 StPO.). Die Vollstreckung des Verweises liegt ausschließlich den Amtsrichtern ob (Anordng. des Kaiserl. Statthalters vom 8. Februar 1886 GBl. S. 5), auch in denjenigen Amtsgerichtsbezirken, in denen durch Verfügung des Ministeriums die Vollstreckung der Freiheitsstrafen der Staatsanwaltschaft belassen ist (Min. Verf. vom 20. Februar 1886 Smlg. Bd. 11 S. 43). Ob der Verweis mündlich oder schriftlich zu erteilen ist, bleibt dem richterlichen Ermessen überlassen (Art. XII d. EG. z. StGB. für Elf.-Lothr.).

3. Anwendung findet ferner gemäß § 1 FPStG. die besonders wichtige Vorschrift des § 56 Abs. 1 StGB.:

„Ein Angeschuldigter, welcher zu einer Zeit, als er das zwölfte, aber nicht das achtzehnte Lebensjahr vollendet hatte, eine strafbare Handlung begangen hat, ist freizusprechen, wenn er bei Begehung derselben die zur Erkenntnis ihrer Strafbarkeit erforderliche Einsicht nicht besaß."

Die Amtsanwälte haben in dem Ermittlungsverfahren gegen Jugendliche, die zur Zeit der Tat das zwölfte, aber nicht das achtzehnte Lebensjahr vollendet hatten, insbesondere auch wenn ein Strafbefehl beantragt werden soll, stets zu prüfen, ob die zur Erkenntnis der Strafbarkeit erforderliche Einsicht vorhanden war. Die Gründe, aus denen das Vorhandensein der letzteren angenommen wird, sind aktenkundig zu machen. (Vgl. §21 Abs. 5 der Geschäftsanweisung für Amtsanwälte vom 15. Juli 1880, Smlg. Bd. 5 S. 240 und Verfügung des Oberstaatsanwalts vom 8. Oktober 1885, Smlg. Bd. 10 S. 231). Auch die mit der unmittelbaren Verfolgung der Feldfrevel betrauten Beamten (Feldhüter, Gendarmen ꝛc.) sollen bei jugendlichen Tätern im Alter von zwölf bis achtzehn Jahren von vornherein ihr Augenmerk darauf richten, ob sie zur Zeit der Tat die erforderliche Einsicht hatten und, was hierüber in Erfahrung gebracht wurde, schon in die Anzeige aufnehmen.

Für den Fall der Freisprechung eines jugendlichen Täters wegen Mangels der erforderlichen Einsicht bestimmt § 56 Abs. 2 StGB.:

„In dem Urteile ist zu bestimmen, ob der Angeschuldigte seiner Familie überwiesen oder in eine Erziehungs- oder Besserungsanstalt gebracht werden soll. In der Anstalt ist er so lange zu behalten, als die der Anstalt vorgesetzte Verwaltungsbehörde solches für erforderlich erachtet, jedoch nicht über das vollendete zwanzigste Lebensjahr."

4. Personen, die bei Begehung der Handlung das 12. Lebensjahr noch nicht vollendet hatten, können gemäß § 55 StGB. wegen derselben strafrechtlich nicht verfolgt werden. Doch sollen die Amtsanwälte gemäß § 4 der erwähnten Geschäftsanweisung die zu ihrer Kenntnis kommenden strafbaren Handlungen solcher Personen geeignetenfalls dem Amtsrichter mitteilen, damit dieser Anlaß zu vormundschaftlichen Maßregeln nehmen kann. Auch die Feldhüter und Gendarmen haben, sofern die Umstände des Falles auf Verwahrlosung oder frühzeitige Verdorbenheit des jugendlichen Täters schließen lassen, Anzeige zu erstatten.

5. §§ 55 und 56 StGB. enthalten nur einen persönlichen Strafausschließungsgrund zugunsten der Jugendlichen allein. Infolgedessen ist der Erwachsene strafbar, der sich an der Zuwiderhandlung des Jugendlichen als Mittäter, Anstifter oder Gehülfe beteiligt, oder der sich in Beziehung auf eine solche Handlung der Hehlerei oder der Begünstigung schuldig gemacht hat.

Haftbarkeit Dritter.

§ 3.

Für die auf Grund dieses Gesetzes (§ 5) dem Verletzten zuzusprechende Buße und für die Kosten, zu denen Personen verurteilt werden, welche unter der Gewalt, der Aufsicht oder im

Dienste eines anderen stehen und zu dessen Hausgenossenschaft gehören, ist letzterer für den Fall des Unvermögens des Verurteilten haftbar zu erklären, und zwar unabhängig von der etwaigen Strafe, zu welcher er selbst auf Grund dieses Gesetzes oder des § 361 Nr. 9 des Strafgesetzbuchs verurteilt wird.

Sind die gesetzlichen Vertreter (Ehemänner, Eltern, Vormünder) der Verurteilten als haftbar zu erklären, so erstreckt sich die Haftbarkeit auch auf die Geldstrafe.

Wird festgestellt, daß der als haftbar in Anspruch Genommene die Tat nicht verhindern konnte, so tritt die Haftbarkeit nicht ein.

Gegen die in Gemäßheit der vorstehenden Bestimmungen als haftbar Erklärten tritt an die Stelle der Geldstrafe eine Freiheitsstrafe nicht ein.

1. § 361 Nr. 9 StGB. bestimmt:

„Mit Haft wird bestraft: wer Kinder oder andere unter seiner Gewalt stehende Personen, welche seiner Aufsicht untergeben sind und zu seiner Hausgenossenschaft gehören, von der Begehung von Diebstählen, sowie von der Begehung strafbarer Verletzungen der Zoll- oder Steuergesetze, oder der Gesetze zum Schutze der Forsten, der Feldfrüchte, der Jagd oder der Fischerei abzuhalten unterläßt. Die Vorschriften dieser Gesetze über die Haftbarkeit für die den Täter treffenden Geldstrafen oder anderen Geldleistungen werden hierdurch nicht berührt.

In den Fällen der Nr. 9 kann statt der Haft auf Geldstrafe bis zu einhundertfünfzig Mark erkannt werden."

Hierüber hinausgehend setzen §§ 3 und 4 eine Haftbarkeit Dritter für bestimmte ihrer Hausgenossenschaft angehörende Personen fest. In den Fällen des § 3 ist die Haftbarkeit nur eine subsidiäre, d. h. sie tritt nur ein, wenn die an erster Stelle verurteilten Personen zur Zahlung unvermögend sind.

2. Voraussetzung der Haftbarkeit ist, daß die an erster Stelle verurteilten Personen der Gewalt, der Aufsicht oder dem Dienste des Mithaftenden unterstehen und zu seiner Hausgenossenschaft gehören, d. h. mit ihm in häuslicher Gemeinschaft wohnen. Unter Gewalt ist nicht nur die elterliche Gewalt des Vaters oder der Mutter und die vormundschaftliche Gewalt zu verstehen, sondern, wie sich aus Abs. 2 ergibt, auch die Gewalt, welche der Ehemann über seine Ehefrau ausübt. Das Bestehen eines solchen allerdings beschränkten Gewaltverhältnisses erkennt auch das BGB. an, indem es in § 1354 dem Mann die Entscheidung in allen das gemeinschaftliche eheliche Leben betreffenden Angelegenheiten zuweist. Infolgedessen haftet der Ehemann auch unter der Herrschaft des BGB. für die Zuwiderhandlungen seiner Ehefrau gegen das FPStG.

Der Ausdruck „gesetzliche Vertreter" in Abs. 2 ist allerdings in Ansehung der Ehefrau ungenau, da weder nach der richtig verstandenen Lehre des Code civil noch vom Standpunkt des BGB. aus der Ehemann gesetzlicher Vertreter seiner Frau ist. (Vgl. LG. Colmar Beschl. 29./8. 1900 Ztschr. Bd. 26 S. 51.)

Volljährige Kinder stehen, auch wenn sie zu der Hausgenossenschaft ihrer Eltern gehören, gemäß § 1626 BGB. nicht unter elterlicher Gewalt. Soweit sie daher nicht zugleich in einem Dienstverhältnis zu den Eltern stehen, haften diese nicht für die Zuwiderhandlungen der volljährigen Kinder gegen das FPSt.G.

Unter Aufsicht stehende Personen im Sinne des § 3 sind beispielsweise Schüler, sofern sie mit ihrem Erzieher in häuslicher Gemeinschaft leben, Zwangszöglinge, die in einer Familie untergebracht sind.

3. Der **Umfang** der Haftbarkeit des Dritten erstreckt sich bei Personen, die seinem Dienste oder seiner Aufsicht unterstellt sind, nur auf die Buße und die Kosten, bei Ehefrauen, minderjährigen Kindern und Mündeln auch auf die Geldstrafe. Dagegen darf gegen ihn gemäß Abs. 4, niemals auf eine subsidiäre Freiheitsstrafe an Stelle der Geldstrafe erkannt werden. Die Haftbarkeit muß in dem Urteil oder in dem amtsgerichtlichen Strafbefehl (§ 41 dieses Ges.) stets ausdrücklich ausgesprochen werden. Von den Rechtsmitteln kann der für haftbar Erklärte unabhängig von dem an erster Stelle Verurteilten für sich Gebrauch machen. Daneben gilt selbstverständlich auch § 340 StPO.

Die Haftbarkeit für den durch eine Zuwiderhandlung gegen das FPStG. entstandenen Schaden nach Maßgabe des obigen Paragraphen besteht nur dann, wenn der Schaden im Strafverfahren als Buße „auf Grund dieses Gesetzes (§ 5)" geltend gemacht wird. Im übrigen sind für die Verpflichtung eines Dritten zum Ersatze des Schadens, den ihm unterstellte Personen durch Zuwiderhandlungen gegen das FPStG. verursacht haben die Vorschriften des BGB. (§§ 831, 832) maßgebend. Ueber den Vorbehalt des Art. 107 EG. z. BGB. zugunsten der landesgesetzlichen Buße in Feldpolizeistrafsachen vgl. die „Einleitende Vorbemerkung" gegen Ende.

4. Der **Grund** der Haftbarkeit liegt darin, daß der Gesetzgeber bei Zuwiderhandlungen der im obigen Paragraphen aufgeführten Personen eine **Fahrlässigkeit** in der Handhabung der Aufsicht seitens der hierzu Berufenen vermutet. Infolgedessen tritt die Haftbarkeit nach der Bestimmung in Abs. 3 nicht ein, wenn festgestellt wird, daß der als haftbar in Anspruch Genommene die Tat, etwa wegen Krankheit oder Abwesenheit, **nicht verhindern konnte**. Dies ist aber nicht etwa so zu verstehen, als wenn den Gewalthaber usw. die Beweislast dafür träfe, daß er die Tat nicht verhindern konnte. Ob dies der Fall ist, hat vielmehr der Richter gemäß § 260 StPO. nach seiner freien aus dem Inbegriff der Verhandlung geschöpften Ueberzeugung zu entscheiden. Auch die mit der Ermittlung und Verfolgung der strafbaren Handlungen betrauten Be-

amten (Feldhüter, Gendarmen und Amtsanwälte) haben in geeigneten Fällen nach dem Vorhandensein von Gründen, aus denen etwa die Tat von dem Gewalthaber rc. nicht verhindert werden konnte, zu forschen. Daß die Tat ohne sein Vorwissen verübt wurde, genügt nicht, um seine Haftbarkeit auszuschließen.

§ 4.

Hatte der Täter noch nicht das zwölfte Lebensjahr vollendet, so wird derjenige, welcher in Gemäßheit des § 3 haftet, zur Zahlung der Geldstrafe, der Buße und der Kosten als unmittelbar haftbar verurteilt.

Dasselbe gilt, wenn der Täter zwar das zwölfte, aber noch nicht das achtzehnte Lebensjahr vollendet hatte und wegen Mangels der zur Erkenntnis der Strafbarkeit seiner Handlung erforderlichen Einsicht freizusprechen ist, oder wenn derselbe wegen eines seine freie Willensbestimmung ausschließenden Zustandes straffrei bleibt.

1. Im Gegensatz zu § 3 ist hier die Haftbarkeit eine **unmittelbare**, da die Person, welche die Zuwiderhandlung begangen hat, gemäß §§ 55, 56, 51 StGB. zur Bestrafung nicht herangezogen werden kann. Im übrigen gelten aber auch hier, wie schon die Erwähnung des § 3 im Gesetzestext angibt, sämtliche in § 3 hervorgehobenen Einschränkungen, insbesondere daß gegen die für haftbar erklärten Personen an Stelle der Geldstrafe nicht auf eine Freiheitsstrafe erkannt werden kann.

2. Zustände, welche die freie Willensbestimmung ausschließen können, sind Bewußtlosigkeit und krankhafte Störung der Geistestätigkeit (§ 51 StGB.). Unter Bewußtlosigkeit kann unter Umständen hochgradige Trunkenheit fallen.

Buße.

§ 5.

In den Fällen einer Zuwiderhandlung gegen dieses Gesetz kann, wenn durch dieselbe ein Schaden entstanden ist, auf Verlangen des Verletzten neben der Strafe auf eine an denselben zu zahlende Buße bis zum Betrage von einhundert Mark erkannt werden.

Eine erkannte Buße schließt die Geltendmachung eines weiteren Entschädigungsanspruchs aus.

Die zu der Buße Verurteilten haften für dieselbe als Gesamtschuldner.

1. Während im Forststg. (§§ 19 u. 23) neben der Strafe nur auf Werterſatz oder Erſatzgeld zu erkennen iſt, kann in dem Feldpolizeiſtrafverfahren zugunſten des durch die Zuwiderhandlung Geſchädigten, ſofern er es verlangt, auf vollen Erſatz des durch die Tat entſtandenen Schadens erkannt werden. Der zuerkannte Betrag darf aber die Summe von einhundert Mark nicht überſteigen. Der Charakter dieſer Buße iſt der gleiche wie im gemeinen Strafrecht (§§ 188, 231 StGB.). Darüber, daß die Vorſchrift des § 5 auch gegenüber den Beſtimmungen des BGB. fortdauernd Geltung beſitzt, vgl. die „Einleitende Vorbemerkung" gegen Ende.

2. Auf die Buße wird nur auf Verlangen des Geſchädigten erkannt. Das Verlangen kann aber nur bis zum Erlaß des Strafbefehls oder im ordentlichen Verfahren bis zur Verkündigung des Urteils erſter Inſtanz geſtellt werden. Das Verlangen kann auch jederzeit zurückgenommen werden, um den Entſchädigungsanſpruch im Wege der Zivilklage zu verfolgen. Die anzeigenden Beamten (Feldhüter, Gendarmen ꝛc.) haben zweckmäßigerweiſe den Geſchädigten darauf aufmerkſam zu machen, daß er den Erſatz des Schadens im Strafverfahren erhalten kann und gegebenenfalls ſein Verlangen und die Höhe desſelben in die Anzeige aufzunehmen (vgl. die für die Anzeigen der Feldhüter vorgeſchriebenen Formulare im Anhang D). Eine beſondere Form iſt für das Verlangen nicht vorgeſchrieben.

3. Gemäß § 41 kann die Feſtſetzung der Buße im Strafbefehl erfolgen. Im übrigen ſind für die Buße, da das Strafverfahren in Feldpolizeiſachen ſich, abgeſehen von den in §§ 39—41 hervorgehobenen Beſonderheiten, nach der StPO. beſtimmt, die Vorſchriften der §§ 443 bis 446 StPO. entſprechend anwendbar.

4. Die Zuerkennung einer Buße unterliegt dem Ermeſſen des Strafrichters. Er kann das Verlangen des Verletzten ablehnen, wenn er findet, daß kein genügender Nachweis über das Vorhandenſein oder die Höhe des behaupteten Schadens erbracht iſt, oder daß aus ſonſtigen Gründen keine Veranlaſſung vorliegt, von der ihm erteilten Befugnis Gebrauch zu machen.

5. Der Strafrichter kann auf die verlangte Buße nur neben der Strafe, alſo nur im Falle eines Strafbefehls oder der Verurteilung des Täters erkennen. Es ſteht ihm ſonach nicht zu, über die Strafverfolgung zunächſt zu entſcheiden, und erſt ſpäter, etwa nach einer vorgängigen etwaigen Beweiserhebung in betreff des erlittenen Vermögensſchadens, in einem nachträglichen Verfahren die Buße ab- oder zuzuſprechen.

6. Die erkannte Buße ſchließt nach Abſatz 2 die Geltendmachung eines weiteren Entſchädigungsanſpruchs aus; die Ablehnung des Antrags würde jedoch eine Zivilentſchädigungsklage nicht ausſchließen, ſollte auch die Ablehnung aus dem Grunde erfolgt ſein, weil der Straf-

richter den Eintritt einer nachteiligen Folge für den Verletzten für nicht erwiesen erachtet, oder von der Strafverfolgung freigesprochen hat, weil die Zuwiderhandlung für ihn nicht erwiesen sei. Der Grundsatz, daß die Urteile der Strafgerichte für die Zivilgerichte nicht bindend sind, findet hier seine volle Anwendung.

7. Absatz 3 überträgt auf die Feldrügesachen die Bestimmungen des § 231 des Strafgesetzbuches über die **Gesamthaftung** der Verurteilten für die Entrichtung der Buße. Diese Gesamthaftung aller zur Buße Verurteilten tritt sonach auch da ein, wo sie im Strafbefehl oder im Urteil nicht ausgesprochen ist, oder wo mehrere Urteile gegen die verschiedenen Beteiligten in verschiedenen Verfahren wegen derselben Zuwiderhandlungen ergangen sind (K. B. S. 400 f.).

8. Die **Vollstreckung** der Buße erfolgt gemäß § 495 StPO. nach den Vorschriften über die Vollstreckung der Urteile der Zivilgerichte (§§ 704 ff. ZPO.). Die Gerichtsschreiber der Amtsgerichte sind durch eine Min. Verf. vom 30. März 1899 (Slg. Bd. 24 S. 319) angewiesen, denjenigen Privatpersonen, denen eine Buße zuerkannt worden ist, hiervon nach Rechtskraft der Entscheidung Kenntnis zu geben. Die Vollstreckung geschieht alsdann durch einen Gerichtsvollzieher, der von dem Berechtigten unmittelbar oder durch Vermittlung des Gerichtsschreibers zu beauftragen ist, auf Grund einer von dem Gerichtsschreiber erteilten, mit der Bescheinigung der Vollstreckbarkeit versehenen beglaubigten Abschrift der Urteilsformel (§ 91 Gesch. Anw. für die Gerichtsvollzieher vom 1. März 1904 — Slg. Bd. 28 S. 286). Soweit die Bußen dem Staate, den Bezirken, den Kreisstraßenbaufonds und den Gemeinden zustehen, werden sie gemäß der vorbezeichneten Verfügung vom 30. März 1899 gleichzeitig mit der Geldstrafe und den Gerichtskosten dem Verkehrssteueramt zur Einziehung überwiesen.

Felddiebstahl und Feldbeschädigung.

§ 6.

Felddiebstähle, Begünstigung und Hehlerei in bezug auf solche, sowie rechtswidrig und vorsätzlich begangene Beschädigungen und Begünstigung in Beziehung auf solche unterliegen den Bestimmungen dieses Gesetzes nur dann, wenn der Wert des Entwendeten oder der angerichtete Schaden zehn Mark nicht übersteigt.

1. Nach obiger Bestimmung unterliegt der **Felddiebstahl** (§ 9) nur dann den Vorschriften dieses Gesetzes, wenn der Wert der entwendeten Gegenstände den Betrag von **zehn Mark** nicht übersteigt. Maßgebend ist für die Anwendung des § 6 auf Felddiebstähle nur der Wert der

entwendeten Sache, nicht der Schaden, der dem Eigentümer durch die Entwendung entsteht. Die Worte „oder der angerichtete Schaden" am Schluß des § 6 beziehen sich also nur auf die dort angeführte vorsätzliche und rechtswidrige Beschädigung, nicht auf die Felddiebstähle (vgl. R. G. Urt. 26/10. 93, Joh. Bd. 14 S. 343 und Daude Bem. zu § 6 pr. F. u. FPG.). Uebersteigt der Wert der entwendeten Gegenstände den Betrag von zehn Mark, so finden auf den Felddiebstahl die Vorschriften der §§ 242 ff. StGB. Anwendung. Wird der Felddiebstahl von mehreren Personen gemeinschaftlich begangen, so ist der Gesamtwert des Entwendeten, nicht aber der Wert des auf den einzelnen entfallenden Anteils entscheidend.

2. In Ansehung der im FPStG. für strafbar erklärten Beschädigungen (vgl. insbesondere § 26) folgt aus obiger Vorschrift in Verbindung mit dem in Bem. 1b zu § 1 Gesagten:

a) Vorsätzlich und rechtswidrig begangene Beschädigungen, die an sich unter § 303 StGB. fallen, unterliegen den Bestimmungen des FPStG., wenn der angerichtete Schaden den Betrag von zehn Mark nicht übersteigt. Andernfalls sind sie nach § 303 StGB. zu bestrafen.

b) Die in diesem Gesetz aufgeführten Beschädigungen sind auch dann strafbar, wenn sie fahrlässig begangen sind.

c) Fahrlässige Beschädigungen sind immer, d. h. auch wenn der angerichtete Schaden den Betrag von zehn Mark übersteigt, nach diesem Gesetz zu bestrafen.

3. Da Begünstigung und Hehlerei in bezug auf einen Felddiebstahl und die Begünstigung in bezug auf eine nach diesem Gesetz strafbare vorsätzliche und rechtswidrige Beschädigung nach § 8 mit der vollen Strafe des Felddiebstahls bezw. der vorsätzlichen Beschädigung bestraft werden, mußten sie auch im obigen Paragraphen neben dem Felddiebstahl und der vorsätzlich und rechtswidrig begangenen Beschädigung genannt werden.

Beihülfe.

§ 7.

Die Beihülfe zu einem Felddiebstahl oder einer vorsätzlichen Beschädigung wird mit der vollen Strafe der Zuwiderhandlung bestraft.

1. Als Gehülfe wird nach § 49 Abs. 1 StGB. bestraft, wer dem Täter zur Begehung eines Verbrechens oder Vergehens durch Rat oder Tat wissentlich Hülfe geleistet hat. Da die meisten nach dem FPStG. strafbaren Handlungen Uebertretungen sind (vgl. oben Bem. 1c zu § 1), würde die zu ihnen geleistete Beihülfe regelmäßig straflos bleiben.

Dies zu verhüten, wird durch die Vorschrift des § 7 bezweckt, wonach die Beihülfe zu einem Felddiebstahl oder zu einer nach dem FPStG. strafbaren vorsätzlichen Beschädigung bestraft werden soll, auch wenn die Tat sich nur als Uebertretung darstellt. Die zu anderen strafbaren Handlungen des FPStG. geleistete Beihülfe bleibt straflos.

2. In § 49 Abs. 2 StGB. ist für den Gehülfen grundsätzlich eine mildere Strafe angedroht als für den Täter. Auch hiervon enthält § 7 eine Ausnahme, indem hier für die Beihülfe zu einem Felddiebstahl und zu einer vorsätzlichen Beschädigung die volle Strafe der Zuwiderhandlung angedroht ist. Dies ist aber nicht etwa dahin zu verstehen, daß in jedem einzelnen Fall der Gehülfe von derselben Strafe getroffen werden soll wie der Täter. Vielmehr bleibt der allgemeine Grundsatz maßgebend, daß jeder Beteiligte nach dem Grade seiner Verschuldung zu bestrafen ist (Begr. S. 16).

3. Die Beihülfe ist nur dann strafbar, wenn sie vorsätzlich zu vorsätzlichem Tun geleistet wird. Fahrlässige Beihülfe zu vorsätzlichem Tun bleibt straflos. Wird fahrlässige Beihülfe zu fahrlässigem Handeln geleistet, so kann sie als fahrlässige Mitverursachung in Betracht kommen. Die Beihülfe muß tatsächlich geleistet sein. Der Versuch der Beihülfe ist nicht strafbar. Dagegen ist auch die zu einer blos versuchten Handlung geleistete Beihülfe, sofern der Versuch strafbar ist, zu bestrafen. Die Beihülfe ist ferner strafbar, wenn sie einem Jugendlichen, dem der persönliche Strafausschließungsgrund des § 55 oder 56 zur Seite steht, geleistet wird (vgl. Bem. 5 zu § 2).

4. Ueber die Anstiftung enthält das FPStG. keine besondere Bestimmung. Es finden deshalb gemäß § 1 die Bestimmungen des StGB. Anwendung. § 48 StGB. bestimmt:

„Als Anstifter wird bestraft, wer einen andern zu der von demselben begangenen strafbaren Handlung durch Geschenke oder Versprechen, durch Drohung, durch Mißbrauch des Ansehens oder der Gewalt, durch absichtliche Herbeiführung oder Beförderung eines Irrtums oder durch andere Mittel vorsätzlich bestimmt hat.

Die Strafe des Anstifters ist nach demjenigen Gesetz festzusetzen, welches auf die Handlung Anwendung findet, zu welcher er wissentlich angestiftet hat."

Die Anstiftung muß hiernach vorsätzlich geschehen sein. Auch der Täter, der bestimmt worden ist, muß vorsätzlich gehandelt haben. Fahrlässige Bestimmung zu fahrlässigem Handeln kann aber unter Umständen fahrlässige Mitverursachung sein. Der bloße Versuch der Anstiftung ist nicht strafbar. Dagegen ist die Anstiftung schon strafbar, wenn sie zu einem strafbaren Versuch der Handlung — nach § 8 ist nur der Versuch des Felddiebstahls strafbar — geführt hat. Davon, daß der Täter zur Verantwortung gezogen werden werden kann, hängt die Strafbarkeit des Anstifters nicht ab. Ist z. B. eine jugendliche Person, welcher der

Strafausschließungsgrund des § 55 oder 56 StGB. zur Seite steht (vgl. Bem. 5 zu § 2), zur Tat bestimmt worden, so ist der Anstifter trotzdem strafbar. Fehlt es dem die Tat Ausführenden überhaupt an der freien Willensbestimmung, wie z. B. einem Blödsinnigen oder sinnlos Betrunkenen, so ist derjenige, der durch seine Beeinflussung den Handelnden als Werkzeug mißbraucht hat, selbst Täter.

Versuch, Begünstigung, Hehlerei.

§ 8.

Der Versuch des Felddiebstahls, die Begünstigung und Hehlerei in Beziehung auf einen solchen, sowie die Begünstigung in Beziehung auf eine nach diesem Gesetz strafbare vorsätzliche Beschädigung werden mit der vollen Strafe des Felddiebstahls, bezw. der vorsätzlichen Beschädigung bestraft.

1. Die für den Begriff des Versuches maßgebende Bestimmung des StGB. lautet in § 43 Abs. 1:

„Wer den Entschluß, ein Verbrechen oder Vergehen zu verüben, durch Handlungen, welche einen Anfang der Ausführung dieses Verbrechens oder Vergehens enthalten, betätigt hat, ist, wenn das beabsichtigte Verbrechen oder Vergehen nicht zur Vollendung gekommen ist, wegen Versuches zu bestrafen."

Nach § 44 StGB. ist der Versuch milder zu bestrafen als die vollendete strafbare Handlung.

Von diesen beiden Bestimmungen enthält § 8 insofern eine Ausnahme, als auch der Versuch des Felddiebstahls, trotzdem dieser in der Regel nur eine Uebertretung darstellt, strafbar ist und zwar mit der vollen Strafe, die für den Felddiebstahl angedroht ist.

Nur der Versuch eines Felddiebstahls ist strafbar. Der Versuch anderer nach diesem Gesetz strafbarer Handlungen bleibt straflos.

Anwendung auf den Versuch des Felddiebstahls findet § 46 Ziff. 1 StGB., wonach der Versuch als solcher straflos bleibt, wenn der Täter die Ausführung der beabsichtigten Handlung aufgegeben hat, ohne daß er an dieser Ausführung durch Umstände gehindert worden ist, welche von seinem Willen unabhängig waren.

2. Der Begriff der Begünstigung ergibt sich aus § 257 Abs. 1 StGB., welcher lautet:

„Wer nach Begehung eines Verbrechens oder Vergehens dem Täter oder Teilnehmer wissentlich Beistand leistet, um denselben der Bestrafung zu entziehen oder um ihm die Vorteile des Verbrechens oder Vergehens zu sichern, ist wegen Begünstigung mit Geldstrafe bis zu sechshundert Mark oder mit Gefängnis bis zu einem Jahre und, wenn er diesen Beistand seines Vorteils wegen leistet, mit Gefängnis zu bestrafen. Die Strafe darf jedoch, der Art oder dem Maße nach, keine schwerere sein, als die auf die Handlung selbst angedrohte."

Nach § 8 FPStG. ist die Begünstigung in Beziehung auf einen Felddiebstahl und eine nach dem FPStG. strafbare vorsätzliche Beschädigung, desgleichen auf einen strafbaren Versuch des Felddiebstahls selbst dann strafbar, wenn diese Handlungen Uebertretungen sind. Die Begünstigung, die in bezug auf eine andere nach diesem Gesetze strafbare Handlung geleistet wird, bleibt straflos.

Die Begünstigung muß wissentlich geleistet sein, d. h. der Begünstiger muß wissen, daß der Begünstigte sich eines Felddiebstahls oder einer vorsätzlichen Beschädigung schuldig gemacht hat, dagegen braucht er nicht die näheren Umstände, unter denen die Tat begangen worden ist, zu kennen. Selbstverständlich aber können diese Umstände, sofern sie nach §§ 11—13 die Strafbarkeit erhöhen, ihm nur insoweit zugerechnet werden, als er sie gekannt hat (§ 59 StGB.), sodaß z. B. derjenige, welcher Begünstigung in bezug auf einen Felddiebstahl im Rückfall geleistet hat, ohne zu wissen, daß der Täter rückfällig war, nur nach § 9 dieses Gesetzes bestraft werden kann. Eine irrtümliche rechtliche Beurteilung, z. B. die Annahme, die Begünstigung sei, weil zu einer Uebertretung geleistet, nicht strafbar, schließt die Strafbarkeit nicht aus (vgl. hierzu RG. Urt. 26./9. 81 E. 5 S. 23).

Eine Begünstigung, um den Täter der Bestrafung zu entziehen, liegt u. a. vor, wenn auf den zur Anzeige verpflichteten Beamten (Feldhüter, Gendarmen) in der Absicht, ihn von der Beanzeigung abzuhalten, eingewirkt wird, oder wenn diesen oder anderen mit der Verfolgung der Tat betrauten Beamten wissentlich falsche Angaben zugunsten des Täters gemacht werden, nach der Rechtsprechung des RG. (Urt. 6./4. 86 E. 14 S. 88) sogar dann, wenn versucht wird, die zur Anzeige eines von Amtswegen zu verfolgenden Delikts entschlossene Person von ihrem Vorhaben abzubringen. Dagegen liegt keine Begünstigung vor, wenn jemand einem zur Anzeige oder Untersuchung der Tat verpflichteten Beamten Auskunft über den Täter oder die Tat zu geben sich überhaupt weigert (RG. Urt. 23/11 83 E. 9 S. 433), oder wenn auf den Antragsberechtigten eingewirkt wird, um ihn von der Stellung des Antrags abzuhalten oder zur Zurücknahme des gestellten Antrags zu veranlassen.

Nicht nur die Begünstigung in der Absicht, den Täter der Verurteilung zu entziehen, sondern auch die Begünstigung, die geleistet wird, um die Vollstreckung zu vereiteln, ist strafbar, so daß sich unter Umständen der Begünstigung schuldig macht, wer die dem Täter auferlegte Geldstrafe bezahlt. Vgl. hierüber RG. Urt. 21./9. 97 E. 30 S. 232.

Die Begünstigung setzt nicht voraus, daß der beabsichtigte Erfolg, den Täter der Bestrafung zu entziehen, wirklich eingetreten ist, ebensowenig aber auch, daß der Täter zur Bestrafung gebracht worden ist. Sie ist vielmehr ein völlig selbständiges Delikt. Ein dem Täter zur Seite stehender persönlicher Strafausschließungsgrund (S. Bem. 5 zu § 2) schützt demgemäß den Begünstiger nicht vor Strafe.

Ein im Entwurf dieses Gesetzes dem § 7 (jetzigen § 8) beigefügter zweiter Absatz, wonach die Bestimmungen des § 257 Abs. 2 und 3 StGB. Anwendung finden sollten, ist von der Kommission gestrichen worden (K. B. S. 400). Demgemäß ist auch die Begünstigung, welche dem Täter oder Teilnehmer von einem Angehörigen gewährt worden ist, um ihn der Bestrafung zu entziehen, nicht straflos (anders § 14 Forststg.).

3. Es gibt zwei Arten von Hehlerei:

a) § 258 StGB. „Wer seines Vorteils wegen sich einer Begünstigung schuldig macht, wird als Hehler bestraft."

b) § 259 StGB. „Wer seines Vorteils wegen Sachen, von denen er weiß oder den Umständen nach annehmen muß, daß sie mittels einer strafbaren Handlung erlangt sind, verheimlicht, ankauft, zum Pfande nimmt oder sonst an sich bringt oder zu deren Absatze bei anderen mitwirkt, wird als Hehler bestraft." (Partiererei.)

Während aber nach dem StGB. die Hehlerei allgemein mit Gefängnis als Mindeststrafe bedroht ist, finden auf die Bestrafung der Hehlerei, die in bezug auf einen Felddiebstahl geleistet worden ist, die milderen Vorschriften dieses Gesetzes Anwendung.

Die Hehlerei ist ein selbständiges Delikt. Persönliche Strafausschließungsgründe in der Person des Felddiebes (S. Bem. 5 zu § 2) machen den Hehler nicht straflos. Auch ist die Hehlerei, die in Beziehung auf einen nur auf Antrag zu verfolgenden Felddiebstahl verübt wird, selbst dann strafbar, wenn wegen der Haupttat ein Antrag nicht gestellt wird.

Bei der zweiten Art von Hehlerei (Partiererei) ist nicht erforderlich, daß der Hehler beim Erwerb der Sachen wußte, daß sie mittels einer strafbaren Handlung erlangt waren. Es genügt vielmehr, daß er zu jenem Zeitpunkt nach den Umständen, unter denen sich der Erwerb vollzog, bei Aufwendung einiger Sorgfalt darüber nicht zweifelhaft sein konnte.

Das Ansichbringen des § 259 StGB. bedeutet einen Wechsel des Gewahrsams. Ein bloßes Mitgenießen von Früchten, die durch einen Felddiebstahl erlangt sind, etwa durch die Familienangehörigen des Diebes, ist infolgedessen der Regel nach keine Hehlerei.

4. Über gewerbs- und gewohnheitsmäßige Hehlerei s. § 13 Ziff. 2 dieses Ges.

Felddiebstahl.

§ 9.

Mit Geldstrafe bis zu einhundert Mark oder mit Haft bis zu vier Wochen wird wegen Felddiebstahls bestraft, wer noch nicht eingeheimste Gartenfrüchte, Feldfrüchte odere andere Boden-

erzeugnisse aus Gartenanlagen aller Art, Weinbergen, Obstanlagen, Baumschulen, Saatkämpen, von Aeckern, Wiesen, Weiden, Plätzen, Gewässern, Wegen oder Gräben entwendet.

1. Der Felddiebstahl unterscheidet sich vom gemeinen Diebstahl (§ 242 StGB.) in drei Beziehungen, die zusammen treffen müssen:

a) in Ansehung des Wertes des Entwendeten,

b) in Ansehung des Gegenstandes und

c) in Ansehung des Ortes, wo er begangen wird.

Die Unterscheidungsmerkmale sind rein äußerlicher Natur. Seinem Wesen nach ist der Felddiebstahl nicht verschieden von dem gemeinen Diebstahl. Infolgedessen müssen neben den erwähnten drei Beziehungen sämtliche Voraussetzungen des gemeinen Diebstahls vorhanden sein, also insbesondere die Absicht der rechtswidrigen Zueignung und das Bewußtsein der Rechtswidrigkeit des Handelns auf seiten des Täters. Wenn daher der Täter in der Vorstellung, durch seine Handlung niemandes Rechte zu verletzen, handelt, z. B. eine Blume, gegen deren Wegnahme der Eigentümer in der Regel nichts einzuwenden hat, vom Felde pflückt, ist er nicht strafbar.

2. Wegen des Wertes s. § 6 dieses Ges.

3. Als Gegenstand des Feldiebstahls nennt § 9 noch nicht eingeheimste Gartenfrüchte, Feldfrüchte oder andere Bodenerzeugnisse. Zum Begriff des Einheimsens genügt weder das bloße Abtrennen vom Boden noch das Verbringen in Haufen auf dem Felde im allgemeinen. Vielmehr ist eine Einheimsung erst mit dem Augenblick eingetreten, in welchem die Feldfrüchte und sonstigen Felderzeugnisse an denjenigen Ort verbracht sind, der zu ihrer Verwahrung bis zu ihrem Verbrauch oder ihrer Verwertung, überhaupt zu einer verhältnismäßig dauernden Aufbewahrung, bestimmt ist. Infolgedessen ist es nur Felddiebstahl, wenn Heu, das bei der Ernte vom Boden getrennt und in Haufen gesetzt ist, oder Kartoffeln, die aus dem Boden genommen und zum Trocknen auf dem Felde ausgebreitet sind, entwendet werden. Dagegen wird wegen gemeinen Diebstahls bestraft, wer Feldfrüchte aus einem zu längerer Aufbewahrung dienenden Ort entwendet, mag dieser in dem Anwesen des Besitzers oder auch, wie es in erntereichen Jahren häufig vorkommt, auf dem Felde, das die Früchte erzeugt hat, in daselbst angelegten Mieten, Schobern oder Gruben sich befinden (vgl. OLG. Urt. 3./11. 90 Ztschr. Bd. 15 S. 539, ferner RG. Urt. 7./2. 82 E. 5 S. 385 und 2./11. 83 E. 9 S. 163 zu § 18 pr. F. u. FPG.).

Garten- und Feldfrüchte fallen unter den weiteren Begriff der Bodenerzeugnisse. Sie unterscheiden sich von den „anderen" Bodenerzeugnissen dadurch, daß sie zum Genusse durch Mensch oder Vieh bestimmt sind. „Andere" Bodenerzeugnisse sind z. B. auch Blumen.

Für den Begriff des Bodenerzeugnisses ist nach dem Gesagten bedeutungslos, ob bereits eine Trennung von Grund und Boden stattgefunden hat. Es macht ferner keinen Unterschied, ob die Pflanze aus Samen, der in den Boden eingelegt wurde, bzw. sog. Wurzelausschlag entsprossen ist, oder ob sie erst als Pflanze in den Boden eingesetzt wurde und sich in ihm weiter entwickelt hat (RG. Urt. 26./10. 82 E. 7 S. 190, Urt. 1./11.'92 E. 23 S. 269 zu § 18 pr. F. u. FPG.). Dagegen ist erforderlich, daß die eingesetzte Pflanze die Eigenschaft einer natürlichen Zugehörigkeit zum Grund und Boden erlangt hat, m. a. W. daß sich in der Pflanze infolge der Einwirkung von Grund und Boden das pflanzliche Leben weiter entwickelt hat (RG. Urt. 16./1. 02 E. 35 S. 67 zu § 18 pr. F. u. FPG.). Infolgedessen sind z. B. Blumenstöcke und Reben, die der Gärtner zur Aufbewahrung vorübergehend in den Boden bloß eingeschlagen hat, um sie bald darauf zu verkaufen, keine Bodenerzeugnisse. Wer sie entwendet, begeht daher einen gemeinen Diebstahl nach § 242 StGB. Oft wird allerdings der Zeitpunkt, zu welchem jene natürliche Verbindung des pflanzlichen Organismus mit dem Grund und Boden eintritt, nur schwer zu bestimmen sein, und zwar nicht bloß bei Pflanzen, sondern noch in erhöhtem Maße bei Samen und Saatfrüchten, z. B. Saatkartoffeln. In solchen zweifelhaften Fällen wird zugunsten des unbefugt Wegnehmenden die ihm vorteilhaftere Annahme, daß nämlich die Verbindung bereits besteht, Platz zu greifen haben (RG. Urt. 1. 11. 92 E. 23 S. 269 zu § 18 pr. F. u. FPG.).

4. Ein Felddiebstahl liegt nur dann vor, wenn die Entwendung der Bodenerzeugnisse aus Gartenanlagen aller Art, Weinbergen, Obstanlagen, Baumschulen, Saatkämpen, von Aeckern, Wiesen, Weiden, Plätzen, Gewässern, Wegen oder Gräben stattgefunden hat. Wie in der „Einleitenden Vorbemerkung" bereits hervorgehoben, beziehen sich die Bestimmungen des Gesetzes über Felddiebstahl *sowohl auf städtische als auch auf ländliche Orte und Feldmarken.* Es ist deshalb ohne Einfluß, ob die bezeichneten Grundstücke und Gewässer sich innerhalb von Städten und anderen bebauten Ortschaften oder auf dem Lande sich befinden. Im *einzelnen* ist zu bemerken:

a) Unter Gartenanlagen sind nicht bloß Gärten im engeren Sinne zu verstehen, sondern alle Orte, die eine gartenähnliche Ausstattung haben. Infolgedessen gehören auch *Kirchhöfe* und *Gräber* hierher, sofern sie nach den konkreten Verhältnissen, nach Ausstattung und Einrichtung sich als Gartenanlagen darstellen (RG. Urt. 27./10. 96 E. 29 S. 138, Urt. 26./10. 82 E. 7 S. 190, Urt. 15./1. 83 R. 5 S. 30, sämtlich zu § 18 pr. F. u. FPG.). Demgemäß fällt die Entwendung von Blumenstöcken und Blumen aus Kirchhöfen und von Gräbern unter obigen Paragraph, und die Entwendung von daselbst wachsenden Bäumen, Frucht- und Ziersträuchern unter § 12 Ziff. 6, sofern der Wert der entwendeten Gegenstände den Betrag von zehn Mark nicht übersteigt.

Unter Umständen kann allerdings eine solche Entwendung aus Kirchhöfen und von Gräbern auch unter § 168 StGB. fallen. Doch ist hierzu eine Zerstörung oder Beschädigung des Grabes selbst, etwa eine durch das Ausgraben der Pflanzen bewirkte erhebliche Zerwühlung oder Einlöcherung des Grabhügels vorausgesetzt. Auch die Entwendung von Pflanzen im Werte bis zu zehn Mark aus öffentlichen Gartenanlagen fällt unter § 9 bezw. § 12 Ziff. 6 dieses Ges., desgleichen die Entwendung aus einem Park, sofern dieser nicht nach forstwirtschaftlichen Betriebsregeln eingerichtet ist, oder wesentlich zur Holznutzung dient, sondern der Verschönerung der Landschaft und der Annehmlichkeit des Publikums gewidmet ist (vgl. Daube Bem. 4 zu § 18 pr. F. u. FPG.).

b) Saatkämpe werden meistens zu den Zwecken der Holznutzung angelegt sein. Entwendungen aus ihnen fallen daher in der Regel unter das Forststg. Doch gibt es auch Saatkämpe, die nicht zu den in § 10 Forststg. bezeichneten Grundstücken gehören. Auf Entwendungen aus solchen bezieht sich § 9 FPStG.

c) Gewässer. Nach ausdrücklicher Bestimmung des § 9 soll auch die Entwendung von Wasserpflanzen als Feldbiebstahl behandelt werden. Dagegen sind Anlandungen an Flußufern nicht angeführt. Sind dieselben aber mit Weiden oder anderen Holzarten besetzt, und dienen sie demgemäß zur Holznutzung, so fällt die Entwendung des Holzes unter § 10 Forststg.

d) Topfpflanzen sind keine Bodenerzeugnisse. Sie empfangen ihre Nahrung aus dem in dem Topf vorhandenen Grund, nicht aus der Erdoberfläche und stellen sich diesen gegenüber als eine völlig selbständige Sache dar. Die Entwendung von Topfpflanzen mit oder ohne Topf fällt daher nnter § 242 StGB. Dabei kommt nicht in Betracht, ob der Blumentopf seinen Standort in dem Garten hat, in dem früher die Pflanze gezogen wurde (RG. Urt 30./7. 04 E. 26 S. 101).

§ 10.

Wer, nachdem er wegen vollendeten oder versuchten Felddiebstahls oder wegen Teilnahme, Begünstigung oder Hehlerei in Beziehung auf einen solchen in Elsaß-Lothringen auf Grund dieses Gesetzes rechtskräftig verurteilt worden ist, innerhalb der nächsten zwei Jahre abermals eine dieser Handlungen begeht, befindet sich im Rückfalle.

Im zweiten oder ferneren Rückfalle befindet sich, wer nach rechtskräftiger Verurteilung wegen ersten, zweiten oder ferneren Rückfalls innerhalb der nächsten zwei Jahre abermals eine der bezeichneten Handlungen begeht.

1. Unter Teilnahme fallen Mittäterschaft, Anstiftung und Beihülfe. Ueber die beiden letzteren Begriffe, vgl. die Bemerkungen zu § 7, über Begünstigung und Hehlerei § 8.

2. Unter rechtskräftiger Verurteilung ist nicht nur die Bestrafung durch Urteil, sondern auch die zufolge eines Strafbefehls zu verstehen. Für den Beginn der zweijährigen Frist ist nicht wie in §§ 245, 264 StGB. maßgebend der Zeitpunkt der völligen oder teilweisen Verbüßung oder des gänzlichen oder teilweisen Erlasses der Strafe, sondern der Zeitpunkt der Rechtskraft des Richterspruchs. Ein Urteil ist rechtskräftig, wenn es durch ein Rechtsmittel nicht mehr angefochten werden kann, ein Strafbefehl ist rechtskräftig, wenn die Einspruchsfrist verstrichen, oder der erhobene Einspruch zurückgenommen ist.

3. Ein zweiter oder fernerer Rückfall im Sinne dieses Gesetzes liegt nur vor, wenn in der früheren richterlichen Entscheidung die Bestrafung wegen ersten, zweiten oder ferneren Rückfalls ausdrücklich ausgesprochen worden ist.

4. Behufs Erleichterung der Feststellung von Rückfällen sind die Amtsanwälte zu bestimmten gegenseitigen Mitteilungen sowie zur besonderen Kenntlichmachung der auf Grund des FPStG. ergangenen Bestrafungen in ihren Strafprozeßlisten verpflichtet. (Vgl. des näheren die Verf. des Oberstaatsanwalts zu Colmar vom 31. Dezember 1888, Smlg. Bd. 13 S. 430 und vom 23. April 1891, Smlg. Bd. 16 S. 118.)

§ 11.

Geldstrafe von fünf bis zu einhundertfünfzig Mark oder Haft tritt für den Felddiebstahl ein:

1. wenn derselbe an einem Sonn- oder Festtage oder in der Zeit von Sonnenuntergang bis Sonnenaufgang begangen ist:
2. wenn der Täter Mittel angewendet hat, um sich unkenntlich zu machen:
3. wenn der Täter dem Feldhüter oder einem anderen zuständigen Beamten oder dem Beschädigten seinen Namen oder Wohnort anzugeben sich geweigert oder falsche Angaben über seinen oder seiner Gehülfen Namen oder Wohnort gemacht oder auf Anrufen der vorstehend genannten Personen, anstatt stehen zu bleiben, die Flucht ergriffen oder fortgesetzt hat;
4. wenn die Tat unter Benutzung von Aexten, Sägen, Messern, Spaten oder ähnlichen Werkzeugen begangen ist;
5. wenn der Täter die Auslieferung der zum Gebrauch bei dem Felddiebstahl bestimmten Werkzeuge verweigert hat;

6. wenn zum Zwecke des Felddiebstahls ein zur Fortschaffung größerer Mengen geeignetes Gerät, Fahrzeug oder Lasttier mitgebracht ist;

ferner, wenn der Felddiebstahl begangen ist:

7. aus einem umschlossenen Raume mittels Einsteigens;
8. an Kien, Harz, Saft, Wurzeln, Rinde oder Trieben stehender Bäume oder Rebstöcke, sofern derselbe nicht als Forstdiebstahl strafbar ist;
9. an Trauben im Innern von Weinbergen;
10. im Rückfalle.

1. In den §§ 11--13 sind die Umstände hervorgehoben, bei deren Vorhandensein der Felddiebstahl im Interesse eines erhöhten Schutzes der Bodenerzeugnisse schärfer zu bestrafen ist. Die erschwerenden Umstände ergeben sich, abgesehen von dem Rückfall, teils aus der Erschwerung der Beaufsichtigung der Grundstücke, teils aus der größeren Schwierigkeit der Entdeckung oder Ergreifung des Täters, teils aus der größeren Gefährlichkeit der strafbaren Handlung. Um dem Richter einen Anhalt für die Bemessung der Strafen zu geben, hat in §§ 11—13 eine dreifache Abstufung der Schärfungsgründe stattgefunden. Zu beachten ist, daß § 14 für sämtliche drei Stufen wiederum das Vorhandensein von Milderungsgründen vorsieht, die in den Fällen der §§ 11 und 12 das Zurückgehen auf die Mindeststrafe von einer Mark und im Falle des § 13 Haft an Stelle von Gefängnis gestatten.

2. Zu den einzelnen Schärfungsgründen des § 11 ist zu bemerken:

Zu Ziff. 1:

Die Frage, ob ein Tag im Sinne dieses Gesetzes ein Festtag ist, beurteilt sich ausschließlich nach tatsächlichen Verhältnissen, nämlich danach, ob ein bestimmter Tag nach den Gebräuchen der in Betracht kommenden Bevölkerungskreise als Festtag gehalten wird. Festtag ist hier nicht gleichbedeutend mit gesetzlichem Feiertag. Der gesetzgeberische Gedanke ist der, daß die Felder an einem Tage, an dem die Bevölkerung und infolgedessen regelmäßig auch die zur Aufsicht über die Felder bestimmten Beamten sich kirchlichen oder anderen festlichen Veranstaltungen widmen, den Angriffen der Frevler eher ausgesetzt sind, als an gewöhnlichen Tagen. Ob ein solcher Festtag vorliegt, kann sich demgemäß nicht nach den durch verschiedenartige Erwägungen veranlaßten Vorschriften anderer Gesetze über äußere Heilighaltung gewisser Tage bestimmen, sondern bemißt sich lediglich nach dem Herkommen und den Gebräuchen der Bevölkerung. An einem Orte, mit ausschließlich oder vorwiegend protestantischer Bevölkerung z. B. ist ein am Karfreitag begangener Felddiebstahl schwerer zu bestrafen, als die gleiche am selben Tage in einer wesentlich nur von Katholiken bewohnten Gegend begangene strafbare Handlung, während andererseits in einer Gemeinde mit vorwiegend katholischer Bevölkerung der Patronstag als Festtag im Sinne dieses Gesetzes zu gelten hat. (Vgl. hierzu die Ausführungen des Unterstaatssekretärs v. Puttkamer und des Berichterstatters Abgeordneten Dr. Gunzert bei der 2. Lesung des Gesetzentwurfes [L. A. S. 379].) Auf die Konfession der Schutzbeamten kommt es bei dieser Frage nicht an. A. M. Daude Bem. 2 zu § 2 pr. F. u. FPG. Wohl aber ist gemäß § 59 StGB. vorausgesetzt, daß der Täter gewußt hat, daß der Tag, an dem die Tat begangen wurde, ein Festtag im Sinne dieses Gesetzes ist.

Zu Ziff. 2:

Mittel, um sich unkenntlich zu machen, sind z. B. Vermummen, Färben des Gesichts, Tragen falscher Bärte.

Zu Ziff. 3:

Zuständiger Beamter ist jeder Beamte, der befugt ist, den Namen oder den Wohnort der in Betracht kommenden Person festzustellen, daher außer dem in Ziff. 3 genannten Feldhüter auch andere Gemeindepolizeibeamte und staatliche Schutzleute, ferner der Gendarm und der Bürgermeister. Unter Namen ist nicht nur der Familiennamen, sondern auch der Vornamen zu verstehen. Ja sogar, wer unter mehreren Vornamen einen falschen Rufnamen angibt, kann sich des obigen Erschwerungsgrundes schuldig machen. (Vgl. RG. Urt. 2/2. 06 Dtsche. Juristenzeitung XI. Jahrg. S. 711.) Häufig wird der Erschwerungsgrund der falschen Namensangabe mit einer Uebertretung nach § 360 Ziff. 8 StGB. zusammentreffen, wonach mit Geldstrafe oder mit Haft bestraft wird, wer sich eines ihm nicht zukommenden Namens einem zuständigen Beamten gegenüber bedient. In diesen Fällen kann selbstverständlich nicht neben der aus § 11 Ziff. 3 verschärften Strafe auf eine Strafe nach § 360 Ziff. 8 erkannt werden.

Bloße Weigerung, den Namen oder Wohnort des Gehülfen anzugeben, ist kein Schärfungsgrund. Nur die falsche Angabe kommt in Betracht. Sie ist auch lediglich Schärfungsgrund und kann nicht etwa noch besonders als Begünstigung des Gehülfen gemäß § 257 StGB. bestraft werden.

Zu Ziff. 4:

Anders wie im Forststg. § 12 Ziff. 4 ist nicht nur der Gebrauch schneidender, sondern auch hauender Werkzeuge ein Schärfungsgrund.

Zu Ziff. 5:

Es genügt die Weigerung der Herausgabe. Aktiver Widerstand ist nach § 113 StGB zu bestrafen.

Zu Ziff. 6:

Die Geräte 2c. müssen zum Zwecke des Feldiebstahls mitgebracht sein. Ihre Verwendung ist kein Schärfungsgrund, wenn sie bloß zufällig, weil sie gerade zur Hand waren, benutzt wurden. Zur Fortschaffung größerer Mengen geeignete Geräte sind auch größere Säcke und Körbe. Als Fahrzeug genügt ein Handkarren oder Kinderwagen. Auch ein Kahn fällt unter den Begriff des Fahrzeugs.

Zu Ziff. 7:

Umschlossener Raum ist ein begrenzter Teil der Erdoberfläche. Die Umschließung muß die Absicht des Eigentümers erkennen lassen, das Eindringen Unberechtigter abzuhalten, hierzu aber auch geeignet sein. (Beispiele: Hecke, Mauer, Graben, Gitter.) Einsteigen ist die Tätigkeit, welche der Zuwiderhandelnde auf nicht ordnungsmäßigem Wege anwendet, um von außen über die Umschließung in den umschlossenen Raum zu gelangen. Doch ist ein Betreten des Inneren erforderlich. Ein bloßes Emporklettern etwa an einer Mauer und Herauslangen von Früchten oder Blumen genügt nicht. (Vgl. des näheren die Kommentare zu § 243 Ziff. 2 StGB.)

Zu Ziff. 8:

Kien sind die besonders harzreichen Holzteile der Kiefer, die zum Feueranzünden, als Beleuchtungsmaterial, zur Herstellung von Teer und Kienruß verwendet werden. Harz ist der von Nadelhölzern an offenen Stellen der Rinde ausgeschwitzte Saft. Die Schärfung tritt nur dann ein, wenn die Entwendung der bezeichneten Holzteile an stehenden Bäumen oder Rebstöcken erfolgt, nicht wenn die Bäume schon gefällt waren oder jene Holzteile sich bereits vorher vom Holz gelöst hatten.

Als Forstdiebstahl sind die bezeichneten Entwendungen zu bestrafen, wenn sie in einem Wald oder auf einem anderen hauptsächlich zur Holznutzung bestimmten Grundstück verübt wurden (§ 10 Forststg.).

Zu Ziff. 9:

Der erschwerende Umstand wird darin gefunden, daß der Täter, welcher sich in das Innere des Weinbergs hineinbegibt, den Augen des Wächters entschwindet und viel schwieriger zu entdecken ist.

Zu Ziff. 10:

Ueber Rückfall s. § 10.

§ 12.

Gefängnisstrafe bis zu drei Monaten tritt ein, wenn der Felddiebstahl begangen ist:

1. unter Mitführung von Waffen:
2. aus einem umschlossenen Raume mittels Einbruchs;
3. dadurch, daß zur Eröffnung der Zugänge eines umschlossenen Raumes falsche Schlüssel oder andere zur ordnungsmäßigen Eröffnung nicht bestimmte Werkzeuge angewendet worden sind;
4. von drei oder mehr Personen in gemeinschaftlicher Ausführung;

5. zum Zweck der Veräußerung des Entwendeten oder daraus hergestellter Gegenstände:
6. durch Wegnahme stehender Bäume, Frucht- oder Ziersträucher, sofern die Entwendung nicht als Forstdiebstahl strafbar ist;
7. von dem Aufseher in dem seiner Aufsicht unterstellten Grundstücke:
8. im zweiten Rückfalle.

Zu Ziff. 1:

Waffen sind zunächst die eigentlichen Stoß-, Hieb- und Schußwaffen. Aber auch andere Werkzeuge, die zur Beibringung bedeutender Körperverletzungen geeignet sind, z. B. Holzknüppel, sind unter Umständen als Waffen anzusehen. Ein gewöhnliches Taschenmesser ist keine Waffe. Zum Begriff des Mitführens genügt, daß der Täter das Bewußtsein hat, eine Waffe bei sich zu haben. Die Absicht, sie anzuwenden, ist nicht erfordert. Sind mehrere Personen an dem Felddiebstahl beteiligt, von denen nur eine Person eine Waffe mit sich führt, so trifft der Schärfungsgrund des § 12 Ziff. 1 auch diejenigen, welche von der Mitführung der Waffe Kenntnis hatten.

Zu Ziff. 2:

Ueber umschlossenen Raum vgl. Bem. zu § 11 Ziff. 7. Einbruch ist die gewaltsame Aufhebung des Zusammenhanges der äußeren Umschließung eines umschlossenen Raumes durch Herstellung einer Oeffnung oder durch Vergrößerung einer bereits vorhandenen Oeffnung (Olshausen Note 16 zu § 243 StGB.). Nach der Rechtsprechung des Reichsgerichts ist nicht erforderlich, daß durch die gewaltsame Aufhebung des Zusammenhanges eine Zerstörung oder Beschädigung der Umschließung verursacht wird, so daß z. B. schon das ohne Beschädigung erfolgende Ausheben eines verschlossenen Torflügels oder das ebenso erfolgende Auseinanderbiegen der Flügel eines verschlossenen Tores Einbruch ist. (Vgl. des näheren die Kommentare zu § 243 Ziff. 2 StGB.)

Zu Ziff. 3:

Ein falscher Schlüssel ist ein Schlüssel, der nicht zur ordnungsmäßigen Eröffnung des Schlosses bestimmt ist. Zur ordnungsmäßigen Eröffnung bestimmt ist derjenige Schlüssel, den der zur Verfügung über den umschlossenen Raum Berechtigte zur Eröffnung des Schlosses dauernd gewidmet hat. „Andere" zur ordnungsmäßigen Eröffnung nicht bestimmte Werkzeuge sind Dietriche, Haken und sonstige Gegenstände, welche den Verschlußmechanismus in Bewegung setzen. (Vgl. des näheren die Kommentare zu § 243 Ziff. 3 StGB.)

Zu Ziff. 4:

Vorausgesetzt ist ein bewußtes Zusammenwirken von drei oder mehr Personen.

Zu Ziff. 6:

Als Forstdiebstahl ist die Entwendung strafbar, wenn sie in einem Wald oder auf einem anderen hauptsächlich zur Holznutzung bestimmten Grundstück verübt wurde (§ 10 Forststg.).

Ueber die Wegnahme der Bäume Frucht- und Ziersträucher aus öffentlichen Anlagen und Kirchhöfen s. Bem. 4a u. 4d zu § 9.

Zu Ziff. 7:

Aufseher im Sinne der obigen Vorschrift sind auch die Feldhüter und andere von der Gemeindebehörde mit der Aufsicht über die Feldmark betraute Personen in Ansehung der ihnen unterstellten Grundstücke. Auf die Form der Bestellung oder die Dauer der Anstellung kommt es nicht an.

Zu Ziff. 8:

Ueber Rückfall s. § 10.

§ 13.

Auf Gefängnisstrafe von einer Woche bis zu einem Jahre ist zu erkennen:

1. wenn im Falle eines Felddiebstahls der Schuldige sich im dritten oder ferneren Rückfalle befindet;
2. wenn die Hehlerei gewerbs- oder gewohnheitsmäßig begangen ist.

1. Ueber Rückfall s. § 10.

2. Ueber Hehlerei im allgemeinen s. Bem. 3 zu § 8. Die Gewerbsmäßigkeit der Hehlerei setzt nicht notwendig eine wiederholte Begehung der strafbaren Handlung des eigenen Vorteils wegen voraus. Es genügt schon ein einziger Fall von Hehlerei, sofern die Absicht des Hehlers darauf gerichtet war, sich durch künftige mehrmalige Wiederholung dieses Vergehens eine Einnahmequelle, d. h. einen wiederkehrenden Erwerb zu verschaffen. Dagegen ist die Gewohnheitsmäßigkeit durch eine mehrmalige Begehung, infolge deren der Hang zur Hehlerei verstärkt ist, bedingt.

Gewerbs- und Gewohnheitsmäßigkeit sind persönliche Eigenschaften im Sinne des § 50 StGB. Haben daher mehrere Personen sich derselben Hehlerei schuldig gemacht, so trifft der Strafschärfungsgrund des § 13 Ziff. 2 nur denjenigen, der gewerbs- oder gewohnheitsmäßig handelte.

§ 14.

Sind mildernde Umstände vorhanden, so kann in den Fällen des § 11 auf Geldstrafe von weniger als fünf Mark, in denen des § 12 auf Geldstrafe bis zu einhundertfünfzig Mark oder auf Haft, in denen des § 13 auf Haft erkannt werden.

S. Bem. 1 zu § 11.

§ 15.

Bei Felddiebstählen (§§ 9 bis 14) finden die Bestimmungen des § 247 des Strafgesetzbuchs entsprechende Anwendung.

Liegen die Voraussetzungen des § 370 Nr. 5 des Strafgesetzbuchs vor, so tritt die Verfolgung nur auf Antrag ein. Die Zurücknahme des Antrags ist zulässig.

1. § 247 St.GB. lautet:

„Wer einen Diebstahl oder eine Unterschlagung gegen Angehörige, Vormünder oder Erzieher begeht, oder wer einer Person, zu der er im Lehrlingsverhältnisse steht, oder in deren häuslicher Gemeinschaft er als Gesinde sich befindet, Sachen von unbedeutendem Werte stiehlt oder unterschlägt, ist nur auf Antrag zu verfolgen. Die Zurücknahme des Antrages ist zulässig.

Ein Diebstahl oder eine Unterschlagung, welche von Verwandten aufsteigender Linie gegen Verwandte absteigender Linie oder von einem Ehegatten gegen den anderen begangen worden ist, bleibt straflos.

Diese Bestimmungen finden auf Teilnehmer oder Begünstiger, welche nicht in einem der vorbezeichneten persönlichen Verhältnisse stehen, keine Anwendung."

a) Angehörige im Sinne des StGB. und daher auch des FPStG. sind nach § 52 StGB.: Verwandte und Verschwägerte auf- und absteigender Linie, Adoptiv- und Pflege-Eltern und -Kinder, Ehegatten, Geschwister und deren Ehegatten, und Verlobte. Vormund ist nur derjenige, dem die Sorge für die Person und das Vermögen eines geschäftsunfähigen oder beschränkt geschäftsfähigen Menschen im ganzen obliegt. Nicht hierher gehört der Gegenvormund oder der Pfleger, dem nur die Sorge für einzelne Angelegenheiten oder einen Kreis von Angelegenheiten übertragen ist. Erzieher ist nur derjenige, dem die Leitung der körperlichen und sittlichen Ausbildung eines anderen im ganzen obliegt, also in der Regel nicht der Lehrer, wohl aber eine Person, der durch die zuständige Behörde ein Kind in Zwangserziehung gegeben wurde. Lehrling ist jeder, welcher bei einem Lehrherrn zur Erlernung eines Gewerbes in Arbeit tritt, ohne Unterschied ob die Erlernung gegen Lehrgeld oder unentgeltliche Hilfeleistung stattfindet, ob der Lehrling Kost und Lohn erhält oder nicht, ob er der häuslichen Gemeinschaft des Lehrherrn angehört oder nicht. Der Begriff des Gesindes ist in der Gesindeordnung vom 26. Juli 1903 (GBl. S. 51) nicht ausdrücklich bestimmt. Nach der Begründung des Entwurfes zur Gesindeordnung und der landesüblichen Auffassung gehören nur solche Personen zu dem Gesinde, die zu landwirtschaftlichen oder häuslichen Diensten niederer Art verpflichtet sind. Ihr Verhältnis zur Dienstherrschaft muß so sein, daß sie dieser jeder Zeit zur Verfügung stehen. Eines Antrages zur Strafverfolgung wegen Felddiebstahls bedarf es nur, wenn, was allerdings die Regel ist, das Gesinde in die häusliche Gemeinschaft der Herrschaft aufgenommen ist.

b) Der Felddiebstahl muß gegen die Angehörigen, gegen den Lehrherrn oder die Dienstherrschaft begangen sein. Wenn daher durch den Felddiebstahl mehrere Personen verletzt sind von denen nur eine in dem bezeichneten Verhältnis zu dem Täter steht, wenn z. B. die entwendete Sache mehreren Personen gehört, die nicht sämtlich Angehörige rc. des Täters sind, so ist der Felddiebstahl auch ohne Antrag zu verfolgen.

c) Bei Felddiebstählen gegen Angehörige, Vormünder und Erzieher tritt die Verfolgung ohne Rücksicht auf den Wert des Entwendeten immer nur auf Antrag ein. Bei Felddiebstählen dagegen, die durch Lehrlinge oder Gesinde begangen wurden, bedarf es eines Antrages bloß dann, wenn es sich um Bodenerzeugnisse von unbedeutendem Wert handelt.

2. § 370 Ziff. 5 StGB lautet, soweit er hier in Betracht kommt:

„Mit .... wird bestraft: 5) wer Nahrungs- oder Genußmittel von unbedeutendem Wert oder in geringer Menge zum alsbaldigen Verbrauch entwendet."

Die Vorschrift des Abs. 2 bezieht sich ebenso wie die des Abs. 1, wo §§ 9—14 ausdrücklich angeführt sind, auch auf die erschwerten Fälle von Felddiebstahl.

Nahrungsmittel sind solche Mittel, welche der Ernährung des menschlichen Körpers dienen. Unerheblich ist, ob diese Mittel, wie die Mehrzahl der Feldfrüchte, erst nach vorheriger Zubereitung genießbar sind oder ohne solche. Daher gehört z. B. auch Brotgetreide hierher (vgl. RG. Urt. 26./2. 86 E. 13 S. 371). Viehfutter ist kein Nahrungsmittel im Sinne des § 370 Ziff. 5. Der Begriff des Genußmittels umfaßt alle Gegenstände, welche — auch ohne der Ernährung zu dienen — durch die menschlichen Organe dem Körper zugeführt und mit dem Genusse verbraucht werden. Daher gehören Blumen und Feuerungsmaterialien nicht hierher.

Der Wert muß ein unbedeutender, die Menge eine geringe sein. Beteiligen sich mehrere an dem Felddiebstahl, so ist der Gesamtwert oder die Gesamtmenge des Entwendeten entscheidend, nicht der nach der Teilung auf den einzelnen Mittäter entfallende Anteil.

Alsbaldiger Verbrauch ist nicht gleichbedeutend mit sofortigem Verbrauch. Die obige Vorschrift findet daher auch dann Anwendung, wenn die Feldfrüchte zunächst nach Hause gebracht werden, um dort eine Zubereitung zu erfahren. Immerhin muß aber die Befriedigung eines augenblicklichen, nicht etwaiger künftiger Bedürfnisse bezweckt sein. Felddiebstähle an Nahrungsmitteln von unbedeutendem Wert gehören infolgedessen dann nicht hierher, wenn ein Ansammeln beabsichtigt ist.

3. Der Antrag muß binnen einer Frist von drei Monaten gestellt werden. Diese Frist beginnt mit dem Tage, seit welchem der zum Antrage Berechtigte von der Handlung und der Person des Täters Kenntnis erlangt hat (§ 61 StGB.). Sind an der Handlung mehrere Personen beteiligt, so beginnt die Antragsfrist allen Beteiligten gegenüber mit dem Tage, an welchem der Antragsberechtigte von der Tat und von der Person auch nur eines Beteiligten Kenntnis erlangt hat. Der Antrag kann nicht geteilt werden. Das gerichtliche Verfahren findet gegen sämtliche an der Handlung Beteiligte (Täter und Teilnehmer), sowie gegen den Begünstiger statt, auch wenn nur gegen eine dieser Personen auf Bestrafung angetragen worden ist (§ 63 StGB.).

Antragsberechtigt ist der Verletzte, d. i. derjenige, in dessen Recht durch die Straftat eingegriffen worden ist, also nicht nur der Eigentümer des Grundstücks, auf dem der Felddiebstahl verübt wurde, sondern z. B. auch der Nießbraucher und der Pächter. Sind mehrere Personen verletzt, so sind sie in der Ausübung ihres Rechts unabhängig von einander. Versäumt daher einer von ihnen die dreimonatliche Frist, so wird hierdurch das Recht der übrigen nicht ausgeschlossen (§ 62 StGB.). Es ist nicht nötig, daß der Verletzte den Antrag persönlich stellt, er kann auch vertreten werden, sei es durch einen von ihm besonders hierzu Beauftragten, zu dessen Legitimation auch mündlicher Auftrag genügt, sei es durch eine mit seiner Vertretung im allgemeinen oder für einen bestimmten Kreis von Angelegenheiten betraute Person, sofern im letzteren Falle die Zuwiderhandlung diese Angelegenheiten betroffen hat. Unter diesem Gesichtspunkt sind z. B. der Nachlaßpfleger und der Gutsverwalter zur Stellung eines Strafantrages befugt. Ist durch die Zuwiderhandlung das Eigentum des Staates oder einer anderen öffentlichen Korporation verletzt worden, so ist zur Antragstellung der nächste mit selbständiger Verwaltung des Eigentums betraute Beamte befugt, z. B. bei Staats-, Bezirks- und Vizinalstraßen — man denke an Felddiebstähle, die an Früchten von Straßenbäumen begangen werden — der Kreisbauinspektor. Minderjährige sind zur selbständigen Stellung des Strafantrags berechtigt, sofern sie das achtzehnte Lebensjahr vollendet haben. Unabhängig von dieser Befugnis hat neben dem Minderjährigen auch der gesetzliche Vertreter das Recht, den Antrag zu stellen. Der gesetzliche Vertreter ist der allein Berechtigte, wenn der Verletzte geschäftsunfähig ist oder das achtzehnte Lebensjahr noch nicht vollendet hat (§ 65 StGB.).

Der Strafantrag muß seinem Inhalt nach die Absicht kundgeben, eine strafrechtliche Verfolgung herbeiführen zu wollen. Was die Form angeht, so kann er bei Gericht oder bei der Staatsanwaltschaft schriftlich oder zu Protokoll erklärt werden. Bei allen anderen Behörden, insbesondere auch Gendarmen, Schutzleuten und Feldhütern, muß er schriftlich angebracht werden (§ 156 Abs. 2 StPO.). Es genügt aber, wenn der Antrag von diesen Behörden niedergeschrieben und von dem Verletzten unterschrieben wird. Die Unterschrift braucht aber nicht notwendig in dem ausgeschriebenen Namen zu bestehen, es genügt z. B. auch Unterkreuzung.

Die Zurücknahme des Antrags ist in allen Fällen, in denen das FPStG. die Verfolgbarkeit von der Antragstellung abhängig macht, ausdrücklich für zulässig erklärt. Sie kann nur erfolgen bis zur Verkündung eines auf Strafe lautenden Urteils (§ 64 StGB.). Einem solchen steht ein amtsrichterlicher Strafbefehl gleich, wenn er durch Ablauf der Einspruchsfrist vollstreckbar geworden ist (§ 450 StPO. und Oppenhoff zu § 64 StGB. Bem. 10). Die rechtzeitige Zurücknahme des Strafantrages gegen einen der an einer Zuwiderhandlung gegen das FPStG. Beteiligten, hat die Einstellung des Verfahrens auch gegen die anderen zur Folge. Im Falle des Abs. 1 der obigen Gesetzesbestimmung hat die Zurücknahme des Strafantrages bei mehreren Mittätern aber nur die Straflosigkeit derjenigen Personen zur Folge, die zu dem Verletzten in dem in § 247 Abs. 1 StGB. bezeichneten Verhältnis stehen. Die Zurücknahme des Strafantrags ist an eine Form nicht gebunden. Sie kann insbesondere auch dem Feldhüter oder einem anderen zur Verfolgung der strafbaren Handlung zuständigen Beamten, bei dem der Antrag gestellt war, mündlich erklärt werden. Doch muß dafür gesorgt werden, daß die Zurücknahme rechtzeitig zur Kenntnis des Gerichts gelangt. Ein Widerruf der Zurücknahmeerklärung ist unzulässig.

§ 16.

Im Falle des § 12 Nr. 1 ist neben der Geldstrafe oder der Freiheitsstrafe die Einziehung der Waffen, welche der Täter bei der Zuwiderhandlung bei sich geführt hat, auszusprechen, ohne Unterschied, ob sie dem Schuldigen gehören oder nicht.

In allen Fällen der §§ 9 bis 14 können die zur Begehung der strafbaren Zuwiderhandlung geeigneten Werkzeuge, welche der Täter bei der Zuwiderhandlung bei sich geführt hat, eingezogen werden, ohne Unterschied, ob sie dem Schuldigen gehören oder nicht. Die Tiere und andere zur Wegschaffung des Entwendeten dienende Gegenstände, welche der Täter bei sich geführt hat, unterliegen nicht der Einziehung.

Ist die Verfolgung oder Verurteilung einer bestimmten Person nicht ausführbar, so kann die Einziehung selbständig erkannt werden.

1. Die obige Vorschrift weicht in mehrfacher Beziehung von der allgemeinen Bestimmung des § 40 StGB. über Einziehung ab. Zunächst

ist die Einziehung nicht bloß dann zulässig, wenn der Felddiebstahl unter den Begriff des Vergehens fällt, sondern schon dann, wenn er bloß eine Uebertretung darstellt. Sie ist ferner zulässig ohne Rücksicht darauf, ob die einzuziehenden Gegenstände dem Schuldner gehören oder nicht. Schließlich muß die Einziehung in Ansehung der Waffen, die der Täter bei sich führte, immer ausgesprochen werden. Dagegen unterliegt die Einziehung der zur Begehung des Felddiebstahls geeigneten Werkzeuge, welche der Täter bei der Zuwiderhandlung bei sich führte, dem Ermessen des Richters. Da aber die Einziehung erfahrungsgemäß ein sehr wirksames Mittel zur Bekämpfung des Felddiebstahls ist, empfiehlt es sich, daß die Amtsanwälle die Einziehung nicht nur da beantragen, wo sie vom Gesetz geboten, sondern auch da, wo sie bloß zugelassen ist. Daß der Täter die Waffen oder Werkzeuge gebraucht hat, ist nicht erforderlich. Es genügt, wenn er sie bei der Begehung der strafbaren Handlung bei sich führte. Tiere und andere zur Wegschaffung des Entwendeten dienende Gegenstände (z. B. Wagen, Körbe, Säcke) unterliegen niemals der Einziehung.

2. Die Einziehung ist in dem Urteil oder dem Strafbefehl (§ 447 Abs. 2 StPO.) auszusprechen. Der einzuziehende Gegenstand ist möglichst genau zu bezeichnen. Soll die Einziehung selbständig erkannt werden, weil die Verfolgung oder Verurteilung einer bestimmten Person nicht ausführbar ist (Abs. 3), so richtet sich das Verfahren nach § 447 ff. StPO. mit dem Unterschiede, daß die Verhandlung und Entscheidung gemäß § 40 FPStG. durch die Amtsgerichte ohne Zuziehung von Schöffen stattfindet. Die Ausführung der von dem Gericht angeordneten Einziehung und die Verwendung der eingezogenen Gegenstände liegt den Verkehrssteuerämtern ob. (Vgl. § 43 der Dienstanweisung für die Verkehrssteuerbehörden vom 10. Dezember 1896 ABl. 1897 S. 9 ff.).

3. Wichtig für die Feldschutzbeamten, die bekanntlich gemäß der Verordnung des Reichskanzlers vom 13 Juni 1879 Hülfsbeamte der Staatsanwaltschaft hinsichtlich der strafbaren Handlungen gegen die Forst-, Feld-, Jagd- und Fischereigesetze sind, ist die Bestimmung des § 94 StPO., wonach Gegenstände, welche der Einziehung unterliegen oder als Beweismittel für die Untersuchung von Bedeutung sein können, in Verwahrung zu nehmen oder in anderer Weise sicher zu stellen sind. Werden diese Gegenstände auf Verlangen des Feldhüters usw. nicht freiwillig herausgegeben, so bedarf es der Beschlagnahme, d. i. der ausdrücklichen Anordnnng der amtlichen Verwahrung oder Sicherstellung. Hierzu ist in der Regel nur der Richter befugt. Doch kann die Anordnung bei Gefahr im Verzug auch von dem Feldschutzbeamten und sonstigen Hülfsbeamten der Staatsanwaltschaft getroffen und nötigenfalls mit Gewalt sofort durchgesetzt werden. Es empfiehlt sich aber, von dieser Befugnis nur in dringenden Fällen und nur in dem Maße, als zur Erreichung der Zweckes unbedingt erforderlich ist, Gebrauch zu machen, z. B. wenn die Vernichtung oder Beiseite-

schaffung eines der Einziehung unterliegenden wichtigen Gegenstandes oder eines wichtigen Beweismittels droht. Erhebt der Betroffene oder im Falle seiner Abwesenheit ein erwachsener Angehöriger ausdrücklich Widerspruch gegen die Beschlagnahme, so soll der Beamte, welcher die Beschlagnahme angeordnet hat, binnen drei Tagen dem zuständigen Amtsrichter Anzeige von der Beschlagnahme erstatten und seine Bestätigung nachsuchen. Das Gleiche gilt, wenn bei der Beschlagnahme weder der davon Betroffene noch ein erwachsener Angehöriger anwesend war. (Vgl. § 98 StPO.)

In allen Fällen hat der den Feldfrevel anzeigende Beamte die der Einziehung unterliegenden Gegenstände in seiner Anzeige möglichst genau zu bezeichnen, damit ihre Auffindung bei späterer Vollstreckung der Einziehung erleichtert ist.

Andere Fälle der Aneignung.

§ 17.

Mit Geldstrafe bis zu dreißig Mark oder mit Haft bis zu acht Tagen wird bestraft, wer unbefugt:

1. Dungstoffe von Äckern, Wiesen, Weiden, Gärten, Obstanlagen oder Weinbergen aufsammelt;
2. fremde, auf dem Felde zurückgelassene Ackergeräte gebraucht.

Im Falle der Nr. 2 tritt die Verfolgung nur auf Antrag ein. Die Zurücknahme des Antrags ist zulässig.

1. Die §§ 17, 18 und 19 behandeln bestimmte Arten von Wegnahmen, die im Gegensatz zu § 9 ff. selbst dann strafbar sind, wenn die in § 242 StGB. normierten Voraussetzungen des gemeinen Diebstahls, welche bekanntlich (vgl. Bem. 1 zu § 9) auch für den Felddiebstahl gelten, nicht zutreffen. Insbesondere wird nicht erfordert, daß der Täter die Absicht hatte, die weggenommenen Gegenstände sich rechtswidrig anzueignen. Ferner ist die Wegnahme unter Umständen selbst dann strafbar, wenn sie an Gegenständen erfolgt ist, die der Eigentümer aufgegeben hatte, oder die keinen Wert haben. Denn es handelt sich in den bezeichneten Gesetzesparagraphen weniger um den Schutz des Privateigentums, als vielmehr um die Wahrung der öffentlichen Ordnung. (Vgl. Begr. S. 19.)

In den Fällen der §§ 17 und 18 ist nur das „unbefugte" Handeln strafbar. Da dieses ein gesetzliches Tatbestandsmerkmal bildet, tritt Strafe auch nur dann ein, wenn der Täter den Mangel seiner Befugnis kannte oder nach den Umständen kennen mußte. (Vgl. § 59 StGB. und Bem. 1 b zu § 1.)

2. Unter Dungstoffen sind alle zum Düngen bestimmten Gegenstände zu verstehen, also neben dem natürlichen Dünger auch Kunstdünger, ferner Straßenkot und Bauschutt, sofern diese zum Zwecke der Verbesserung des Bodens auf die Grundstücke gebracht sind.

3. Unter dem Aufsammeln ist nur das allmähliche Zusammentragen des auf Äckern, Wiesen, Weiden etc. zerstreut liegenden Düngers zu verstehen. Die Wegnahme einer größeren Quantität aufgeschütteten Düngers wird regelmäßig als Diebstahl nach § 242 StGB. zu bestrafen sein (K. B. S. 402, vgl. auch RG. Urt. 16./12. 90 E. 21 S. 245 zu § 25 pr. F. u. FPG.).

4. Die Strafbarkeit des Gebrauches auf dem Felde zurückgelassener Ackergeräte wird selbstverständlich durch die ausdrückliche oder stillschweigende Einwilligung des Eigentümers ausgeschlossen; es liegt dann kein unbefugter Gebrauch vor. (K. B. S. 403.)

5. Über Strafantrag vgl. Bem. 3 zu § 15.

§ 18.

Mit Geldstrafe bis zu zehn Mark oder mit Haft bis zu drei Tagen wird bestraft, wer unbefugt:

1. das auf oder an Grenzrainen, Wegen, Triften oder an oder in Gräben wachsende Gras oder sonstige Viehfutter oder in Teichen oder sonst auf Privateigentum Schilf abschneidet oder abrupft;
2. von Bäumen, Sträuchern oder Hecken Laub abpflückt oder Zweige abbricht, insofern dadurch ein Schaden entsteht.

Die Verfolgung tritt nur auf Antrag ein. Die Zurücknahme des Antrags ist zulässig.

1. In Ziff. 1 handelt es sich um Wegnahme von Bodenerzeugnissen, die nur geringen Wert haben und regelmäßig nicht zum ordnungsmäßigen Ertrag im wirtschaftlichen Sinne gerechnet werden, da die Bodenfläche, auf der sie wachsen, meist anderen Zwecken als der Fruchtgewinnung dient. Werden die Gegenstände in größeren Mengen entwendet, und sind auch im übrigen die Voraussetzungen des gemeinen oder Felddiebstahls, insbesondere die Absicht der rechtswidrigen Zueignung und das Bewußtsein der Rechtswidrigkeit gegeben, so fallen diese Entwendungen unter § 242 StGB. oder § 9 dieses Gesetzes. Das Abrupfen und Abschneiden von Gras und Viehfutter auf Wiesen fällt niemals unter § 18, sondern unter die härtere Strafvorschrift des § 9, sofern natürlich auch die übrigen Voraussetzungen dieses Paragraphen gegeben sind.

2. Nur das Abpflücken von Laub und das Abbrechen von Zweigen fällt unter die milde Strafvorschrift des § 18 Ziff. 2. Alle anderen

Beschädigungen von Bäumen, Sträuchern und Hecken, insbesondere auch das Abschneiden sind, sofern nicht § 304 StGB. zur Anwendung kommt, nach § 26 Ziff. 3 u. 4 dieses Gesetzes zu bestrafen.

Das Abpflücken von Laub und Abbrechen von Zweigen bleibt straflos, wenn dadurch ein Schaden nicht entstanden ist.

3. Über Strafantrag vgl. Bem. 3 zu § 15.

Nachlese.

§ 19.

Mit Geldstrafe bis zu fünfzig Mark oder mit Haft bis zu vierzehn Tagen wird bestraft, wer:

1. an Orten, welche dem Gebrauch der Nachlese nicht unterworfen sind, oder auf eingefriedigten Grundstücken, oder einem Beschlusse des Gemeinderats zuwider in Weinbergen die Nachlese ausübt;
2. an Orten, welche dem Gebrauch der Nachlese unterworfen sind, dieselbe unter Nichtbeachtung der für die Ausübung gegebenen ortspolizeilichen Anordnungen oder vor vollständig vollendeter und abgeräumter Ernte oder in der Zeit zwischen Sonnenuntergang und Sonnenaufgang ausübt.

1. Die Nachlese ist eine ihrer juristischen Natur nach den beschränkten persönlichen Dienstbarkeiten des BGB. nahestehende Belastung von Grundstücken zugunsten der Armen, wonach diesen gestattet ist, die nach vollständig vollendeter und abgeräumter Ernte auf den Feldern zurückgelassenen Bodenerzeugnisse einzusammeln und sich anzueignen (Vgl. auch Dalloz Rép. V° commune 833 und droit rural 101 ff.). Sie hat ihren Ursprung in den durch das mittelalterliche Recht übernommenen Vorschriften der mosaischen Gesetzgebung (Kap. 23 Vers 22 des 3. und Kap. 24 Vers 19 und 21 des 5. Buches Mose). Ihr Gebrauch hat sich in Frankreich unter dem Schutz königlicher Edikte und lokaler Reglements trotz des häufigen Widerstrebens der Grundeigentümer bis zur Revolution beinahe überall erhalten. Das Ruralgesetz vom Jahre 1791 hat in Tit. II Art. 21 und 22 die bestehenden Gebräuche anerkannt. In Elsaß-Lothringen scheint der Gebrauch der Nachlese an den weitaus meisten Orten zu herrschen. Wo sie besteht, ist sie auch durch das BGB. unberührt geblieben. (Vgl. Art. 115 u. 184 EG. z. BGB.)

Das deutsche Wort „Nachlese" umfaßt die in der französischen Sprache mit drei Einzelausdrücken bezeichneten Tätigkeiten des glanage d. i. des Auflammelns von Ähren oder anderen Ackererzeugnissen, insbesondere auch Kartoffeln, des râtelage d. i. des Zusammenrechens von Gräsern und Heu auf abgemähten Wiesen und des grapillage d. i. des Abpflückens von Trauben, die bei der Weinlese übersehen oder sonst zurückgelassen worden sind.

2. Voraussetzung jeder Berechtigung zur Nachlese ist gemäß § 19 Ziff. 1, daß sie an dem Orte, wo sie ausgeübt wird, von Alters her gebräuchlich ist. Sie ist ferner unbedingt verboten auf eingefriedigten Grundstücken. Die Art der Einfriedigung ist nicht, wie in dem hinsichtlich des Weidgangs und der Koppelhütung noch geltenden Art. 6 Abschnitt 4 Titel I des Ruralgesetzes vorgeschrieben. Es genügt daher jede Art von Umschließung des Grundstücks, um dieses von der Nachlese zu befreien. Schließlich kann die Nachlese durch Beschluß des Gemeinderats in Weinbergen auch an den Orten, wo sie bisher gebräuchlich war, absolut verboten werden. Um der Strafsanktion des § 19 Ziff. 1 teilhaftig zu werden, muß der Gemeinderatsbeschluß verkündet werden. Es genügt nicht, wie Bruck Bem. 12 zu § 16 Gem.-O. anzunehmen scheint, daß der Bürgermeister eine das Verbot der Nachlese in Weinbergen enthaltende ortspolizeiliche Verordnung „mit Zustimmung des Gemeinderats" erläßt. Letzteres trifft zu für die von Bruck im gleichen Zusammenhang berührte Verordnung über die Schließung der Weinberge und die Erntezeit (§§ 38, 48 FPStG. und § 368 Ziff. 1 StGB.). Vielmehr muß der Gemeinderatsbeschluß als solcher verkündet werden. So ist durch das von Bruck a. a. O. angezogene Urteil des OLG. vom 2. Mai 1885 (Ztschr. Bd. 11 S. 23 ff.) in Ansehung des gleich gelagerten Falles der Regelung des Weidgangs durch den Gemeinderat ausdrücklich entschieden worden. Demgemäß ist der Beschluß des Gemeinderats durch den Bürgermeister auch dann zu verkünden, wenn er selbst im Gemeinderat gegen das Verbot der Nachlese gestimmt hat. Er bedarf nicht der Genehmigung der Aufsichtsbehörde. Es ist aber zweckmäßig, daß der Bürgermeister den Beschluß des Gemeinderats wie die von ihm selbst erlassenen Polizeiverordnungen der Aufsichtsbehörde entsprechend der Vorschrift des § 16 Gem.-O. vorlegt.

3. Ziff. 2 des § 19 bestimmt über die Art, in welcher die Nachlese an den Orten, die ihrem Gebrauch unterworfen sind, ausgeübt werden darf. Demnach ist die Nachlese untersagt vor vollständig vollendeter und abgeräumter Ernte und in der Zeit zwischen Sonnenuntergang und Sonnenaufgang. Außerdem müssen bei Ausübung der Nachlese die auf sie bezüglichen ortspolizeilichen Anordnungen beobachtet werden. Zuständig zum Erlaß dieser ortspolizeilichen Anordnungen sind die Bürgermeister, auch in Straßburg und Mülhausen, da in diesen Städten durch die Verordnung vom 28. Februar 1873 bezw. 28. Mai 1889 die Feldpolizei den Bürgermeistern überlassen ist, dagegen nicht in Metz, wo gemäß § 14 Abs. 2 des Gesetzes vom 30. Dezember 1871 mangels abweichender besonderer Bestimmung auch die Feldpolizei dem Polizeidirektor zusteht.

Durch die in Ziff. 2 erwähnten ortspolizeilichen Anordnungen können die schädlichen Auswüchse, zu denen unter Umständen die Ausübung der Nachlese, insbesondere in der Nähe großer Städte oder Industriezentren, führt, in wirksamer Weise beseitigt werden. So können die Bürgermeister anordnen, daß nur Gemeindeeinwohner und unter diesen nur Greise, Kinder und andere arbeitsunfähige Personen die Nachlese ausüben dürfen, ja sogar die Verstattung zur Nachlese von einer jeder einzelnen Person zu erteilenden besonderen Erlaubnis abhängig machen (vgl. Dalloz Rép. V[v] droit rural 106). Die Ortspolizeibehörde kann ferner die Zeit, zu welcher die Nachlese beginnen darf, und ihre Dauer festsetzen, auch die Werkzeuge bestimmen, die gebraucht werden dürfen, und das Mitbringen von Werkzeugen überhaupt verbieten. Die Anordnungen der Ortspolizeibehörden dürfen aber natürlich nicht so weit gehen, daß dadurch die Ausübung der Nachlese tatsächlich aufgehoben wird. Gegen sämtliche auf die Ausübung der Nachlese bezüglichen Bestimmungen der Ortspolizeibehörden steht der Beschwerdeweg an die Aufsichtsbehörde gemäß § 71 Gem.-O. offen.

Alte Gebräuche, durch welche die Ausübung der Nachlese eingeschränkt wird, genießen nicht des strafrechtlichen Schutzes des § 19. Sie werden eines solchen erst dann teilhaftig, wenn sie in die Form von Ortspolizeiverordnungen gekleidet sind.

Die Eigentümer der der Nachlese unterworfenen Grundstücke sind nach dem FPStG. während deren Ausübung an keinerlei Beschränkung in der Benutzung ihres Eigentums gebunden. Die Bestimmung des Art. 22 Tit. 2 des Ruralgesetzes, wonach Grundeigentümer und Weideberechtigte, um die Nachlese nicht zu stören, nach vollendeter Ernte zwei volle Tage warten mußten, bevor sie Weidevieh auftreiben durften, ist dem FPStG. unbekannt und durch § 51 aufgehoben.

Weidefrevel.

§ 20.

Mit Geldstrafe bis zu fünfzig Mark oder mit Haft bis zu vierzehn Tagen wird bestraft, wer unbefugt auf einem Grundstück Vieh weidet.

1. Ein Weidefrevel liegt nach dem Gesetz vor:

a) wenn das Weiden an Orten ausgeübt wird, wo der Zuwiderhandelnde kein Weiderecht besitzt (§§ 20 und 21),

b) wenn Vieh oder Hausgeflügel unbeaufsichtigt herumläuft (§§ 22 und 24),

c) wenn der Weideberechtigte sich den Anordnungen nicht fügt, welche in bezug auf die Gemeindeweide, den Weidgang und die Koppelhütung erlassen sind (§ 25),

d) wenn das Vieh auf dem Marsche über fremde Grundstücke getrieben wird (§ 29 Ziff. 2).

Der Schutz der Grundstücke gegen Weidefrevel hat in diesem Gesetz eine weitere Verstärkung noch dadurch erfahren, daß dem Beschädigten in einzelnen Fällen, nämlich wenn Vieh oder Hausgeflügel unbeaufsichtigt auf einem fremden Grundstück herumläuft oder wenn fremdes Vieh über Grundstücke getrieben wird (Fälle b und d), das Recht zur Pfändung des Viehs zusteht (vgl. des näheren §§ 42 ff.). Ueber das Tötungsrecht gegenüber Tauben s. § 24 Abs. 2.

2. Unter **Vieh** sind in § 20 nur solche Tiere verstanden, welche nach den Regeln des landwirtschaftlichen Betriebs durch bestimmte Personen gehütet oder dauernd unter Aufsicht gehalten werden, also Rindvieh, Pferde, Esel, Schweine, Schafe, Ziegen, aber auch Federvieh, sofern dieses zur Weide getrieben wird, was insbesondere bei Gänsen häufig der Fall ist.

3. Die Strafe trifft die das Vieh hütende Person (s. aber auch § 23), wenn sie die Tiere auf Grundstücken, wo dieselben nicht geweidet werden dürfen, vorsätzlich oder fahrlässigerweise weiden läßt. Der Tatbestand ist aber erst erfüllt, wenn ein wirkliches Abweiden stattgefunden hat. Die im § 14 Abs. 2 pr. F. u. FPG. getroffene Bestimmung, wonach die Strafe verwirkt ist, sobald das Vieh die Grenzen des Grundstücks, auf welchem es nicht geweidet werden darf, überschritten hat, ist durch die Landesausschußkommission ausdrücklich abgelehnt worden (K. B. S. 403).

4. **Unbefugt** ist das Weiden da, wo dem Eigentümer des Viehs überhaupt kein Recht zum Weiden zusteht, oder wo ein bestehendes Weiderecht (z. B. eine Dienstbarkeit) überschritten, etwa mittels einer größeren Anzahl Vieh als gestattet, oder zu einer verbotenen Zeit ausgeübt wird. Kein unbefugtes Weiden im Sinne des § 20 liegt vor, wenn das Vieh auf die Gemeindeweide oder die dem Weidgang oder der Koppelhütung unterliegenden Grundstücke getrieben wird. Läßt der Eigentümer oder der Hüter des Viehs die für die Benutzung der Gemeindeweide oder die Ausübung der bezeichneten Weiderechte geltenden Anordnungen außer acht, so macht er sich nach § 25 strafbar.

Da das unbefugte Handeln in § 20 ein gesetzliches Tatbestandsmerkmal ist, muß der Täter **wissen**, daß ein Recht zum Weiden nicht bestand. Dem Wissen steht die durch Fahrlässigkeit verursachte Unkenntnis gleich (vgl. Bem. 1 b zu § 1).

5. Das unbefugte Weiden **auf allen Arten von Grundstücken** ist untersagt, daher auch das Weiden auf Wegen, Straßenrändern, Böschungen von Straßen, Wasserläufen und Eisenbahnen, sowie in Straßengräben. Die auf den Schutz dieser Anlagen abzielenden besonderen Bestimmungen des französischen Rechts (arrêt vom 16. Dezember 1759, betr. das Abweiden der Ränder großer Straßen, Art. 2 des Ges. vom 15. Juli 1845, betr. die Eisenbahnen, sowie Nr. 14 des Art. 372 der in Gemäßheit des

Ministerialzirkulars vom 21. Juli 1854 erlassenen Vizinalwegeordnungen) sind durch § 20 aufgehoben.

§ 21.

Geldstrafe von fünf bis zu einhundertfünfzig Mark oder Haft tritt ein, wenn der Weidefrevel (§ 20) begangen ist

1. auf Grundstücken, deren Betreten durch Warnungszeichen verboten ist;
2. auf eingefriedigten Grundstücken;
3. auf bestellten Äckern oder auf Wiesen, welche mit künstlichen Be- oder Entwässerungsanlagen versehen, oder auf welchen solche Anlagen in der Ausführung begriffen sind, in Gärten, Baumschulen, Weinbergen, Graben- oder Kanalböschungen;
4. in der Zeit von Sonnenuntergang bis Sonnenaufgang;
5. nachdem der Täter im Laufe der dem Frevel vorausgegangenen zwei Jahre in Elsaß-Lothringen auf Grund dieses oder des vorhergehenden Paragraphen rechtskräftig verurteilt worden war.

1. Die größere Strafbarkeit der hier hervorgehobenen Weidefrevel findet ihren Grund darin, daß unter Ziff. 1 das Beweiden der Grundstücke gegen das erkennbare Verbot des Besitzers vorausgesetzt ist, die eingefriedigten (Ziff. 2) und die unter Ziff. 3 aufgeführten Grundstücke eines höheren Schutzes bedürfen, die Weide zur Nachtzeit die Entdeckung des Frevlers erschwert und der Rückfall bei Weidefrevel als ein erschwerender Umstand sich darstellt.

2. Zu den einzelnen Erschwerungsgründen ist zu bemerken:

Zu Ziff. 1:

Jeder Eigentümer oder Besitzer ist ohne weiteres befugt, das Betreten seines Grundstücks durch Warnungszeichen zu verbieten (§ 903 BGB.). Es bedarf daher keiner behördlichen Anerkennung des Warnungszeichens, um den Erschwerungsgrund des § 21 Ziff. 1 hervorzurufen. Die äußerlichen Merkmale, welche als Warnungszeichen dienen können, sind nach dem Ortsgebrauch verschieden. Es kommen in Betracht: Tafeln mit oder ohne Aufschriften, Gräben, Kreuze, Schlagbäume und Strohwische. In allen Fällen ist erforderlich, daß der Zuwiderhandelnde die Bedeutung des äußeren Merkmals als eines Warnungszeichens erkannt hat oder nach Lage der Verhältnisse hat erkennen müssen. Die unbefugte Fortnahme oder Beschädigung der Warnungszeichen wird nach § 26 Ziff. 2 bestraft.

Zu Ziff. 2:

Jede Art von Umschließung des Grundstücks genügt. Die Einfriedigung braucht nicht die besonderen in Art. 6 Abschn. IV Tit. I des Ruralgesetzes von 1791 bezeichneten Eigenschaften zu haben, da diese Bestimmung im wesentlichen nur noch für die Ausübung des Weidgangs und der Koppelhütung Geltung hat.

Zu Ziff. 3:

Als bestellt ist der Acker anzusehen, sobald er besäet oder bepflanzt ist; das bloße Pflügen des Ackers genügt nicht. Nur die im Gesetzestext näher bezeichneten Kunstwiesen genießen den besonderen in der erhöhten Strafbarkeit liegenden Schutz. Das Weiden auf gewöhnlichen Wiesen ist nach § 20 strafbar. Nicht erforderlich ist, daß die künstlichen Be- oder Entwässerungsanlagen vollendet sind. Es genügt schon, daß solche Anlagen in der Ausführung begriffen sind.

Zu Ziff. 5:

Ueber Rückfall s. die Bemerkungen zu § 10. Eine weitere Schärfung der Strafe im zweiten und ferneren Rückfall wie beim Felddiebstahl findet beim Weidefrevel nicht statt.

Nachlässigkeit des Hirten.

§ 22.

Mit Geldstrafe bis zu zehn Mark oder mit Haft bis zu drei Tagen wird der Hirt bestraft, welcher das ihm zur Beaufsichtigung anvertraute Vieh ohne Aufsicht oder unter der Aufsicht einer hierzu untüchtigen Person läßt.

1. Der Hirt macht sich nach obiger Gesetzesvorschrift strafbar:
   a) wenn er das Vieh ohne Aufsicht,
   b) wenn er es unter der Aufsicht einer hierzu untüchtigen Person läßt.

„Die Tüchtigkeit der von dem Hirten bestellten Person zu beurteilen, bleibt in den einzelnen Zuwiderhandlungsfällen dem Ermessen des Gerichts überlassen. Es wird diese Beurteilung von der Art und Zahl der dem Hirten anvertrauten Viehstücke, von der örtlichen Lage des Weidelandes, der Lage und Bestellungsart der an das Weideland angrenzenden Felder, der Schwierigkeit der Abkehr des Viehs vom nachbarlichen Boden und dergl. abhängig sein" (Daube zu § 12 pr. F. u. FPG.).

2. Die Strafe des § 22 ist lediglich dadurch bedingt, daß das Vieh ohne Aufsicht oder ohne geeignete Aufsicht gelassen ist. Daß eine Beschädigung fremder Grundstücke oder etwa ein Abweiden stattgefunden habe, ist nicht erfordert.

Haftbarkeit für den Hirten.

§ 23.

Für die Geldstrafe, die Buße und die Kosten, zu welchen der Hirt auf Grund der §§ 20, 21, 22 verurteilt wird, ist der

Besitzer der Viehs nach Maßgabe der §§ 3 und 4 auch dann haftbar, wenn jener nicht zu seiner Hausgenossenschaft gehört.

1. § 23 enthält eine über §§ 3 und 4 hinausgehende Bestimmung bezüglich der Haftbarkeit für die Straftat eines andern. Sie ist durch Art. 107 EG. z. BGB. gegenüber den Vorschriften des letzteren über die Haftung für fremdes Verschulden aufrecht erhalten worden. Die Haftbarkeit erstreckt sich auf die Geldstrafe, die Buße und die Kosten und tritt auch dann ein, wenn der Hirt nicht zu der Hausgenossenschaft des Viehbesitzers gehört. Im übrigen gelten die Vorschriften der §§ 3 und 4 auch hier. Demgemäß ist die Haftung nur eine subsidiäre, d. h. sie tritt nur im Fall des Unvermögens des Hirten ein, es sei denn, daß die Voraussetzungen des § 4 gegeben sind. Der Viehbesitzer haftet nicht, wenn er die Tat nicht verhindern konnte. Ferner tritt an die Stelle der Geldstrafe, für welche der Viehbesitzer für haftbar erklärt ist, eine Freiheitsstrafe nicht ein. Die Festsetzung der Haftbarkeit kann im Strafbefehl erfolgen (§ 41).

2. Der Zweck der gesetzlichen Bestimmung ist, bei der gewöhnlichen Mittellosigkeit der Hirten und ihrer Abhängigkeit von ihrem Brotherrn die Verfolgung der Geldstrafe, der Buße und der Kosten gegen denjenigen sicher zu stellen, welcher den Vorteil von dem Vieh, mittels welchen der Weidefrevel begangen wurde, hat. Infolgedessen ist unter „Viehbesitzer" derjenige zu verstehen, welcher das Vieh zu eigenem Vorteil nutzt, also der besitzende Eigentümer, der Nießbraucher, der Pächter und der Nutzungspfandgläubiger.

Selbstverständlich bleiben neben der Bestimmung des § 23, welcher lediglich die im Strafverfahren geltend zu machende Verantwortlichkeit betrifft, die Vorschriften des Bürgerlichen Rechts über die Haftung für Tierschaden und fremdes Verschulden in Geltung (§§ 831 ff. BGB.).

Mehrere Besitzer eines Viehstücks haften als Gesamtschuldner (§ 840 BGB.). Wird der Weidefrevel mittels Viehs begangen, das zu einer das Vieh mehrerer Besitzer umfassenden Herde gehört, so haften nicht wie nach § 74 pr. F. u. FPG. die sämtlichen an der Herde beteiligten Viehbesitzer, sondern nur die Besitzer derjenigen Stücke, welche übergetreten sind, diese aber als Gesamtschuldner (s. auch § 5 Abs. 3 dieses Ges.).

Unbeaufsichtigtes Vieh, Tauben.

§ 24.

Mit Geldstrafe bis zu zehn Mark oder mit Haft bis zu drei Tagen wird bestraft, wer

1. Vieh oder Hausgeflügel außerhalb eingefriedigter Grundstücke ohne gehörige Aufsicht herumlaufen läßt, sofern nicht

nach den Umständen anzunehmen ist, daß die Gefahr einer Beschädigung Dritter nicht vorliegt;

2. Tauben zur Saat- oder Erntezeit innerhalb des durch die Ortspolizei bestimmten und öffentlich bekannt gemachten Termins nicht eingeschlossen hält.

Tauben, welche der Vorschrift unter Nr. 2 zuwider auf einem fremden Grundstück betroffen werden, gelten als schädliches Wild im Sinne des § 2 des Jagdpolizeigesetzes vom 7. Mai 1883 (Gesetzbl. S. 57).

1. Durch das Herumlaufenlassen von Vieh und Hausgeflügel, insbesondere Hühnern, ohne gehörige Aufsicht außerhalb eingefriedigter Grundstücke werden Belästigungen und Zwistigkeiten jeglicher Art verursacht, auch häufig Veranlassung zu Beschädigung gegeben. Infolgedessen stellt § 24 Ziff. 1 dieses Herumlaufenlassen als eine mildere Art des Weidefrevels unter Strafe, die jedoch nicht eintritt, wenn nach den Umständen anzunehmen ist, daß die Gefahr einer Beschädigung Dritter nicht vorliegt. Aus der Form dieses einschränkenden Zusatzes ist nicht etwa zu schließen, daß den Zuwiderhandelnden eine Beweislast für das Vorhandensein der strafausschließenden Umstände trifft. Vielmehr haben die Staatsanwaltschaft und der Richter, sofern nach Lage der Verhältnisse Anlaß gegeben ist, von Amtswegen nach dem Vorhandensein solcher Umstände zu forschen (vgl. auch § 260 StPO.).

Welche Aufsicht als eine genügende anzusehen ist, ist eine Frage tatsächlicher Natur und in jedem Falle nach den obwaltenden Umständen, insbesondere auch den örtlichen Verhältnissen und Wirtschaftsgebräuchen zu beurteilen.

2. Nur das Herumlaufenlassen von Vieh und Hausgeflügel ist nach § 24 Ziff. 1 strafbar. Das Herumschwimmenlassen von Tieren in Gewässern fällt nicht unter diese Vorschrift, da jegliche Anhaltspunkte dafür fehlen, daß das Gesetz hier etwas anderes im Auge hatte, als die Felder und Bodenerzeugnisse gegen herumlaufendes Vieh und Hausgeflügel zu schützen, insbesondere nicht ersichtlich ist, daß das Gesetz, wie allerdings an anderen Stellen geschehen (§§ 30, 31, 35), hier auch den Schutz der Gewässer beabsichtigt hat (vgl. OLG. Urt. 21. 11. 92, Ztschr. Bd. 18 S. 182 ff. und KG. Urt. 13. 2. 02, Joh. Bd. 24 S. 60 zu § 11 pr. F. u. FPG.). Doch kann das Herumschwimmenlassen von Tieren in Gewässern nach anderen gesetzlichen Bestimmungen strafbar sein. In Betracht kommt § 38 des Fischereigeś. vom 2. Juli 1891 (GBl. S. 69), wonach es verboten ist, Enten, Gänse, Schwäne oder andere dem Fischlaich schädliche Wasservögel in die Laich- und Hegeplätze zu setzen oder auslaufen zu lassen.

Ueber das Recht zur Pfändung aufsichtsloser Viehs und Hausgeflügels s. §§ 42 ff.

3. Tauben sind der Landwirtschaft in der Saat- und Erntezeit besonders gefährlich, da sie häufig und mitunter scharenweise in die Felder einfallen und die offen liegenden Saatstoffe und Bodenerzeugnisse verzehren. Infolgedessen können durch ortspolizeiliche Verordnung während der Saat- und Ernteperioden Sperrzeiten bestimmt werden, innerhalb deren die Tauben eingeschlossen zu halten sind. Zuständig zum Erlaß der ortspolizeilichen Verordnungen ist der Bürgermeister, in Metz der Polizeidirektor (vgl. Bem. 3 zu § 19). Wer während der angeordneten Sperrzeiten seine Tauben nicht eingeschlossen hält, verfällt der Strafe des § 24. Außerdem aber gelten nach Abs. 2 Tauben, die während der Sperrzeit auf fremden Grundstücken betroffen werden, als schädliches Wild im Sinne des § 2 des Jagdpolizeigesetzes vom 7. Mai 1883. Nach dieser Bestimmung ist es nicht als Ausübung der Jagd anzusehen, wenn Eigentümer, Besitzer oder Pächter schädliches Wild auf ihren Ländereien vertilgen. Demgemäß haben diese Personen das Recht, die während der Sperrzeiten nicht eingeschlossenen Tauben zu töten. Die Mittel, mit welchen die Vertilgung erfolgen darf, werden durch das Ministerium bestimmt. Die auf die Tauben bezügliche Ministerialverordnung ist am 31. Dezember 1888 erlassen worden (s. Anhang B). Hervorzuheben ist, daß die Anwendung von Schußwaffen nur mit Erlaubnis des Kreis- oder Polizeidirektors statthaft ist. Die Erlaubnis wird nur unter der Voraussetzung erteilt, daß die Tauben scharenweise in ein Grundstück einfallen.

Uebrigens ist dadurch, daß in § 24 Abs. 2 FPStG. in Verbindung mit § 2 Jagdpolizeiges. den Eigentümern, Besitzern und Pächtern das Recht zur Vertilgung der der Vorschrift des Abs. 2 zuwider nicht eingeschlossenen Tauben gegeben ist, das Recht des Jagdpächters zur Erlegung dieser Tiere nicht ausgeschlossen. Vielmehr kann neben den vorhin bezeichneten Personen auch der Jagdpächter Tauben, die er während der Sperrzeit in seinem Revier antrifft, töten, und zwar auch ohne besondere Erlaubnis unter Anwendung der Schußwaffe. (Vgl. OLG. Urt. 27./7. 91 Ztschr. Bd. 17 S. 70.)

Das Recht zur Tötung von Tauben, die während der Sperrzeit auf fremden Grundstücken angetroffen werden, ist durch Art. 130 EG. z. BGB. aufrecht erhalten worden.

Besondere Vorschriften gelten zufolge des Reichsgesetzes vom 28. Mai 1894 zugunsten der Brieftauben. (S. Anhang C.)

Weideberechtigungen.

§ 25.

Mit Geldstrafe bis zu dreißig Mark oder mit Haft bis zu acht Tagen wird bestraft, wer bei Ausübung der Gemeindeweide, des Weidganges (vaine pâture) und der Koppelhütung (parcours)

1. den darüber erlassenen Polizeiverordnungen zuwiderhandelt;
2. Ziegen, welche nicht in einer Herde geführt werden, anders als angebunden weiden läßt;
3. Zuchtstiere mit der Gemeindeherde austreibt.

1. § 25 betrifft die polizeiliche Regelung der Weideberechtigungen. Letztere sind teils öffentlich-rechtlicher, teils privatrechtlicher Natur. Zum öffentlichen Recht gehören die Nutzungsrechte der Gemeindeangehörigen an der Gemeindeweide (Allmend). Ueber die Art ihrer Ausübung beschließt der Gemeinderat gemäß §§ 54 Ziff. 2, 76 Ziff. 13 Gem.-O. Privatrechtlichen Charakter dagegen haben das Recht des Weidgangs und der Koppelhütung. Weidgang (vaine pâture) ist das Recht der Angehörigen einer Gemeinde, ihr Vieh innerhalb des Gemeindebannes auch auf fremden Grundstücken weiden zu lassen. Die Koppelhütung (parcours) ist das Recht der Angehörigen mehrerer Gemeinden oder Ortschaften, ihr Vieh auch auf ihnen nicht gehörigen Grundstücken weiden zu lassen. Letzteres Recht kommt nur noch sehr selten vor, bald als einseitiges, bald als wechselseitiges. (Vgl. Begründung des Entwurfs der Gem.-O. in Verhandlungen des Landesausschusses 1894 Bd. 1 Vorlage 5 S. 67.)

Das Ruralgesetz vom 6. Oktober 1791 hat im vierten Abschnitt des ersten Titels die Rechte des Weidgangs und der Koppelhütung, die ebenso wie die Rechte an der Gemeindeweide in ihrem wirtschaftsgeschichtlichen Ursprung auf die Zeit zurückgehen, zu der noch kein Privateigentum an Grund und Boden bestand, insoweit aufrecht erhalten, als sie auf besonderem Titel, auf Gesetz oder Gewohnheit oder auf unvordenklichem Ortsgebrauch beruhen. Die juristische Natur dieser Rechte ist zweifelhaft. Es dürfte richtig sein, sie nicht als Servituten, wie das Ruralgesetz sie bezeichnet, sondern als Nutzungsrechte nach Art der beschränkten persönlichen Dienstbarkeiten aufzufassen. Subjekte der Rechte sind die Grundbesitzer (Eigentümer, Nießbraucher, Pächter) und die Vieh besitzenden seßhaften Familienhäupter der einen oder mehrerer Gemeinden. Sie bilden eine Art von Gemeinschaft, aus welcher der einzelne jedoch nicht ohne weiteres austreten kann. Die Zahl des Viehs, das jeder auftreiben darf, bestimmt sich grundsätzlich nach der Ausdehnung der Ländereien des Berechtigten. Doch kann auch jedes seßhafte Familienhaupt, welches weder Eigentümer noch Nutzungsberechtigter ist, bis zu sechs Stück Wollvieh und eine Kuh mit ihrem Kalb weiden lassen. (Art. 13 und 14 IV. Abschn. Tit. I des Ruralgesetzes.) Das BGB. hat diese Rechte, wo sie bestehen, und die sie betreffenden Rechtsnormen in Art. 184, 115 EG. aufrecht erhalten. (Vgl. zu dem Gesagten Kisch S. 714 ff.)

Dem Weidgang und der Koppelhütung sind nicht unterworfen Grundstücke von bestimmter Kulturart und, sofern das Recht sich nicht auf einen besonderen Titel gründet, solche Grundstücke, die in der durch das

Gesetz bestimmten Weise eingefriedigt sind (vgl. über diese Beschränkungen Bem. 2 zu § 49). Die Einfriedigung eines Grundstücks vermindert zugleich aber auch das Recht des Eigentümers zum Auftrieb von Vieh im Verhältnis zu der eingefriedigten Fläche (Art. 5, 7 und 16 a. a. O.). Zeitlich beschränkt ist das Recht dadurch, daß es auf Wiesen niemals vor dem ersten Grasschnitt und auf besäten oder mit einem Erzeugnis bestandenen Grundstücken nicht eher als nach vollendeter Ernte ausgeübt werden darf (Art. 9 u. 10 a. a. O.).

Im übrigen bestimmen über den Umfang und die Art der Ausübung die örtlichen Gewohnheiten und Gebräuche. Aber auch der Gemeinderat derjenigen Gemeinde, auf deren Bann das Recht ruht, hat gemäß § 56 Ziff. 13 Gem.-O. in Ansehung dieser Rechte eine Reglementierungsbefugnis, wodurch jene Gewohnheiten und Gebräuche festgelegt und selbständig abgeändert werden können. Vgl. jedoch auch § 49 dieses Gesetzes. Der Gemeinderat bezeichnet insbesondere die Grundstücke oder Grundstücksarten, die der Weide unterworfen sind, bestimmt nach dem oben hervorgehobenen Verhältnis die Zahl des Viehs, das jeder auftreiben darf (z. B. so und soviel Stück auf einen Morgen), die Jahreszeit, zu welcher die Weide stattfindet, die Stunden, zu denen sie verboten ist (z. B. von Sonnenuntergang bis Sonnenaufgang). Er kann auch den einzelnen Arten von Tieren bestimmte Bezirke zuweisen und einzelne Tierarten wegen des besonders großen Schadens, den sie beim Weiden verursachen, ganz ausschließen (vgl. Dalloz Rép. V° commune 820, 823 f.). Niemals aber dürfen die Anordnungen des Gemeinderats so weit gehen, daß dadurch die Ausübung des Weiderechts tatsächlich verhindert ist.

Die Beschlüsse des Gemeinderats über den Weidgang sind auch für die Grundeigentümer in Ansehung der Ausübung der Weide auf ihren eigenen Grundstücken verpflichtend, so daß ein Eigentümer zu der Zeit, zu welcher der Weidgang stattfindet, nicht mehr Tiere, als ihm gestattet, auf seinen dem Weidgang unterworfenen Grundstücken weiden lassen darf. (Vgl. Dalloz Rép. V° commune 808 ff., OLG. Urt. 23./11. 89 Ztschr. Bd. 15 S. 117 ff.)

Grundsätzlich wird der Weidgang und die Koppelhütung in einer gemeinsamen Herde unter einem Gemeindehirten ausgeübt. Nach Art. 12 u. 15 IV. Abschn. Tit. I des Ruralgesetzes kann aber auch jeder Eigentümer oder Nutzungsberechtigter die ihm gestattete Anzahl Vieh unter einem besonderen Hirten weiden lassen (vgl. LG. Metz Urt. 22./10. 90 Ztschr. Bd. 16 S. 180). Dagegen ist es nicht zulässig, daß zwei oder mehrere Viehbesitzer ihre Tiere zu einer besonderen von der Gemeindeherde getrennten Herde vereinigen, da hierbei leicht Ueberschreitungen der dem einzelnen Viehbesitzer zur Weide gestatteten Höchstzahl von Viehstücken vorkommen können (Dalloz Rép. commune 814).

2. Die Zuständigkeit des Gemeinderats zur Regelung der Ausübung von Weidgang und Koppelhütung sowie zur Beschlußfassung über

die Art der Nutzung der Gemeindeweide (Allmend) ist unabhängig von der Polizeiverordnungsgewalt des Bürgermeisters. Damit die Beschlüsse des Gemeinderats aber den Charakter von Polizeiverordnungen gewinnen, damit m. a. W. Zuwiderhandlungen gegen sie gemäß § 25 Ziff. 1 dieses Gesetzes strafbar werden, ist erforderlich, daß sie verkündet werden. Zur Verkündung ist der Bürgermeister verpflichtet. Die Verkündung aber muß ergeben, daß es sich um Polizeiverordnungen des Gemeinderats handelt, da in Ansehung der Ausübung der Nutzungsrechte an Gemeindegütern und der Ausübung des Weidgangs und der Koppelhütung nur dem Gemeinderat und nicht dem Bürgermeister die Reglementierungsgewalt zusteht. Der Beschluß des Gemeinderats muß daher als solcher und nicht als bürgermeisterliche Verordnung verkündet werden. (Vgl. OLG. Urt. 2./5. 85 Ztschr. Bd. 11 S. 23 ff. und die daselbst in der Anmerkung des Einsenders angeführten Stellen aus Dalloz Rép. V° commune, ferner Block dictionnaire V° organisation communale 243, OLG. 17./10. 85 bei Franz, Rechtsprechung des OLG. Colmar in Strafsachen Nr. 137 und oben Bem. 1 zu § 19.)

Die Beschlüsse des Gemeinderats über die Gemeindeweide unterliegen in kleinen Gemeinden, sofern die Nutzung neu eingeführt oder die Grundsätze verändert werden, nach welchen die Ausübung bisher erfolgt ist, gemäß § 76 Ziff. 13 Gem.-O. der Genehmigung der Aufsichtsbehörde. Die Beschlüsse über die Ausübung des Weidgangs und der Koppelhütung bedürfen keiner Genehmigung Doch ist es zweckmäßig, daß der Bürgermeister die sämtlichen hierher gehörigen Reglements wie andere Ortspolizeiverordnungen entsprechend § 16 Gem.-O. der Aufsichtsbehörde mitteilt.

Natürlich ist es dem Bürgermeister als Inhaber der örtlichen Polizeigewalt unbenommen, den Reglements des Gemeinderats über die Ausübung der Gemeindeweide, des Weidgangs und der Koppelhütung Vorschriften allgemein feldpolizeilicher Natur anzufügen. So z. B. kann er im Anschluß an ein solches Reglement das Beweiden der Gräben und Böschungen der Gemeindewege verbieten, und zwar selbst dann, wenn etwa der Weidgang seit unvordenklichen Zeiten auf den Gemeindewegen ausgeübt worden ist (OLG. Urt. 3./1. 90 Ztschr. Bd. 15 S. 174). Ferner kann er zur Verhütung ansteckender Krankheiten unter den Tieren auf Grund der gesetzlichen Bestimmungen über die Abwehr und Unterdrückung von Viehseuchen Vorschriften erlassen, welche eine Einschränkung der Ausübung der Weideberechtigungen zur Folge haben. (Vgl. auch Dalloz Rép. V° commune 794, 827 und 830.)

3. Die örtlichen Gebräuche und Gewohnheiten, welche die Ausübung der Gemeindeweide, des Weidgangs und der Koppelhütung zum Gegenstand haben, fallen, obschon sie im übrigen von wesentlicher Bedeutung sind, an sich noch nicht unter den strafrechtlichen Schutz des § 25. Hierzu ist erforderlich, daß sie in einer Polizeiverordnung des Gemeinderats festgestellt werden (vgl. OLG. Urt. 21./5. 87 Ztschr. Bd. 12 S. 245).

4. Die Polizeiverordnungen des Gemeinderats über die Ausübung der Weideberechtigungen sind nicht nur für die Eigentümer, Nutzungsberechtigten und anderen Personen, deren Berechtigung sie zum Gegenstande haben, verbindlich, sondern auch für alle diejenigen Personen, welche für jene oder in deren Auftrag Vieh zur Weide führen. Daher verfallen insbesondere auch Hirten, die jenen Polizeiverordnungen zuwiderhandeln, der in § 25 bestimmten Strafe. (OLG. Urt. 23./11. 88 Ztschr. Bd. 15 S. 117.)

5. Ziff. 2 des § 25 verbietet, daß Ziegen, die nicht in einer Herde geführt werden, bei Ausübung der Gemeindeweide, der Koppelhütung und des Weidgangs anders als angebunden geweidet werden. Die Vorschrift rechtfertigt sich aus der besonderen Natur der Ziegen, die, sofern sie einzeln und unangebunden weiden, die Anpflanzungen, welche sich in der Nähe der Weide befinden, in besonders hohem Grade gefährden.

6. Durch Ziff. 3 des § 25 ist verboten, Zuchtstiere mit der Gemeindeherde auszutreiben. Die Vorschrift ist mit Rücksicht auf die öffentliche Sicherheit und Schamhaftigkeit sowie im Hinblick darauf getroffen worden, daß schlechte Zuchtstiere, die mit den Kühen auf die Weide getrieben werden, die Nachzucht in der Gemeinde verschlechtern können. Das Verbot bezieht sich nur auf die Gemeindeherde. Wer sich eine eigene Herde mit einem Zuchtstier hält, darf diesen mit den Kühen austreiben (LA. S. 387 ff.).

Beschädigung.

§ 26.

Mit Geldstrafe bis zu einhundertfünfzig Mark oder mit Haft wird bestraft, wer unbefugt,

1. abgesehen von den Fällen des § 305 des Strafgesetzbuchs, öffentliche Wege jeder Art oder fremde Privatwege oder deren Zubehörungen beschädigt;
2. abgesehen von den Fällen des § 274 Nr. 2 des Strafgesetzbuchs und des § 59 des Katastergesetzes vom 31. März 1884 (Gesetzbl. S. 59), Steine, Pfähle, Tafeln, Stroh- oder Hegewische, Hügel, Gräben oder ähnliche zur Abgrenzung, Absperrung oder Vermessung von Grundstücken oder Wegen dienende Merk- oder Warnungszeichen, desgleichen Merkmale, die zur Bezeichnung eines Wasserstandes bestimmt sind, sowie Wegweiser fortnimmt, vernichtet, umwirft, beschädigt oder unkenntlich macht;
3. natürliche oder künstliche Einfriedigungen, Geländer, oder die zur Sperrung von Wegen oder Eingängen in einge-

friedigte Grundstücke dienenden Vorrichtungen beschädigt oder vernichtet;

4. abgesehen von den Fällen des § 304 des Strafgesetzbuchs, Bäume, Sträucher, Pflanzen oder Feldfrüchte, die zum Schutze von Bäumen, Reben oder Hopfen dienenden Pfähle oder sonstigen Vorrichtungen beschädigt.

1. Die in §§ 26 ff. hervorgehobenen Beschädigungen werden nur dann nach dem FPStG. bestraft, wenn sie entweder aus Fahrlässigkeit oder zwar vorsätzlich verübt worden sind, aber nicht einen zehn Mark übersteigenden Schaden verursacht haben (§ 6). Anderenfalls unterliegen sie der Strafbestimmung des § 303 StGB. Auf die in §§ 304, 305 StGB. hervorgehobenen schweren Fälle der vorsätzlich und rechtswidrig begangenen Sachbeschädigung findet das FPStG. überhaupt keine Anwendung (S. Ziff. 1 u. 4 des § 26). Soweit die Beschädigungen nach dem FPStG. strafbar sind, bedarf es keines Strafantrages.

2. §§ 26 und 27 ersetzen in ihrer Beziehung auf Wege die Verbote des Art. 372 der auf Grund des Ministerialrundschreibens vom 21. Juli 1854 erlassenen Vizinalwegeordnungen, soweit dies nicht bereits durch § 366 Ziff. 9 und § 370 Ziff. 2 StGB. und § 20 dieses Gesetzes geschehen ist. Neben den angeführten gesetzlichen Bestimmungen ist für den Schutz der Wegepolizei von besonderer Bedeutung noch § 366 Ziff. 10 StGB., wonach mit Geldstrafe bis zu sechzig Mark oder mit Haft bis zu vierzehn Tagen bestraft wird, wer die zur Erhaltung der Sicherheit, Bequemlichkeit und Ruhe auf den öffentlichen Wegen, Straßen, Plätzen oder Wasserstraßen erlassenen Polizeiverordnungen übertritt. (Vgl. Anhang A.)

3. §§ 26 und 27 dienen nur dem Schutze von Feldgrundstücken, Feldwegen einschließlich der Landstraßen und von im Feldbann errichteten Einfriedigungen, nicht aber beziehen sie sich auf die im Inneren von Ortschaften belegenen Grundstücke. So die ständige Rechtsprechung des OLG., welche sich auf die Entstehungsgeschichte der angeführten gesetzlichen Bestimmungen und in Ansehung der Wege insbesondere den Umstand stützt, daß §§ 26 und 27 an Stelle der in Art. 372 der Vizinalwegeordnungen vom Jahre 1854 enthaltenen Verbote getreten sind, die nur die Wege auf dem Lande (voirie rurale), nicht aber die öffentlichen Verkehrswege in Städten oder Dörfern (voirie urbaine ou municipale) zum Gegenstande hatten. (Vgl. OLG. Urt. 28./2. 91 Ztschr. Bd. 16 S. 280 ff., Beschl. 2./5. 91 ebenda S. 469, Urt. 15./12. 03 ebenda Bd. 29 S. 588, ferner Dufour, traité général du droit administratif Bd. 3 Nr. 324, Ducrocq, cours de droit administratif Nr. 333.) Im übrigen kommt aber die in strafrechtlicher Beziehung wertlose Unterscheidung zwischen Straßen der grande und der petite voirie und der voirie vicinale hier nicht in Betracht (s. aber Bem. 4 zu § 1). Den

öffentlichen Wegen sind die dem Täter nicht gehörigen Privatwege gleichgestellt (§ 26 Ziff 1: „öffentliche Wege jeder Art oder fremde Privatwege").

4. Nur unbefugtes Handeln ist nach § 26 strafbar. Nicht unbefugt handelt, wer auf seinem Grundstück berechtigterweise Handlungen vornimmt, welche auf das Nachbargrundstück oder den vorbeiführenden Weg schädigend einwirken (Vgl. einen solchen Fall in OLG. Urt. 8./11. 89 Ztschr. Bd. 15 S. 171.) Nach § 59 StGB. muß der Täter wissen, daß ihm kein Recht zu der Handlung, die den Schaden verursacht, zusteht. Seinem Wissen steht die durch Fahrlässigkeit verursachte Unkenntnis gleich (vgl. Bem. 1b zu § 1).

5. Zu den einzelnen Bestimmungen des § 26 ist zu bemerken:

Zu Ziff. 1:

§ 305 StGB. lautet:

„Wer vorsätzlich und rechtswidrig .... eine gebaute Straße, welche fremdes Eigentum ist, ganz oder teilweise zerstört, wird mit Gefängnis nicht unter einem Monat bestraft.

Der Versuch ist strafbar."

Fahrlässige Beschädigungen einer gebauten Straße, die fremdes Eigentum ist, werden nach § 26 Ziff. 1 bestraft.

Zu Ziff. 2:

§ 274 Ziff. 2 StGB. lautet:

„Mit Gefängnis, neben welchem auf Geldstrafe bis zu dreitausend Mark erkannt werden kann, wird bestraft, wer einen Grenzstein oder ein anderes zur Bezeichnung einer Grenze oder eines Wasserstandes bestimmtes Merkmal in der Absicht, einem anderen Nachteil zuzufügen, wegnimmt, vernichtet, unkenntlich macht, verrückt oder fälschlich setzt."

§ 59 des Katastergesetzes lautet:

„Wer unbefugt Grenzsteine oder andere Grenzmarken setzt, wieder aufrichtet, herausnimmt oder in ihrer Lage verändert, wird mit Geldstrafe bis zu einhundertfünfzig Mark oder mit Haft bestraft, sofern nicht die Bestimmungen der §§ 274 Nr. 2 und 370 Nr. 1 des Reichsstrafgesetzbuches Platz greifen."

Dem Täter muß die Eigenschaft der Steine, Pfähle rc. als Grenz- und Warnungszeichen rc. bekannt gewesen sein. Dem steht gleich, wenn er diese Eigenschaft zwar nicht kannte, nach Lage der Verhältnisse aber hätte kennen müssen (Vgl. KG. Urt. 12./2. 94 Joh. 15 S. 310 zu § 30 pr. F. u. FPG.)

Die in der Absicht rechtswidriger Zueignung erfolgte Wegnahme oder Beschädigung von Grenzzeichen, Wegweisern, Einfriedigungen u. dgl. ist kein bloßer Feldfrevel, sondern wird als Diebstahl oder Diebstahlsversuch nach § 242 StGB. bestraft. (Vgl. RG. Urt. 11./1 95 E. 26 S. 367 zu § 30 pr. F. u. FP.G.)

Zu Ziff. 3:

Unter natürlichen Einfriedigungen sind insbesondere die Hecken zu verstehen. Wegen des Abpflückens von Laub und des Abbrechens von Zweigen aus Hecken s. § 18 Ziff. 2.

Zu Ziff. 4:

§ 304 StGB., soweit derselbe hier in Betracht kommt, lautet:

„Wer vorsätzlich und rechtswidrig .... Gegenstände, welche zum öffentlichen Nutzen, oder zur Verschönerung öffentlicher Wege, Plätze oder Anlagen dienen, beschädigt oder zerstört, wird mit Gefängnis bis zu drei Jahren oder mit Geldstrafe bis zu eintausendfünfhundert Mark bestraft.

Neben der Gefängnisstrafe kann auf Verlust der bürgerlichen Ehrenrechte erkannt werden.

Der Versuch ist strafbar."

Hiernach sind Beschädigungen von Bäumen, Sträuchern und Pflanzen, die zum öffentlichen Nutzen oder zur Verschönerung öffentlicher Wege, Plätze oder Anlagen dienen, sowie Beschädigungen der zum Schutze der vorerwähnten Bäume rc. dienenden Pfähle oder sonstigen Vorrichtungen nach § 304 StGB. zu bestrafen. Dabei ist aber ferner vorausgesetzt, daß die Beschädigungen vorsätzlich und rechtswidrig geschehen sind, und daß durch sie die Bäume rc. gerade in derjenigen Beziehung minder tauglich werden, vermöge welcher sie zu den durch die Strafdrohung des § 304 StGB. besonders geschützten gehören. Infolgedessen können geringfügige Beschädigungen der in § 304 aufgeführten Gegenstände doch nach dem FPStG. zu ahnden sein, z. B. das Abpflücken von Blumen aus Anlagen oder von den Grabhügeln eines Kirchhofs (s. auch Bem. 4a zu § 9). Andererseits kann unter Umständen das Abpflücken einer einzigen Blume aus einer öffentlichen Anlage den Tatbestand des § 304 StGB. erfüllen, wenn z. B. eine besonders kostbare, nur selten blühende Pflanze, die für sich allein zur Verschönerung der Anlage erheblich beiträgt, ihres Blütenschmucks beraubt wird. (Vgl. RG. Urt. 30./12. 81 E. 5 S. 318, Urt. 29./11. 83 E. 9 S. 219.)

Das bloße Abpflücken von Laub und Abbrechen von Zweigen aus Bäumen, Sträuchern und Hecken ohne Anwendung von Werkzeugen fällt, sofern nicht der Tatbestand des § 304 StGB. erfüllt ist, unter die mildere Strafvorschrift des § 18 Ziff. 2 FPStG., ist demgemäß nur auf Antrag strafbar und bleibt straflos, wenn dadurch ein Schaden nicht entstanden ist.

6. Bezüglich des Strafmaßes hebt der Kommissionsbericht (K. B. S. 407) besonders hervor, daß eine nachdrückliche Strafe gerechtfertigt sei, wenn die Beschädigung aus reiner Bosheit geschehen ist.

Verunreinigung von Wegen.

§ 27.

Mit Geldstrafe bis zu fünfzig Mark oder mit Haft bis zu vierzehn Tagen wird bestraft, wer

1. abgesehen von den Fällen des § 366 Nr. 9 des Strafgesetzbuchs, auf öffentlichen Wegen oder fremden Privatwegen oder deren Zubehörungen unbefugt und ohne daß ein Bedürfnis besteht, Erde, Steine, Holz, Schutt, Dünger oder andere Materialien lagert oder Stroh, Laub, Streu ausbreitet oder auf solche Wege aufgesammelte Steine wirft oder auf andere Weise dieselben verunreinigt oder ihre Benutzung erschwert;
2. über die unter Nr. 1 genannten Wege mit nicht aufgehobenem Pfluge fährt;
3. auf den unter Nr. 1 genannten Wegen die zur Bezeichnung der Fahrbahn gelegten Steine, Faschinen oder sonstigen Zeichen entfernt oder in Unordnung bringt.

1. Im allgemeinen vgl. Bemerkungen 1 bis 3 zu § 26.

Die §§ 27 und 28 stellen bestimmte schadendrohende Handlungen unter Strafe, ohne Rücksicht darauf, ob ein Schaden wirklich eingetreten ist. § 27 verbietet gewisse Handlungen, welche der Ordnung auf Wegen, die nicht der alleinigen Willkür des Täters unterstehen, zuwider sind.

2. Zu Ziff. 1 ist zu bemerken:

§ 366 Ziff. 9 StGB. (s. Anhang A) droht demjenigen mit Strafe, der auf öffentlichen Wegen, Straßen, Plätzen oder Wasserstraßen Gegenstände, durch welche der freie Verkehr gehindert wird, aufstellt, hinlegt oder liegen läßt. Darüber hinausgehend bestimmt § 27 Ziff. 1 FPStG., daß, ohne Rücksicht darauf, ob ein Verkehrshindernis tatsächlich bewirkt wurde, strafbar ist, wer auf öffentlichen Wegen oder fremden Privatwegen unbefugt die oben bezeichneten Gegenstände lagert oder ausbreitet rc. Die Strafe tritt aber nicht ein, wenn ein Bedürfnis zu der Lagerung rc. bestand. Bedürfnis ist nicht gleichbedeutend mit dringender Not (cas de nécessité absolue des Art. 372 Ziff. 2 der auf Grund des Rundschreibens vom 21. Juli 1854 erlassenen Vizinalwegeordnungen). Ein Bedürfnis kann z. B. schon bestehen, wenn ein Grundstück keine Zufahrtswege besitzt, und aus diesem Grunde Gegenstände, welche zur Bewirtschaftung des Grundstücks notwendig sind, z. B. Dünger, vorübergehend auf dem in der Nähe vorbeiführenden Wege gelagert werden. (Vgl. OLG. Urt. 8./3. 92, Ztschr. Bd. 17 S. 371.) Auch in solchen Fällen darf aber der Verkehr auf dem öffentlichen Wege nicht gehindert sein, da sonst die Strafvorschrift des § 366 Ziff. 9 StGB. eingreift. Keinesfalls ist aus § 27 Ziff. 1 FPStG. zu folgern, daß jemand im Fall eines Bedürfnisses das Recht hätte, Materialien auf einem öffentlichen Wege oder fremden Privatwege zu lagern. Die Verpflichtung, die Materialien wieder wegzuräumen, besteht natürlich fort, und die Polizei hat selbstverständlich das Recht, den Betreffenden zum Wegräumen anzuhalten. (Vgl. die Ausführungen des Regierungskommissars in LA. S. 390.)

Störungen des Besitzes.

§ 28.

Mit Geldstrafe bis zu dreißig Mark oder mit Haft bis zu acht Tagen wird bestraft, wer unbefugt,

1. abgesehen von den Fällen des § 366 Nr. 7 des Strafgesetzbuchs, Steine, Scherben, Schutt oder Unrat auf Grundstücke wirft oder in dieselben bringt;
2. Leinwand, Wäsche oder ähnliche Gegenstände zum Bleichen, Trocknen oder zu anderen derartigen Zwecken ausbreitet oder niederlegt;
3. Gruben auf fremden Grundstücken anlegt oder auf fremden Teichen Löcher ins Eis haut;
4. in fremde Gärten oder andere Grundstücke über Mauern, Hecken oder Zäune einsteigt oder durch solche einbricht;
5. die zur Sperrung von Wegen oder Eingängen in eingefriedigte Grundstücke dienenden Vorrichtungen öffnet oder offen stehen läßt;
6. mit Steinen oder anderen Gegenständen in fremde Bäume wirft oder schlägt.

1. § 28 faßt einige ordnungswidrige Handlungen zusammen, welche die ruhige Benutzung des ländlichen Eigentums stören. Hervorzuheben ist, daß nur das unbefugte Handeln strafbar ist (vgl. Bem. 1 b zu § 1).

2. Im einzelnen ist zu bemerken:

Zu Ziff. 1:

Nach § 366 Ziff. 7 StGB. (Anhang A) ist strafbar, wer Steine oder andere harte Körper oder Unrat .... gegen fremde Häuser, Gebäude oder Einfriedigungen oder in Gärten oder in eingeschlossene Räume wirft. Nach § 26 Ziff. 1 FPStG. ist das Werfen oder Einbringen von Steinen, Scherben, Schutt und Unrat in fremde Grundstücke ganz allgemein strafbar, auch wenn diese keine Gärten und nicht eingeschlossen sind. Unter den Grundstücken sind aber hier nicht die Wege zu verstehen, da für diese in § 27 Ziff. 1 eine besondere Bestimmung gegeben ist. (Vgl. auch § 366 Ziff. 9 StGB. im Anhang A.)

Zu Ziff. 2:

Wird durch unbefugtes Ausbreiten oder Niederlegen von Wäsche ꝛc. auf öffentlichen Wegen ꝛc. der freie Verkehr gehindert, so greift § 366 Ziff. 9 StGB. ein.

Zu Ziff. 3:

Die Bestimmung betrifft ohne Rücksicht auf den Zweck der Handlung das unbefugte Anlegen von Gruben auf fremden Grundstücken und das unbefugte Schlagen von Löchern in das Eis fremder Teiche, wie es häufig zur Erleichterung des Fisch- und Froschfangs vorkommt. Sind diese Handlungen zum Zweck des unberechtigten Jagens oder Fischens geschehen, so greifen die Bestimmungen der §§ 292, 293, 370 Ziff. 4 StGB. ein.

Zu Ziff. 4:

Bestraft wird das unbefugte Einsteigen oder Eindringen in fremde eingefriedigte Grundstücke an sich, auch wenn der Nachweis des Hausfriedensbruchs oder des Versuches einer anderen strafbaren Handlung, z. B. des Diebstahls, nicht erbracht werden kann (vgl. auch § 29 Ziff. 1).

Zu Ziff. 5:

Durch Ziff. 5 soll verhindert werden, daß die Vorrichtungen, welche zur Sperrung von Wegen oder eingefriedigten Grundstücken dienen, durch bösen Willen oder Nachlässigkeit von Passanten geöffnet oder offen stehen gelassen werden. Auch wenn das Oeffnen ein befugtes ist, kann das Offenstehenlassen nach dem Hindurchgehen unbefugt und damit strafbar sein. Strafbar ist sogar derjenige, welcher eine derartige Vorrichtung, die er offen gefunden hat, nach dem Hindurchgehen unbefugt offen stehen läßt.

Zu Ziff. 6:

Wird durch das Werfen oder Schlagen in fremde Bäume eine Beschädigung derselben verursacht, so tritt § 26 Ziff. 4, unter Umständen § 304 StGB. ein.

Betreten von Grundstücken.

§ 29.

Mit Geldstrafe bis zu zehn Mark oder Haft bis zu drei Tagen wird bestraft, wer:

1. abgesehen von den Fällen des § 123 des Strafgesetzbuchs, von einem Grundstücke, auf dem er ohne Befugnis sich

befindet, auf die Aufforderung des Berechtigten sich nicht entfernt;

2. abgesehen von den Fällen des § 368 Nr. 9 des Strafgesetzbuchs, unbefugt über Grundstücke reitet, karrt, fährt, Vieh treibt, Holz schleift, den Pflug wendet oder über Aecker, deren Bestellung vorbereitet oder in Angriff genommen ist, geht.

Die Verfolgung tritt nur auf Antrag ein. Die Zurücknahme des Antrags ist zulässig.

Im Falle der Nr. 2 bleibt der Zuwiderhandelnde straflos, wenn er durch die schlechte Beschaffenheit eines an dem Grundstücke vorüberführenden und zum gemeinen Gebrauch bestimmten Wegs oder durch ein anderes auf dem Wege befindliches Hindernis zu der Uebertretung genötigt ist, oder wenn durch das Wenden des Pfluges ein Schaden nicht entstanden ist.

1. § 123 StGB., soweit er hier in Betracht kommt, lautet:

„Wer in die Wohnung, in die Geschäftsräume oder in das befriedete Besitztum eines anderen oder in abgeschlossene Räume, welche zum öffentlichen Dienst bestimmt sind, widerrechtlich eindringt, oder wer, wenn er ohne Befugnis darin verweilt, auf die Aufforderung des Berechtigten sich nicht entfernt, wird wegen Hausfriedensbruches mit Gefängnis bis zu drei Monaten oder mit Geldstrafe bis zu dreihundert Mark bestraft.

Die Verfolgung tritt nur auf Antrag ein."

Abgesehen von der Wohnung, den Geschäftsräumen und den abgeschlossenen Räumen, welche zum öffentlichen Dienst bestimmt sind, ist nach § 123 StGB. das widerrechtliche Eindringen und unbefugte Verweilen nur auf befriedetem Besitztum strafbar. Befriedetes Besitztum sind aber nicht nur solche Grundstücke, die eingehegt sind, sondern auch Grundstücke, die ohne besondere Einhegung in derartig erkennbarem Zusammenhang mit Wohnungen oder Geschäftsräumen stehen, daß die „Befriedung" der letzteren sich auch auf jene erstreckt (vgl. RG. Urt. 3./12 89 E. Bd. 20 S. 150).

Hierüber hinausgehend, bestimmt § 23 Ziff. 1 FPStG., daß derjenige, der ohne Befugnis auf einem nicht schon durch das StGB. geschützten, fremden Grundstück verweilt und sich auf die Aufforderung des Berechtigten nicht entfernt, strafbar ist. Inwieweit, abgesehen von § 123 StGB., das unbefugte Betreten (Eindringen) fremder Grundstücke strafbar ist, beurteilt sich nach § 368 Ziff. 9 StGB. und der Ziff. 2 des obigen § 29.

2. Ohne Befugnis weilt auf fremdem Grundstück jeder, dem nicht zufolge der Erlaubnis des Berechtigten oder zufolge eines Privat- oder öffentlichen Rechtes das Verweilen gestattet ist. Ein Beamter, der auf einem Grundstück eine Amtshandlung vornimmt oder aus Anlaß einer

solchen dort weilt, befindet sich nicht unbefugt auf demselben. Nach Beendigung der Amtshandlung kann das fernere Verweilen aber zu einem unbefugten werden.

3. Berechtigt zur Aufforderung zum Verlassen des Grundstücks ist derjenige, dem die Verfügungsgewalt über das Grundstück zusteht oder dessen Vertreter. Eine einmalige Aufforderung, der nicht nachgekommen wird, genügt, um den Tatbestand des § 29 Ziff. 1 zu erfüllen.

4. Nach § 368 Ziff. 9 (S. Anhang A) ist u. a. strafbar das unbefugte Gehen, Fahren, Reiten und Viehtreiben über Gärten, Weinberge, nicht abgeerntete Wiesen, bestellte Aecker und solche Grundstücke, welche mit einer Einfriedigung versehen sind, oder deren Betreten durch Warnungszeichen untersagt ist.

Darüber geht § 29 Ziff. 2 FPStG. in zwei wesentlichen Punkten hinaus:

a) das unbefugte Fahren, Reiten und Viehtreiben — daneben sind auch das unbefugte Karren, Holzschleifen und Pflugwenden genannt — ist auf allen Grundstücken verboten;

b) das Gehen ist außer über die im § 368 Ziff 9 StGB. genannten Grundstücke auch über Aecker, deren Bestellung vorbereitet oder in Angriff genommen ist, verboten. Eine Vorbereitung oder Inangriffnahme der Bestellung eines Ackers ist anzunehmen, wenn dieser bereits gepflügt oder geegt ist. Nach den besonderen Wirtschaftsregeln kann aber auch unter anderen Voraussetzungen schon eine Vorbereitung oder Inangriffnahme anzunehmen sein.

5. Auch nach Ziff. 2 des § 29 wird ein unbefugtes Gehen, Reiten 2c. vorausgesetzt. Die Anwendbarkeit der Bestimmung ist daher auch abgesehen von den Fällen des Abs. 3 da ausgeschlossen, wo eine öffentlichrechtliche Befugnis zum Betreten des Grundstücks oder ein Privatrecht oder eine Erlaubnis des Berechtigten vorliegt. Unter den öffentlichrechtlichen Befugnissen ist insbesondere hervorzuheben das Recht der Beamten, in rechtmäßiger Ausübung einer dienstlichen Verrichtung oder aus Anlaß einer solchen fremde Grundstücke zu betreten. S. ferner § 17 des Gesetzes über das Postwesen des Deutschen Reiches vom 28. Oktober 1871 (RGBl. S. 347), welcher lautet: „In besonderen Fällen, in denen die gewöhnlichen Postwege gar nicht oder schwer zu passieren sind, können die ordentlichen Posten, die Extraposten, Kuriere und Estafetten sich der Neben- und Feldwege, sowie der ungehegten Wiesen und Aecker bedienen, unbeschadet jedoch des Rechtes der Eigentümer auf Schadensersatz." Nach Dambach (Postgesetz 5. Aufl. 1892 S. 115) findet diese Bestimmung auch Anwendung zugunsten der zu Fuß gehenden Briefträger und Postboten. Zu den Privatrechten, die das Betreten gestatten, gehört u. a. auch das Recht des Jagdpächters. Vgl. ferner Bem. 1a zu § 1.

Bei Beurteilung der Frage, ob ein unbefugtes Handeln vorliegt, sind auch bestehende Gebräuche, bei denen ein Einverständnis der Beteiligten anzunehmen ist, zu beachten. So ist es vielfach gebräuchlich, daß im Winter, wo kein Schaden verursacht wird, Erde und Dünger über fremde Grundstücke gefahren werden, ohne daß dies in der Regel für strafbar erachtet werden kann.

6. Die Bestimmung des Abs. 3, wonach im Falle der Ziff. 2 Straflosigkeit eintritt, wenn der Zuwiderhandelnde durch die schlechte Beschaffenheit eines an dem Grundstück vorbeiführenden Weges oder durch ein anderes auf dem Wege befindliches Hindernis zu der Uebertretung genötigt ist, gilt nicht, wenn der schlecht beschaffene oder sonst in seiner Gangbarkeit beeinträchtigte Weg im Eigentum des Zuwiderhandelnden steht, oder dieser zum Unterhalt des Weges verpflichtet ist. Der Umstand, daß man auf sein von allen Seiten eingeschlossenes Grundstück nicht gelangen kann, ohne ein fremdes Grundstück zu betreten, schließt die Strafbarkeit des Betretens ebenfalls nicht aus (K. B. S. 408). Der Eigentümer des eingeschlossenen Grundstücks hat vielmehr nur das Recht, gemäß § 917 BGB. von den Nachbarn die Einräumung eines Notweges zu verlangen.

7. Gänzlich unabhängig von der auf der Bestimmung des Absatz 3 oder anderen besonderen Rechtsgründen beruhenden ausnahmsweisen Straflosigkeit des in § 29 bedrohten Verhaltens ist die Frage der Schadensersatzpflicht. Diese ist ausschließlich nach dem BGB. zu beurteilen. Ein Zusammenhang zwischen Schaden und Strafe besteht nur in Ansehung des Pflugwendens auf fremdem Grundstück, indem dieses straflos bleibt, wenn ein Schaden nicht entstanden ist.

8. In sämtlichen Fällen des § 29 tritt Strafverfolgung nur ein, wenn Strafantrag gestellt ist. Hierüber vgl. Bem. 3 zu § 15.

Störung der Wasserverhältnisse.

§ 30.

Mit Geldstrafe bis zu einhundertfünfzig Mark oder mit Haft wird bestraft:

1. wer durch Hinderung oder Aenderung des natürlichen und regelmäßigen Wasserablaufs ein fremdes Grundstück unter Wasser setzt oder Versumpfung desselben herbeiführt oder einem fremden Grundstück oder einer fremden zum Wasserbezug berechtigten Anlage in schädigender Weise das Wasser entzieht;
2. wer, abgesehen von den Fällen der §§ 321 und 326 des Strafgesetzbuchs, unbefugt das zur Bewässerung von Grundstücken dienende Wasser ableitet, oder Gräben, Wälle, Rinnen oder andere zur Ab- und Zuleitung des Wassers dienende Anlagen herstellt, verändert, beschädigt, öffnet, verstopft oder beseitigt.

1. §§ 30 und 31 betreffen die Wasserpolizei. Sie haben ihren Platz im FPStG. gefunden, weil es zur Zeit der Einführung dieses noch kein die wasserrechtlichen Bestimmungen in Elsaß-Lothringen zusammenfassendes Gesetz gab und weil es u. a. auch einem Bedürfnis der Landwirtschaft entsprach, die wenigen landesrechtlichen Strafbestimmungen zum Schutze der Wasserrechtsverhältnisse zu präzisieren und zu ergänzen. Seither ist das öffentliche Wasserrecht in seinen wichtigsten Beziehungen durch das Gesetz betr. Wasserbenutzung und Wasserschutz vom 2. Juli 1891 (GBl. S. 82) geregelt worden. In § 51 des letzteren Gesetzes sind die §§ 30 und 31 des FPStG. ausdrücklich aufrecht erhalten worden. § 31 hat aber durch § 46 des Wassergesetzes eine bedeutsame Ergänzung erfahren (S. u.).

Da die §§ 30, 31 die Wasserpolizei im allgemeinen zum Gegenstand haben, dienen sie selbstverständlich nicht nur dem Schutze der Landwirtschaft, sondern ebenso demjenigen der Industrie und des Gewerbes, überhaupt jedem von der Rechtsordnung anerkannten Interesse an den Wasserverhältnissen.

Die Beurteilung der Frage, ob eine strafbare Verletzung der wasserpolizeilichen Vorschriften vorliegt, hängt sehr häufig von der Beantwortung wasserbautechnischer Vorfragen ab. Infolgedessen sind die Beamten der Staatsanwaltschaft angewiesen worden, sich bei allen diesbezüglichen Anzeigen vor Schlüssigmachung über die Erhebung der öffentlichen Klage mit den Beamten der Wasserbauverwaltung in Verbindung zu setzen. Das Nähere hierüber enthält die Ministerialverfügung vom 12. Mai 1892 (Smlg Bd. 17 S. 400 ff.).

2. § 30 Ziff. 1 stellt die unberechtigte Aenderung oder Hinderung des natürlichen Wasserablaufs unter Strafe. Demgemäß fällt hierunter das Zuwiderhandeln gegen das Verbot des § 59 Abs. 2 AG. z. BGB., wonach weder der Eigentümer des höher liegenden Grundstücks noch der Eigentümer des niedriger liegenden Grundstücks Vorrichtungen treffen darf, durch welche der infolge der natürlichen Bodenverhältnisse stattfindende Abfluß des Regen-, Schnee-, Eis-, Sicker- und Quellwassers zum Nachteil des Nachbarn geändert wird. Ferner gehört hierher auch die das Maß des § 46 AG. z. BGB. überschreitende Benutzung des Wassers aus natürlichen Wasserläufen, die weder mit Schiffen noch mit zusammengebundenen Flößen befahren werden können. (Vgl. Molitor, Bemerkung zu §§ 46, 47 AG. z. BGB. und Kisch S. 688.) Vorausgesetzt wird aber, daß durch die Hinderung oder Aenderung des Wasserablaufs ein fremdes Grundstück unter Wasser gesetzt oder Versumpfung desselben herbeigeführt wird, oder daß einem fremden Grundstück oder einer fremden zur Wasserbenutzung berechtigten Anlage in schädigender Weise das Wasser entzogen wird.

3. In Ziff. 2 des § 30 wird mit Strafe bedroht, wer unbefugt durch irgendwelche Veranstaltungen künstliche Bewässerungsanlagen herstellt oder irgendwie verändert oder das zur Bewässerung von Grundstücken dienende Wasser ableitet. Die Strafbarkeit ist unabhängig davon, ob ein Reglement über die Verteilung des Wassers (§ 31) besteht. (OLG. Urt. 8./10. 92 Ztschr. Bd. 18 S. 108.)

Neben der unbefugten Herstellung, Veränderung, Beschädigung und Beseitigung der Bewässerungsanlagen ist auch das unbefugte Oeffnen und Verstopfen derselben ausdrücklich mit Strafe bedroht.

4. § 321 StGB. lautet:

„Wer vorsätzlich Wasserleitungen, Schleusen, Wehre, Deiche, Dämme oder andere Wasserbauten, oder Brücken, Fähren, Wege oder Schutzwehre, oder dem Bergwerksbetriebe dienende Vorrichtungen zur Wasserhaltung, zur Wetterführung oder zum Ein- und Ausfahren der Arbeiter zerstört oder beschädigt, oder in schiffbaren Strömen, Flüssen oder Kanälen das Fahrwasser stört und durch eine dieser Handlungen Gefahr für das Leben oder die Gesundheit anderer herbeiführt, wird mit Gefängnis nicht unter drei Monaten bestraft.

Ist durch eine dieser Handlungen eine schwere Körperverletzung verursacht worden, so tritt Zuchthausstrafe bis zu fünf Jahren und, wenn der Tod eines Menschen verursacht worden ist, Zuchthausstrafe nicht unter fünf Jahren ein."

§ 326 StGB. lautet:

„Ist eine der in den §§ 321 bis 324 bezeichneten Handlungen aus Fahrlässigkeit begangen worden, so ist, wenn durch die Handlung ein Schaden verursacht worden ist, auf Gefängnis bis zu einem Jahre und, wenn der Tod eines Menschen verursacht worden ist, auf Gefängnis von einem Monat bis zu drei Jahren zu erkennen."

Die Strafbarkeit nach § 30 Ziff. 2 ist nicht bedingt durch die Entstehung eines Schadens oder einer Gefahr für das Leben oder die Gesundheit anderer.

5. Nur unbefugtes Handeln ist nach § 30 strafbar. Nicht strafbar ist daher, wer zu der Handlung ein Recht besaß oder wer sich gegen schädliche, von einer Bewässerungsanlage ausgehende Einwirkungen, die er zu dulden nicht verpflichtet ist, in berechtigter Weise wehrt, sollte hierdurch auch eine Beeinträchtigung der Anlage selbst verursacht werden. (Vgl. OLG. Urt. 8./11. 89 Ztschr. Bd. 15 S. 171.)

§ 31.

Mit Geldstrafe bis zu dreißig Mark oder mit Haft bis zu acht Tagen wird bestraft, wer den Festsetzungen der Verleihungsurkunde oder der Polizeiverordnung über die Wasserberechtigung der Triebwerke und anderer Anlagen an Gewässern, insbesondere über die Stauhöhe oder die Stau- beziehungsweise Entnahmezeiten, oder den zur Verhütung von Ueberschwemmungen, über Verteilung des Wassers nicht schiff- oder flößbarer Gewässer zwischen Wässerungsberechtigten oder zwischen Landwirtschaft und Gewerbe oder über die Unterhaltung oder Ausräumung solcher Gewässer erlassenen Polizeiverordnungen zuwiderhandelt.

Ist durch die Zuwiderhandlung eine der im § 30 Nr. 1 bezeichneten Folgen herbeigeführt worden, so ist auf Geldstrafe von zehn bis zu einhundertfünfzig Mark oder auf Haft zu erkennen.

1. S. Bemerkung 1 zu § 30.

2. Durch die Strafbestimmung des § 31 soll zunächst die Befolgung derjenigen Festsetzungen gewährleistet werden, welche gegenüber dem Besitzer eines Triebwerkes oder einer anderen Anlage von Gewässern in der Verleihungsurkunde oder bei der polizeilichen Genehmigung der Anlage über die Benutzung des Wassers, insbesondere über die Stauhöhe oder die Stau- bezw. Entnahmezeiten getroffen worden sind. Es werden demgemäß sowohl die Triebwerke mit sog. gesetzlicher Existenz (vgl. hierüber Leoni-Mandel II S. 208, Huber S. 196, Jacob und Fecht S. 105, Kisch S. 692) als auch diejenigen Triebwerke und Anlagen betroffen, welche auf Grund der polizeilichen aus Gründen des öffentlichen Interesses jederzeit widerruflichen Genehmigung errichtet worden sind. Für die Erteilung der letzteren Genehmigung sind jetzt §§ 1 ff. des Wassergesetzes von 1891 maßgebend. Hinsichtlich des früheren Rechts vgl. Huber S. 64 ff. Hervorzuheben ist, daß § 31 nur die Zuwiderhandlungen bezüglich der **Benutzung** des Wassers **mittels der Anlage** mit Strafe bedroht. Ist diese ohne Genehmigung oder entgegen den Bedingungen der Genehmigung errichtet worden oder werden an ihr genehmigungswidrige Aenderungen angebracht, so greifen die Strafbestimmung des § 42 Ziff. 1 Wassergesetzes und bei Neuanlagen für Wassertriebwerke die Bestimmungen der §§ 147 Ziff. 2, 16, 23 und 25 der Gew.-O. Platz.

§ 31 FPStG. schützt weiter die Polizeiverordnungen, welche die im öffentlichen Interesse notwendige Verteilung des Wassers in nicht schiff- oder flößbaren Gewässern und die Unterhaltung oder Ausräumung solcher Gewässer zum Gegenstande haben. Neben älteren Verordnungen (vgl. Huber S. 67 ff.) kommen in ersterer Hinsicht die auf Grund der §§ 9 und 10 des Wassergesetzes erlassenen Verordnungen des Statthalters und des Bezirkspräsidenten, in letzterer Hinsicht die Verordnungen in Betracht, welche das Ministerium, der Bezirkspräsident oder die mit der Ueberwachung der Unterhaltungsarbeiten betrauten Behörden gemäß

§§ 23, 24 und 29 des angeführten Gesetzes zu erlassen zuständig sind. Des näheren vgl. die Bemerkungen von Jacob und Fecht zu den bezeichneten Gesetzesparagraphen.

Schließlich fallen unter § 31 die Zuwiderhandlungen gegen die Benutzungsordnungen für Be- und Entwässerungsanlagen, welche gemäß § 5 des Gesetzes betr. Abänderung der Gesetzgebung hinsichtlich des Wasserrechts vom 11. Mai 1877 (GBl. S. 12) durch die Syndikatsgenossenschaften oder in Ausübung von Dienstbarkeiten durch die Beteiligten oder durch den Bezirkspräsidenten an Stelle der Genossenschaften oder der Beteiligten erlassen worden sind. Die Strafbestimmung in Abs. 4 des bezeichneten § 5 ist durch § 31 FPStG. außer Kraft gesetzt worden. (Vgl. Jacob und Fecht S. 129 Anm.)

3. In Ergänzung des § 31 FPStG. bestimmt § 46 Wassergesetzes:

„In den Fällen des § 31 des Feldpolizeistrafgesetzes vom 9. Juli 1888 (GBl. S. 73) ist die Strafe gegen den Besitzer der Anlage oder, wenn dieser die Leitung des Betriebes nicht persönlich ausübt, gegen den Betriebsleiter festzusetzen, sofern nicht nachgewiesen wird oder aus den Umständen sich ergibt, daß die strafbare Handlung gegen seine ausdrückliche Anordnung erfolgt ist. Auf Haft kann gegen den Besitzer oder Betriebsleiter nur erkannt werden, wenn er die strafbare Handlung selbst begangen oder ausdrücklich angeordnet hat."

Hiernach ist der Werkbesitzer oder, wenn dieser die Leitung des Betriebes nicht persönlich ausübt, der Betriebsleiter auch dann strafbar, wenn er selbst die in der Verleihungsurkunde oder Polizeiverordnung verbotene Handlung nicht vorgenommen hat. Sein strafrechtliches Verschulden wird in der ungenügenden Ueberwachung des Betriebes oder seiner Angestellten gefunden und stellt sich als Fahrlässigkeit dar. Von der Strafe bleibt er frei, wenn die strafbare Handlung gegen seine ausdrückliche Anordnung erfolgt ist. Eine Beweislast trifft den Betriebsleiter oder Werkbesitzer in dieser Beziehung selbstverständlich nicht. Vielmehr ist gegebenenfalls von Amts wegen nach Umständen zu forschen, welche die Strafbarkeit ausschließen (vgl. auch § 260 StPO.). Ueberdies tritt nach allgemeinem Rechtsgrundsatz Straflosigkeit dann ein, wenn die Nichtbefolgung der einschlägigen Vorschriften durch absolute Verhinderung des Werkbesitzers oder Betriebsleiters infolge höherer Gewalt verursacht worden ist. (Vgl. Oppenhoff Bem. 9, 10 zum 29. Abschnitt des StGB.) Ueber den Begriff des Betriebsleiters vgl. OLG. Urt. 22. 12. 93 u. 22. 1. 94 Ztschr. Bd. 19 S. 278 u. 326.

Die strafrechtliche Haftung des Werkbesitzers oder Betriebsleiters ist eine unmittelbare. Der Angestellte, welcher die Handlung vorgenommen hat, ist nur dann strafbar, wenn er dies gegen ausdrückliche Anordnung des Werkbesitzers oder Betriebsleiters getan hat. Doch ist gegen den Betriebsleiter, der die strafbare Handlung nicht selbst begangen und nicht ausdrücklich angeordnet hat, nur auf Geldstrafe zu erkennen. Im Falle der Uneinbringlichkeit ist allerdings auch diese gemäß § 28 StGB. in Haft umzuwandeln.

4. Abs. 2 droht eine schwerere Strafe für den Fall an, daß durch eine der im Abs. 1 bezeichneten Handlungen eine Unterwassersetzung oder Versumpfung eines fremden Grundstückes herbeigeführt worden ist, oder daß einem fremden Grundstück oder einer fremden zum Wasserbezug berechtigten Anlage in schädigender Weise das Wasser entzogen worden ist.

Steinbrüche 2c.

§ 32.

Mit Geldstrafe bis zu einhundertfünfzig Mark oder mit Haft wird bestraft, wer, abgesehen von den Fällen des § 367 Nr. 12 des Strafgesetzbuchs, den Anordnungen der Behörden zuwider es unterläßt:

1. Steinbrüche, Lehm-, Sand-, Kies-, Mergel-, Kalk- oder Tongruben, Bergwerksschachte, Schürflöcher oder die durch Stockroden entstandenen Löcher, zu deren Einfriedigung oder Zuwerfung er verpflichtet ist, einzufriedigen oder zuzuwerfen;
2. Oeffnungen, welche er in die von der Behörde bezeichneten Eisflächen gemacht hat, durch deutliche Zeichen zur Warnung vor Annäherung zu verwahren.

1. Nach § 367 Ziff. 12 StGB. (S. Anhang A) wird bestraft, wer an Orten, an welchen Menschen verkehren, Oeffnungen dergestalt unverdeckt oder unverwahrt läßt, daß daraus Gefahr für andere entstehen kann.

Der § 32 erweitert diese Bestimmung namentlich zum Schutz des Weideviehs und zur Vermeidung von Unglücksfällen und bestraft das bloße Unterlassen der Einfriedigung und Zuwerfung ohne Rücksicht darauf, ob dadurch eine Gefährdung von Menschen oder Tieren herbeigeführt worden ist. Die Bestrafung ist jedoch davon abhängig, daß einer Anordnung der Behörde zuwidergehandelt worden ist. (K. B. S. 408.)

2. Zum Erlaß der „Anordnungen der Behörden" sind zuständig: in Ansehung der Bergwerke, unterirdischen Steinbrüche und Tagebaue auf Eisenerze das Ministerium als Oberbergbehörde und der Bergmeister als Bergbehörde nach Maßgabe der §§ 172 ff. des Berggesetzes vom 16. Dezember 1873 (GBl. S. 397);

in Ansehung der offenen Steinbrüche das Ministerium, die Kreisdirektoren und im Falle dringender Gefahr der Bürgermeister und der Gewerbeaufsichtsbeamte. Vgl. des näheren § 173 Abs. 2 des Berggesetzes vom 16. Dezember 1873, §§ 77 ff. der Verordnung des Oberpräsidenten vom 7. September 1879 (Möllersche Sammlung Bd. III

S. 1031 und Bekanntmachung des Ministeriums vom 18. November 1892 (ABl. S. 404);

im übrigen die Bezirkspräsidenten und Bürgermeister bezw. Polizeidirektoren.

Die Strafbestimmung des § 32 Ziff. 1 schützt nicht nur die auf Grund der soeben bezeichneten gesetzlichen Bestimmungen erlassenen allgemeinen Polizeiverordnungen, sondern auch die den einzelnen Eigentümern oder Unternehmern gegenüber erlassenen Einzelverfügungen der zuständigen Behörden (arrêtés individuels et spéciaux). Sie tritt, soweit sie die Einfriedigung und Zuwerfung der Bergwerkschächte, Schürflöcher und Steinbrüche betrifft, an Stelle der Strafandrohung der §§ 183, 184 des Berggesetzes vom 16. Dezember 1873, die im übrigen selbstverständlich in Kraft bleiben, und des § 87 der Verordnung vom 7. September 1879, welcher auf § 184 des Berggesetzes beruht.

Anordnungen allgemeiner Natur, welche die Einfriedigung der in § 32 Ziff. 1 aufgeführten Anlagen betreffen, sind enthalten:

in Ansehung der Bergwerke in §§ 1 und 10 der Allg. Bergpolizeiverordnung für Elf.-Lothr. vom 3. Mai 1905 (ABl. A S. 189).

in Ansehung der Steinbrüche in §§ 13, 15 und 69 der Verordnung des Oberpräsidenten vom 7. September 1879 (Möllersche Sammlung III S. 1027 ff.),

in Ansehung des Tagebaues auf Eisenerze in § 13 der Verordnung des Ministeriums vom 6. Oktober 1906 (ABl. A S. 173).

3. Die Ziffer 2 des § 32 dient nicht dem Schutze von Polizeiverordnungen, wodurch allgemein angeordnet ist, daß Oeffnungen von Eisflächen zur Warnung vor Annäherung zu verwahren sind, sondern setzt voraus, daß die Behörde (regelmäßig die Ortspolizeibehörde) die Eisfläche, deren Oeffnungen zu verwahren sind, im einzelnen ausdrücklich bezeichnet hat (K. B. S. 408).

4. Ob jemand nach den Bestimmungen des § 32 strafrechtlich verantwortlich ist, beurteilt sich nach den tatsächlichen Verhältnissen des einzelnen Falles. Das Gebot zur Einfriedigung, Zuwerfung und Verwahrung richtet sich nicht ausschließlich an den Eigentümer der Oeffnung, vielmehr an denjenigen, welcher tatsächlich die Verfügung über die Oeffnung hat, somit allerdings in vielen Fällen an den Eigentümer selbst, außerdem aber auch an den Besitzer oder Verwalter, der nicht Eigentümer ist. (Vgl. RG. Urt. 23./2. 82 E. 6 S. 64, Urt. 23./11. 86 E. 15 S. 58 zu § 367 Ziff. 12 StGB. und § 29 pr. F. u. FPG.). Wichtig auch in strafrechtlicher Beziehung ist für die Bergwerke der § 70 des Berggesetzes vom 16. Dezember 1873, wonach die Personen, welche die Leitung oder Beaufsichtigung des Betriebes übernommen haben, für die Innehaltung der Betriebspläne sowie für die Befolgung aller im Gesetze enthaltenen oder auf Grund desselben ergangenen Vorschriften und Anordnungen verantwortlich sind.

Bienenstöcke.

§ 33.

Mit Geldstrafe bis zu fünfzig Mark oder mit Haft bis zu vierzehn Tagen wird bestraft, wer Bienenstöcke den polizeilichen Verordnungen zuwider aufstellt.

Jeder Eigentümer hat das Recht, Bienenstöcke in beliebiger Zahl zu halten. An und für sich kann er sie auf seinem Eigentum auch nach Belieben aufstellen. Da indessen durch Bienenstöcke unter Umständen die öffentliche Sicherheit gefährdet werden kann, wenn sie z. B. in unmittelbarer Nähe von Straßen, Wegen, Plätzen und besonders Spielplätzen gehalten werden, gibt § 47 Abs. 1 FPStG. dem Bezirkspräsidenten die Befugnis, Anordnungen über die Aufstellung der Bienenstöcke zu treffen. Die Zuwiderhandlung gegen die bezirkspolizeilichen Anordnungen wird im obigen Paragraphen mit Strafe bedroht. Die Ortspolizeibehörde kann, wie aus § 47 hervorgeht, keine Polizeiverordnungen über die Aufstellung von Bienenstöcken erlassen (vgl. auch LA. S. 394 und S. 549).

Ueber die zwangsweise Durchführung der von dem Bezirkspräsidenten angeordneten Maßregeln vgl. § 47 Abs. 2.

Tote Tiere.

§ 34.

Mit Geldstrafe bis zu dreißig Mark oder mit Haft bis zu acht Tagen wird bestraft, wer

1. **den das Vergraben gefallenen oder getöteten Viehs betreffenden Polizeiverordnungen zuwiderhandelt oder in Ermangelung solcher es unterläßt, derartiges ihm gehöriges Vieh innerhalb vierundzwanzig Stunden nach eingetretenem Tode auf seinem Eigentum oder an dem von der Ortspolizei bezeichneten Orte so zu vergraben, daß dasselbe mindestens ein Meter hoch mit Erde bedeckt ist;**
2. **tote Maulwürfe, Katzen, Hunde, Haufen toter Mäuse oder andere derartige tote Tiere, welche von ihm getötet sind oder ihm gehören, im Freien liegen läßt;**
3. **verscharrte Tiere oder Tierknochen ohne Erlaubnis der Ortspolizei ausgräbt.**

1. Nach § 47 Abs. 1 sind die Ortspolizeibehörden und die Bezirkspräsidenten befugt, innerhalb des Rahmens ihrer Zuständigkeit besondere polizeiliche Verordnungen über das Vergraben toter Tiere zu erlassen.

In Ermangelung solcher Verordnungen ist gefallenes oder getötetes Vieh (Rindvieh, Pferde, Esel, Schafe, Schweine, Ziegen und dergl.) von dem Eigentümer innerhalb vierundzwanzig Stunden nach eingetretenem Tode auf seinem Eigentum oder an dem von der Ortspolizeibehörde bezeichneten Orte so zu vergraben, daß dasselbe mindestens ein Meter hoch mit Erde bedeckt ist.

2. Um die Gefahr zu beseitigen, welche sich für die öffentliche Sicherheit aus dem im Freienliegenlassen von toten Tieren, die nicht unter den Begriff Vieh fallen, ergibt, bedroht Ziff. 2 denjenigen mit Strafe, welcher tote Maulwürfe, Katzen, Hunde, Haufen toter Mäuse oder andere derartige Tiere, welche von ihm getötet sind oder ihm gehören, im Freien liegen läßt.

3. Unter Ortspolizeibehörde ist hier der Bürgermeister, in den Städten Straßburg, Metz und Mülhausen der Polizeidirektor zu verstehen. Trotzdem nämlich in Straßburg und Mülhausen die Feldpolizei gemäß den Verordnungen vom 28. Februar 1873 bzw. vom 28. Mai 1889 den Bürgermeistern zusteht, unterliegen die in § 34 bezeichneten Gegenstände der Regelung des Polizeidirektors, da für die Bestimmungen des § 34 nicht feldpolizeiliche, sondern wesentlich gesundheitspolizeiliche Gründe maßgebend waren (vgl. LA. S. 439). Selbstverständlich findet auch § 34 nicht nur in ländlichen Orten und in Feldgemarkungen, sondern auch in Städten Anwendung.

Über die zwangsweise Durchführung der angeordneten Maßnahmen vgl. § 47 Abs. 2.

Verunreinigung von Gewässern.

§ 35.

Mit Geldstrafe bis zu fünfzig Mark oder mit Haft bis zu vierzehn Tagen wird bestraft, wer unbefugt:

1. in Gewässern Felle aufweicht oder reinigt;
2. abgesehen von den Fällen des § 366 Nr. 10 des Strafgesetzbuchs, Gewässer verunreinigt;
3. das Wasser in Feldbrunnen oder in sonst zum öffentlichen Gebrauch bestimmten Wasserbehältern verunreinigt.

Dieselbe Strafe trifft denjenigen, welcher:

4. den polizeilichen Verordnungen zuwiderhandelt, durch welche das Hineinwerfen toter Tiere in Gewässer verboten ist;
5. den Beschlüssen des Bezirkspräsidenten über die Dauer des Röstens des Leins oder Hanfs oder über die Orte, an denen dasselbe gestattet ist, oder über die in § 17 Nr. 2 der Verordnung, betreffend die Fischereipolizei vom 12. Februar 1883 (GBl. S. 5), bezeichneten Maßregeln zuwiderhandelt.

1. In § 35 sind unter „Gewässer" alle Arten von Gewässern zu verstehen. Insbesondere besteht kein Unterschied zwischen privaten und öffentlichen Gewässern. § 366 Ziff. 10 StGB. (Anhang A), der in Ziff. 2 des § 35 angezogen ist, hat dagegen nur die zum Schutz des Verkehrs auf den öffentlichen Wasserstraßen erlassenen Polizeiverordnungen zum Gegenstand.

2. Nur das unbefugte Handeln ist strafbar. Nicht unbefugt handelt, wem das Einführen von verunreinigenden Stoffen in Gewässern erlaubt ist. Dies ist von Wichtigkeit insbesondere für Ziff. 2 des § 35. Häufig nämlich ist dem Inhaber eines gewerblichen oder landwirtschaftlichen Betriebes durch die Verwaltungsbehörde gestattet, die Abwässer in einen Wasserlauf abzuleiten. Eines unbefugten Handelns macht sich in solchem Falle der Inhaber der Anlage solange nicht schuldig, als er den in dieser Hinsicht von der Verwaltungsbehörde getroffenen Anordnungen und Auflagen nachkommt. (Vgl. OLG. Urt. 2./5. 99 u. 11./7. 99 Ztschr. Bd. 24 S. 501 u. 540.)

Das Ministerium hat im Interesse der Erhaltung des Fischbestandes auf Grund des § 29 Abs. 2 des Gesetzes, betr. die Fischerei vom 2. Juli 1891 (GBl. S. 69), unter dem 17. Juli 1892 eine Verordnung (ABl. A S. 343) erlassen, welche allgemein die Maßregeln bestimmt, die bei Ableitung der den Fischen schädlichen Stoffe und Abfälle aus Fabriken und sonstigen gewerblichen und landwirtschaftlichen Betrieben zu beobachten sind. Die erforderlichen Anordnungen im einzelnen Falle erlassen die Bezirkspräsidenten auf Grund derselben gesetzlichen Bestimmung. Die Zuwiderhandlung gegen die Anordnungen des Ministeriums und der Bezirkspräsidenten werden nach § 45 Fischereigesetzes bestraft.

3. Eine Verunreinigung der Gewässer ist nur dann anzunehmen, wenn durch die eingeführten Stoffe das Wasser derartig verändert wird, daß der gemeingewöhnliche Gebrauch desselben unmöglich gemacht wird, daß also das Wasser zum Trinken, Waschen und Kochen unbrauchbar gemacht, oder daß die im Wasser vorkommenden Pflanzen oder Tiere geschädigt werden, oder die Schiffahrt auf demselben erschwert wird.

4. Durch Ziff. 4 werden die Polizeiverordnungen geschützt, welche im Interesse der öffentlichen Gesundheitspflege das Werfen toter Tiere, wie Katzen oder Hunde, in Gewässer verbieten. Die Polizeiverordnungen werden gemäß § 47 von der Ortspolizeibehörde und für den Bereich eines ganzen Bezirks oder mehrerer Gemeinden von dem Bezirkspräsidenten erlassen. Unter Ortspolizeibehörde sind die Bürgermeister, in den Städten Straßburg, Metz und Mülhausen aber die Polizeidirektoren zu verstehen. (Vgl. Bem. 3 zu § 34.) Für den Bezirk Oberelsaß ist auf Grund der Ziff. 4 eine Verordnung erlassen am 22./11. 01 (ABl. A S. 331), für den Bezirk Unterelsaß am 8./2. 02 (ABl. A S. 43).

5. Ziff. 5 ist aufgehoben durch § 54 des Fischereigesetzes und ersetzt durch die Strafbestimmung des § 45 des gleichen Gesetzes in Verbindung

mit § 29 Abs. 3 ebenda, wo es heißt: „Die Bezirkspräsidenten haben ferner zu beschließen über die Dauer des Röstens des Leins und Hanfs und die Bezeichnung derjenigen Wasserläufe und Orte, an welchen diese Arbeit mit dem geringsten Nachteil für die Fische stattfinden kann."

Belegenlassen von Tieren.

§ 36.

Mit Geldstrafe bis zu zehn Mark oder mit Haft bis zu drei Tagen wird bestraft, wer den ortspolizeilichen Verordnungen über die Beschaffenheit der Orte, an welchen das Belegen von Pferden oder Rindvieh vorgenommen werden darf, zuwiderhandelt, oder in Ermangelung solcher innerhalb einer Ortschaft das Belegen von Pferden oder Rindvieh auf öffentlichen Straßen oder Plätzen vornehmen läßt.

Nach § 36 ist aus Gründen der öffentlichen Schamhaftigkeit das Belegenlassen von Pferden und Rindvieh innerhalb einer Ortschaft auf öffentlichen Straßen und Plätzen verboten. Außerdem ist es der Ortspolizeibehörde (vgl. Bem. 3 zu § 34) überlassen, im Wege der Verordnung besondere Anforderungen an die Beschaffenheit der Orte, an welchen das Belegen vorgenommen werden darf, zu stellen.

Nützliche und schädliche Tiere und Pflanzen.

§ 37.

Mit Geldstrafe bis zu einhundertfünfzig Mark oder mit Haft wird bestraft, wer, abgesehen von den nach § 368 Nr. 2 des Strafgesetzbuchs und von den nach anderen besonderen Gesetzen strafbaren Fällen, den zum Schutz nützlicher oder zur Vernichtung schädlicher Tiere oder Pflanzen erlassenen Polizeiverordnungen zuwiderhandelt.

1. In § 368 Ziff. 2 StGB. (Anhang A) wird mit Strafe bedroht, wer das durch gesetzliche oder polizeiliche Anordnungen gebotene Raupen unterläßt.

2. Besondere Gesetze zum Schutze nützlicher und zur Vertilgung schädlicher Tiere sind: das Reichsgesetz, betr. den Schutz von Vögeln, vom 22. März 1888 (RGBl. S. 111), das elsaß-lothringische Gesetz zur Ausführung des Reichsgesetzes, betr. den Schutz von Vögeln, vom 2. Juli 1890 (GBl. S. 47) — über das Verhältnis beider Gesetze zu einander und zum Jagdpolizeigesetz vom 7. Mai 1883 (GBl. S. 57) vgl. Leoni-

Mandel II S. 251 und Halley, Jagd- und Vogelschutz S 84 —, ferner das Reichsgesetz, betr. die Bekämpfung der Reblaus vom 6. Juli 1904 (RGBl. S. 261), dazu die Bekanntmachung des Reichskanzlers vom 7. Juli 1905 (RGBl. S. 690).

3. Zum Erlaß der in § 37 und in § 368 Ziff. 2 StGB. vorgesehenen Polizeiverordnungen sind die Ortspolizeibehörden und die Bezirkspräsidenten nach Maßgabe des § 47 Abs. 1 zuständig. Ortspolizeibehörde in Ansehung dieser Verordnungen sind die Bürgermeister — auch in Straßburg und Mülhausen, da diese Verordnungen wesentlich feldpolizeilichen Charakter tragen (vgl. Bem. 3 zu § 31) — in Metz aber der Polizeidrektor, da die Verordnung des Oberpräsidenten vom 11. August 1872 betr. die polizeilichen Funktionen der Gemeindeverwaltung der Stadt Metz die Feldpolizei nicht dem Bürgermeister überträgt.

An bezirkspolizeilichen Verordnungen, die unter dem Schutze der Strafbestimmung des § 37 stehen, sind zu nennen:

für das Oberelsaß: die BPV., betr. die Vertilgung der Mistel, vom 11. Januar 1893 (ABl. B S. 10), die BPV., betr. das Abraupen; vom 29. November 1901 (ABl. A S. 334).;

für das Unterelsaß: die BPV., betr. die Vertilgung der Kleeseide, vom 20. April 1889 (ABl. S. 108), die BPV., betr. die Vertilgung von Misteln, vom 28. Oktober 1892 (ABl. B S. 298), die BPV., betr. die Vertilgung von Disteln, vom 28. März 1898 (ABl. B S. 107).;

für Lothringen: die BPV., betr. die Vertilgung der Disteln, vom 4. April 1889 (ABl. S. 95), die BPV., betr. die Vertilgung der Kleeseide, vom gleichen Tage (ABl. S. 95), die BPV., betr. die Vertilgung der Mistel, vom 25. Juni 1890 (ABl. S. 192).

Ueber die zwangsweise Durchführung der angeordneten Maßnahmen vgl. § 47 Abs. 2.

Erntebann.

§ 38.

Mit Geldstrafe bis zu zehn Mark oder mit Haft bis zu drei Tagen wird bestraft, wer, abgesehen von dem Falle des § 368 Nr. 1 des Strafgesetzbuchs, den ortspolizeilichen Verordnungen über die Erntezeit zuwiderhandelt.

§ 368 Ziff. 1 StGB. (Anhang A) bedroht denjenigen mit Strafe, der den polizeilichen Anordnungen über die Schließung der Weinberge zuwiderhandelt. Die Zuwiderhandlung gegen andere polizeiliche Anordnungen über die Erntezeit wird nach § 38 dieses Gesetzes bestraft. Ueber die Zuständigkeit zum Erlaß der Verordnungen über den Erntebann und darüber, inwieweit solche Verordnungen zulässig sind, bestimmt § 48.

Zweiter Titel.

Strafverfahren und Pfändung.

Schöffengerichte. Berufungskammern.

§ 39.

Für die Zuwiderhandlungen gegen dieses Gesetz sind die Schöffengerichte zuständig.

Die Hauptverhandlung kann in den Fällen der §§ 12, 13 ohne Anwesenheit des Angeklagten erfolgen. Ueber das Rechtsmittel der Berufung verhandeln und entscheiden in diesen Fällen die Strafkammern in der Besetzung von drei Mitgliedern einschließlich des Vorsitzenden.

1. Nach § 3 Abs. 3 des EG. zur StPO. können die Landesgesetze anordnen, daß Forst- und Feldrügesachen durch die Amtsgerichte in einem besonderen Verfahren, sowie ohne Zuziehung von Schöffen verhandelt und entschieden werden können. Das FPStG. hat anders wie das Forststgez. (§§ 63 bis 78, teilweise abgeändert durch das Ges. v. 23. Mai 1888 GBl. S. 45) von diesem Vorbehalt zugunsten der Landesgesetzgebung in den §§ 39—41 nur in sehr beschränktem Maße Gebrauch gemacht. Insbesondere hat es davon abgesehen, die Zuständigkeit nach Vorgang des Forststgez. grundsätzlich den Amtsgerichten ohne Zuziehung von Schöffen zu übertragen. Entscheidend dafür war die Erwägung, daß gerade Feldrügesachen sich zur Entscheidung durch sach- und ortskundige Laien vorzugsweise eignen.

2. Abs. 1 des § 39 enthält eine Besonderheit nur für die Fälle des § 13, da für alle übrigen Fälle die Zuständigkeit des Schöffengerichts sich schon aus § 27 Ziff. 1 u. 2 GVG. ergibt.

3. Nach § 231 Abs. 1 StPO. kann beim Ausbleiben des Angeklagten nur dann zur Hauptverhandlung geschritten werden, wenn die den Gegenstand der Untersuchung bildende Tat nur mit Geldstrafe, Haft oder Einziehung allein oder in Verbindung mit einander bedroht ist. Absatz 2 Satz 1 des § 39 dehnt dies auf die Fälle der §§ 12 und 13 aus, in denen Gefängnisstrafe angedroht wird. In allen Fällen, wo die Hauptverhandlung ohne Anwesenheit des Angeklagten stattfinden kann, muß der Angeklagte in der Ladung auf die Zulässigkeit dieses Verfahrens ausdrücklich hingewiesen werden (§ 231 Abs. 2 StPO).

Was in Abs. 2 Satz 2 für die Fälle der §§ 12 und 13 bestimmt ist, folgt für alle anderen Fälle bereits aus § 77 GVG. Ueber die Revision entscheidet das Oberlandesgericht gemäß § 123 Ziff. 2 GVG.

Einziehung.

§ 40.

Ueber die Einziehung im Falle des § 16 Absatz 3 verhandeln und entscheiden die Amtsgerichte ohne die Zuziehung von Schöffen.

Nach § 40 StGB. ist die Einziehung nur bei Verbrechen und Vergehen zulässig. Nach § 477 StPO. ist, wenn auf die Einziehung von Gegenständen selbständig erkannt werden soll, das Gericht zuständig, welches für den Fall der Verfolgung einer bestimmten Person zuständig sein würde. Das FPStG. läßt in § 16 die Einziehung auch bei Uebertretungen zu und überträgt in den Fällen, wo auf die Einziehung *selbständig* erkannt werden kann, die Entscheidung den Amtsgerichten, die ohne die Zuziehung von Schöffen entscheiden. Das Verfahren richtet sich im übrigen nach §§ 477 ff. StPO.

Strafbefehl.

§ 41.

Die Verurteilung zur Haftbarkeit in Gemäßheit der §§ 3, 4, 23, sowie zur Buße kann in der Form des Strafbefehls erfolgen.

In allen Feldpolizeistrafsachen kann, da sie zur Zuständigkeit der Schöffengerichte gehören, gemäß § 447 StPO. die Strafe in der Form des *Strafbefehls* festgesetzt werden. Auf eine Freiheitsstrafe kann dabei jedoch nur bis zur Dauer von sechs Wochen erkannt werden. Auch eine etwa verwirkte Einziehung kann nach § 447 Abs. 2 StPO. im Strafbefehl festgesetzt werden, es sei denn, daß auf sie selbständig in dem besonderen Verfahren der §§ 477 ff. StPO. zu erkennen ist. (Vgl. oben § 40.) Ueber § 447 StPO. hinausgehend, bestimmt § 41 FPStG., daß auch die Verurteilung zur *Haftbarkeit* in *Gemäßheit* der §§ 3, 4, 23, sowie zur *Buße* in der Form des Strafbefehls erfolgen kann.

Pfändung.

§ 42.

Wird aufsichtsloses oder fremdes durchziehendes Vieh auf einem Grundstück betroffen, auf welchem es nicht geweidet werden darf, oder Hausgeflügel auf einem fremden Grundstück, dem es Schaden zufügen kann, so kann dasselbe auf der Stelle oder in unmittelbarer Verfolgung sowohl von dem Feldhüter als auch von dem Beschädigten oder von solchen Personen gepfändet werden, welche die Aufsicht über das Grundstück führen oder zu der Familie, zu den Dienstleuten oder zu den auf dem Grundstück beschäftigten Arbeitsleuten des Beschädigten gehören.

Die gepfändeten Tiere sind dem Bürgermeister der Gemeinde, in deren Bann die Pfändung vorgenommen wurde, zur Anordnung der Verwahrung zuzuführen.

1. Nach § 229 BGB. handelt, wer zum Zwecke der Selbsthülfe eine Sache wegnimmt, nicht widerrechtlich, wenn obrigkeitliche Hülfe nicht rechtzeitig zu erlangen ist und ohne sofortiges Eingreifen die Gefahr besteht, daß die Verwirklichung des Anspruchs vereitelt oder wesentlich erschwert werde. Unabhängig von diesen Voraussetzungen geben §§ 42 ff. FPStG. ein bestimmt geartetes, im Wege der Selbsthülfe zu verwirklichendes Pfändungsrecht, das zufolge Art. 89 EG. z. BGB. auch unter der Herrschaft des neuen Rechts bestehen geblieben ist.

Dasselbe dient nur zum Schutze gegen Weidefrevel. Sein Zweck ist in erster Linie die Sicherung des Beweises. Infolgedessen richtet es sich nur gegen aufsichtsloses (§ 24 Ziff. 1) und fremdes durchziehendes (§ 29 Ziff. 2) Vieh, sowie gegen Hausgeflügel, das auf einem fremden Grundstück, dem es Schaden zufügen kann, betroffen wird (§ 24 Ziff. 1). Die Bestimmung hinsichtlich des fremden durchziehenden Viehs verfolgt insbesondere den Zweck, dem Grundstückseigentümer den Beweis seines Schadensersatzanspruchs gegen den unbekannten Eigentümer von Vieh, das unter einem fremden Treiber zum Markte getrieben wird oder von diesem heimkehrt, zu sichern. Neben der Beweissicherung dient das Pfändungsrecht dem beschädigten Eigentümer aber auch zur Befriedigung seines Schadensersatzanspruchs. (S. § 43.)

2. Die Pfändung kann nicht nur von dem Feldhüter oder dem Beschädigten, sondern auch von solchen Personen vorgenommen werden, welche die Aufsicht über das Grundstück führen oder zu der Familie, zu den Dienstleuten oder zu den auf dem Grundstück beschäftigten Arbeitsleuten des Beschädigten gehören.

3. Die Festnahme des Viehs kann sowohl auf dem Grundstück, auf welchem der Weidefrevel begangen wurde, als auch in unmittelbarer Verfolgung auf Grundstücken anderer erfolgen. Das Betreten dieser zum Zwecke der Pfändung ist, weil nicht rechtswidrig, auch nicht strafbar (Vgl. Bem. 1a zu § 1.)

4. Ueber den Umfang der Pfändung bestimmt das Gesetz nichts. Infolgedessen kann der Beschädigte auch mehrere Tiere, die im Weidefrevel betroffen wurden, ja sogar eine ganze Herde pfänden. Es liegt aber in der Natur der Sache, daß regelmäßig nicht mehr Stücke ge-

pfändet werden, als erforderlich ist, um den Beweis zu sichern und den entstandenen Schaden zu decken.

5. Der Bürgermeister, dem die Tiere nach der Pfändung zuzuführen sind, hat deren Verwahrung anzuordnen. Es steht nichts im Wege, daß die gepfändeten Tiere dem Beschädigten selbst zur Verwahrung übergeben werden.

6. Bei anderen Feldfreveln als den im § 42 Abs. 1 bezeichneten Weidefreveln, insbesondere beim Felddiebstahl, ist die Pfändung nicht zulässig. Zur Sicherung des Beweises dient in solchen Fällen die obrigkeitliche Beschlagnahme, für welche §§ 94 ff. StPO. maßgebend sind. (Vgl. Bem. 3 zu § 16.)

7. Das Gesetz sieht davon ab, nach Vorbild des § 17 pr. F. u. FPG. besondere Strafbestimmungen aufzustellen, welche die Vereitelung einer Pfändung, den Widerstand gegen eine solche, die Wegnahme gepfändeter Sachen und die unrechtmäßige Pfändung betreffen. Es geschieht dies in der Erwägung, daß die bestehenden Strafvorschriften über Widerstandsleistung gegen Beamte, Erpressung, Bedrohung, Wegnahme von Sachen, an denen der Besitzer ein Pfand- oder ein Zurückbehaltungsrecht hat, nicht bloß pfändenden Feldhütern sondern auch pfändenden Privatpersonen strafrechtlichen Schutz in genügendem Maße gewähren und daß einer besonderen Strafvorschrift gegen unrechtmäßige Pfändungen praktische Bedeutung kaum beizumessen ist.

§ 43.

Die rechtmäßigerweise gepfändeten Tiere haften für den entstandenen Schaden, die Buße und für alle durch die Pfändung und Verwahrung sowie die Schadensfeststellung verursachten Kosten.

1. Die gepfändeten Tiere haften nur für den entstandenen Schaden, bzw. wenn der Schaden gemäß § 5 im Strafverfahren geltend gemacht wird, für die Buße, ferner für alle durch die Pfändung und Verwahrung sowie die Schadensfeststellung verursachten Kosten, dagegen nicht für die Geldstrafe. Anders im Forststg. § 59.

2. Durch den Zusatz „rechtmäßigerweise" wird klar gestellt, daß die Pfandrechte des Beschädigten nur infolge einer dem § 42 entsprechenden Pfändung zur Wirkung gelangen. Eine Pfändung kann unrechtmäßig stattgefunden haben, wenn sie nicht von einer nach § 42 berechtigten Person oder wenn sie nicht wegen eines der in § 42 bezeichneten Weidefrevel vorgenommen worden ist.

§ 44.

Die Pfändung ist binnen 24 Stunden dem Amtsrichter anzuzeigen, welcher die bekannten Beteiligten davon sofort in Kenntnis zu setzen hat, falls dieselben bei der Pfändung nicht zugegen waren.

Ist die Anzeige an den Amtsrichter unterlassen, so kann der Gepfändete die Pfandstücke zurückverlangen. Der Pfändende hat in diesem Falle keinen Anspruch auf den Ersatz der durch die Pfändung entstandenen Kosten.

1. Zuständig ist der Amtsrichter desjenigen Bezirks, innerhalb dessen die Pfändung stattgefunden hat. Seine Bekanntmachungen und Entscheidungen sind den Beteiligten zu eröffnen, Abwesenden schriftlich mitzuteilen. Gegen die Entscheidungen (§§ 45, 46) findet eine Beschwerde nicht statt, da das Feldpolizeistrafgesetz die Zulässigkeit einer solchen nicht vorsieht und bei der Eigenart des Verfahrens weder die Bestimmungen der Zivilprozeßordnung noch die der Strafprozeßordnung anwendbar erscheinen. A. M. scheint der K. B. S. 412 zu sein.

2. Nur das Pfandrecht und der Anspruch auf Ersatz der durch die Pfändung entstandenen Kosten, nicht der Schadensersatzanspruch selbst, gehen im Falle der Unterlassung der Anzeige unter. Der Anspruch auf Rückgabe der gepfändeten Tiere ist gegebenenfalls im Wege des Zivilprozesses zu verfolgen.

§ 45.

Der Amtsrichter kann die gepfändeten Tiere freigeben, wenn zur Deckung der Beträge, für welche sie haften (§ 43), Sicherheit geleistet wird.

Die Sicherheitsleistung ist nach der Bestimmung des Amtsrichters durch Hinterlegung in barem Geld oder in Wertpapieren oder durch Pfandbestellung oder mittels Bürgschaft geeigneter Personen zu bewirken.

1. Wird der Anordnung des Amtsrichters, die Tiere freizugeben, nicht Folge geleistet, so ist zur Durchsetzung des Anspruchs auf Rückgabe wiederum nur der Weg des Zivilprozesses zulässig. Die Anordnung ist kein vollstreckbarer Titel.

2. Werden bares Geld oder Wertpapiere zum Zwecke der Sicherheit geleistet, so hat die Hinterlegung gemäß § 1 des Ges., betr. das Hinterlegungswesen und den Geschäftskreis der Staatsdepositenverwaltung, vom 1. November 1899 (GBl. S. 143) bei der Kasse der Staatsdepositen-

verwaltung oder, sofern außerhalb Straßburgs bares Geld hinterlegt wird, bei den Verkehrssteuerämtern zu geschehen, welche gemäß der Verordnung des Ministeriums vom 23. Dez. 1899 (ABl. S. 396) Hinterlegungen für Rechnung der Staatsdepositenverwaltung entgegenzunehmen haben.

§ 46.

Werden die gepfändeten Tiere nicht innerhalb fünf Tagen nach der Pfändung in Anspruch genommen, so kann der Amtsrichter die Versteigerung durch Gerichtsvollzieher oder durch den Bürgermeister des Verwahrungsortes anordnen. Dieselbe ist von dem beauftragten Beamten 24 Stunden vor dem Beginne und, sofern der Eigentümer der gepfändeten Tiere ermittelt ist, jedenfalls auch an dem Wohnsitz desselben bekannt zu machen.

Die Kosten der Verwahrung und der Versteigerung sind durch den Amtsrichter festzusetzen, und von dem Erlöse vorweg zu nehmen, der Ueberschuß wird zurückbehalten, bis über die Straftat rechtskräftig erkannt ist, und hinterlegt.

1. Die Versteigerung durch den Bürgermeister empfiehlt sich der Kostenersparnis halber insbesondere dann, wenn es sich um Tiere von nur geringem Werte, z. B. Gänse oder Hühner, handelt.

2. Die Hinterlegung hat sofort nach der Versteigerung und der Vorwegnahme der bezeichneten Kosten zu erfolgen. Hinterlegungsstelle sind die Kasse der Staatsdepositenverwaltung und außerhalb Straßburgs die Verkehrssteuerämter. Die Herausgabe des hinterlegten Betrages an die Berechtigten erfolgt in Gemäßheit der §§ 8 ff. des in Bem. 2 zu § 45 angeführten Gesetzes.

Dritter Titel.

Polizeiliche Verordnungen.

Bienenstöcke, totes Vieh, Unkraut ꝛc.

§ 47.

Zum Erlaß der in §§ 34, 35 Nr. 4 und 37 dieses Gesetzes und der in § 368 Nr. 2 des Strafgesetzbuchs vorgesehenen Polizeiverordnungen sind die Ortspolizeibehörden und für den Bereich des ganzen Bezirks oder mehrerer Gemeinden die Bezirkspräsidenten, zum Erlaß der in § 33 vorgesehenen Verordnungen die Bezirkspräsidenten zuständig.

Die genannten Behörden sind befugt, die angeordneten Maßregeln auf Kosten der Säumigen zur Ausführung zu bringen und die Kosten nach den Vorschriften über die Eintreibung öffentlicher Gefälle im Verwaltungswege beizutreiben.

1. §§ 47—49 treffen Anordnungen über die Zuständigkeit zum Erlaß von Polizeiverordnungen bezüglich solcher Angelegenheiten, deren Regelung das Gesetz der polizeilichen Verordnung überläßt. Nicht erwähnt sind hier diejenigen Gegenstände, hinsichtlich deren die Zuständigkeit in der Strafvorschrift selbst angegeben ist, wie z. B. in §§ 19 u. 36, oder hinsichtlich deren nach Lage der Gesetzgebung Zweifel über die Zuständigkeit nicht bestehen können, wie in §§ 31, 32 oder in § 25, soweit in letzterer Vorschrift die Reglements der Gemeinderäte über die Ausübung der Gemeindeweide, des Weidgangs oder der Koppelhütung in Betracht kommen.

Ueber das Polizeiverordnungsrecht im allgemeinen vgl. Leoni-Mandel II S. 105 und Bruck S. 82. Besonders wichtig ist, daß die Polizeiverordnungen gehörig verkündet werden müssen. Ueber den Nachweis der Verkündigung vgl. Ministerialverfügung vom 19. Dezember 1887 (ABl. S. 273 und Bruck S. 84).

In ihrem Wortlaut muß die Polizeiverordnung bezeugen, daß sie in der gesetzmäßigen Weise, insbesondere unter der gesetzmäßigen Mitwirkung derjenigen Organe zu Stande gekommen ist, an deren irgendwie geartete Tätigkeit das Gesetz den Erlaß der Polizeiverordnung knüpft. So müssen, wie bereits oben Bem. 2 zu § 25 hervorgehoben worden, die von dem Gemeinderat zu erlassenden und von dem Bürgermeister zu verkündenden Reglements über die Ausübung der Gemeindeweide, des Weidgangs und der Koppelhütung sich schon äußerlich als Polizeiverordnungen des Gemeinderats darstellen. Wenn der Erlaß einer bürgermeisterlichen Polizeiverordnung an die Zustimmung des Gemeinderats geknüpft ist, wie in § 48, oder wenn vor Erlaß einer bezirkspolizeilichen Verordnung das Gutachten des Bezirkstags einzuholen ist, wie in § 49, so muß der Text der Polizeiverordnung bezeugen, daß diese Erfordernisse erfüllt sind.

Darüber, ob eine Polizeiverordnung im einzelnen Falle anwendbar ist, entscheiden die Gerichte. Dabei kann sich die Frage aufwerfen, ob die anzuwendende Polizeiverordnung gültig ist. Das Prüfungsrecht der Gerichte beschränkt sich in dieser Hinsicht darauf, festzustellen, ob die Polizeiverordnung nicht etwa nach ihrem Inhalt gegen bestehende Gesetze verstößt, sowie festzustellen, ob sie formell gültig ist. In letzterer Beziehung darf der Richter aber nur prüfen, ob die Polizeiverordnung in der gehörigen Form und von der zuständigen Behörde verkündet worden ist, sowie ob sich die Verordnung äußerlich als gesetzmäßig zu Stande gekommen darstellt. Wenn dies der Fall ist, darf der Richter nicht nachprüfen, ob auch tatsächlich das Zustandekommen ein gesetzmäßiges war, ob z. B. der Gemeinderat tatsächlich einer Verordnung des Bürgermeisters über den Erntebann (§ 48) zugestimmt hat. (Vgl. Leoni-Mandel I S. 160, II S. 109, OLG. Urt. 6./5. 99 Ztschr. Bd. 24 S. 506.) Unter keinen Umständen darf der Richter die Zweckmäßigkeit einer Polizeiverordnung in Frage ziehen. (Vgl. auch Bruck S. 84.)

2. Ortspolizeibehörde im Sinne des § 47 ist der Bürgermeister, in Straßburg, Metz und Mülhausen gemäß § 14 Abs. 2 u. 3 des Gesetzes vom 30. Dezember 1871, die Einrichtung der Verwaltung betreffend, der Polizeidirektor. Wo es sich aber um Verordnungen rein feldpolizeilichen Charakters handelt, ist zum Erlaß derselben in Straßburg und Mülhausen ebenfalls der Bürgermeister zuständig, da in diesen beiden Städten auf Grund von § 14 Abs. 2 des Gesetzes vom 30. Dezember 1871 durch Verordnung vom 28. Februar 1873 bzw. vom 28. Mai 1889 die Feldpolizei den Bürgermeistern übertragen worden ist. Rein feldpolizeilichen Charakter haben von den im § 47 angeführten Polizeiverordnungen nur die unter die Strafsanktion des § 37 und des § 368 Nr. 2 StGB. gestellten Verordnungen über den Schutz nützlicher oder die Vertilgung schädlicher Tiere und Pflanzen. (Vgl. ferner § 48.) Die in §§ 34 und 35 Ziff. 4 vorgesehenen Verordnungen beruhen wesentlich auf gesundheitspolizeilichen Erwägungen. (Vgl. LA. S. 439 und die Bemerkungen zu den einzelnen angeführten Paragraphen.)

Der Bezirkspräsident ist zum Erlaß einer Polizeiverordnung über die in §§ 34, 35 Ziff. 4 und 37 bezeichneten Gegenstände nicht nur für den Bereich des ganzen Bezirks, sondern auch dann zuständig, wenn das Bedürfnis zum Erlaß solcher Verordnungen in einer Mehrheit von Gemeinden hervorgetreten ist. Das Gleiche wird auch in Ansehung der in § 33 vorgesehenen Verordnungen, zu deren Erlaß die Ortspolizeibehörde überhaupt nicht befugt ist, zu gelten haben. (Vgl. Leoni-Mandel II S. 108.)

3. Nach Abs. 2 sind die Ortspolizeibehörden und die Bezirkspräsidenten befugt, die angeordneten Maßnahmen, gleichgültig, ob sie im einzelnen Falle durch Ortspolizei- oder Bezirkspolizeiverordnung angeordnet sind, auf Kosten des Säumigen, den außerdem die Strafe für die Uebertretung trifft, vornehmen zu lassen. Die Kosten sind nach den Vorschriften über die Eintreibung öffentlicher Gefälle im Verwaltungswege beizutreiben. (Vgl. auch Gem.-O. § 69.) Demgemäß erfolgt die Beitreibung auf Grund eines Zwangsbefehls, der vom Steuerempfänger oder bei Gemeinden, deren Gefälle durch einen besonderen Gemeinderechner erhoben werden, von diesem ausgestellt und gemäß § 6 des Gesetzes, betr. die Ausführung der Civilprozeßordnung ꝛc., vom 13 November 1899 (GBl. S.157) vom Amtsrichter des Bezirks, worin der Aussteller seinen dienstlichen Wohnsitz hat, für vollstreckbar erklärt wird. Einwendungen, welche den durch den Zwangsbefehl festgestellten Anspruch betreffen, sind von dem Schuldner im Wege der Klage vor dem Landgericht, in dessen Bezirk der Aussteller des Zwangsbefehls seinen diensilichen Wohnsitz hat, zu erheben (§ 7 des angezogenen Ausf.-Ges.). Im einzelnen ist das Zwangsverfahren über die Beitreibung öffentlicher Gefälle im Verwaltungswege durch die auf Grund des § 7a des erwähnten Ausf.-Ges. (eingefügt durch Art. 5 des Gesetzes vom 13. Februar 1905 GBl. S. 3) erlassene Verordnung des Statthalters vom 26. Mai 1905 (ABl. S. 241) geregelt.

Erntebann.

§ 48.

Die in § 38 dieses Gesetzes und in § 368 Nr. 1 des Strafgesetzbuchs vorgesehenen Polizeiverordnungen werden vom Bürgermeister mit Zustimmung des Gemeinderats erlassen. Dabei kann für eine bestimmte, dem Beginn der Weinlese vorausgehende Zeit die Befugnis der Eigentümer zum Betreten ihrer Weinberge geregelt, insbesondere auf bestimmte Tage und Stunden eingeschränkt werden.

Abgesehen von den die Weinberge betreffenden Bestimmungen sind Verordnungen über die Erntezeit nur in Gemeinden zulässig, wo dieselben bisher üblich waren, und nur zum Zwecke der Regelung der Zu- und Abfahrt.

1. In Abs. 2 des heute noch geltenden Art. 1 Abschn. 5 Titel 1 des Ruralgesetzes vom 6. Oktober 1791 ist als Grundsatz ausgesprochen, daß jedem Eigentümer freisteht, seine Ernte, welcher Art sie auch sei, mit jedem Werkzeug und insbesondere zu jeder Zeit, welche ihm an-

steht, zu bewerkstelligen. Hiervon ist in Abs. 3 eine Ausnahme für die Weinernte gemacht, indem bestimmt wird, daß es in Gegenden, wo bisher der Gebrauch bestand, den Beginn der **Weinernte** alljährlich durch behördliche Ankündigung (ban de vendange, Weinlesebann) festzusetzen, hierbei für nicht eingefriedigte Weinberge auch in Zukunft sein Bewenden haben soll. Das Bedürfnis, für die Weinberge eine gemeinsame Erntezeit zu bestimmen, beruht darauf, daß durch frühzeitiges Herbsten einzelner der Ruf des Weines einer ganzen Gegend gefährdet wird. Außerdem wird durch eine gemeinsame Erntezeit die Beaufsichtigung und der Schutz der Weinberge gegen vorzeitiges Eindringen von Personen, welche die Nachlese ausüben wollen, erleichtert. Außer dem Weinlesebann erkannte die Rechtsprechung der französischen Gerichte im Anschluß an Art. 475 Ziff. 1 Code pénal, wo die Zuwiderhandlung gegen den ban de vendange et **autres bans** mit Strafe bedroht war, behördliche Anordnungen über die Erntezeit auch in Ansehung anderer landwirtschaftlicher Bodenerzeugnisse, insofern solche Anordnungen in der betreffenden Gegend von alters hergebracht waren, insbesondere den Heuerntebann, als zu Recht bestehend an. Die Befugnis zum Erlaß des Erntebannes stand dem Bürgermeister auf Grund der Art. 10 Ziff. 1 und Art. 11 Ziff. 1 des Gesetzes vom 18. Juli 1837 zu.

Dieser Rechtszustand ist durch das Feldpolizeistrafgesetz in mehrfacher Beziehung geändert worden:

a) **Zuständig** zum Erlaß des Erntebannes ist der **Bürgermeister mit Zustimmung des Gemeinderats**. Dies gilt auch für diejenigen Gemeinden, in denen die örtliche Polizeiverwaltung einem Polizeidirektor übertragen ist. Die vom Bürgermeister zu erlassende Polizeiverordnung muß in ihrem Wortlaut bezeugen, daß der Gemeinderat seine Zustimmung erteilt hat. Stellt sich aber demgemäß eine Polizeiverordnung über die Erntezeit äußerlich als gesetzmäßig zu Stande gekommen dar, so hat der Richter nicht das Recht, weiter zu prüfen, ob die Zustimmung auch tatsächlich erteilt worden ist. (Vgl. Bem. 1 zu § 47 und OLG. Urt. 6./5. 99 Ztschr. Bd. 24 S. 506.) Die Ortspolizeiverordnung über die Erntezeit braucht nicht wie nach dem Ruralgesetz alljährlich neu erlassen zu werden. Rechtlich steht vielmehr nichts im Wege, den Beginn der Weinlese dauernd für eine Reihe von Jahren festzusetzen. (Vgl. OLG. Urt. 4./7. 95 Ztschr. Bd. 21 S. 102.) Doch ist es praktischer, alljährlich unter Berücksichtigung der jeweiligen Ernteverhältnisse den Zeitpunkt von neuem zu bestimmen.

b) Nach der früheren Rechtsprechung des OLG. war es zweifelhaft, ob das Recht zum Erlaß des Erntebanns, außer dem Recht, den Beginn der Weinernte festzusetzen, auch die weitere Befugnis enthalte, die **Schließung der Weinberge** vor dem Beginn der Ernte anzuordnen. (Vgl. OLG. Urt. 29./1. 87 Ztschr. Bd. 12 S. 207.) Durch Abs. 1 Satz 2 ist dem Bürgermeister ausdrücklich die Befugnis verliehen, nach erfolgter Zustimmung des Gemeinderats für eine bestimmte dem Beginn der Weinlese vorausgehende Zeit die Befugnis der Eigentümer zum Betreten ihrer Weinberge zu regeln, insbesondere auf bestimmte Tage und Stunden einzuschränken. Dagegen darf das Betreten auch nach dem FPStG. nicht vollständig untersagt werden. (Vgl. OLG. Urt. vom 14. Juli 1895 Ztschr. Bd. 21 S. 102.) Die Bestimmung des Ruralgesetzes, wonach der Erntebann nicht für eingefriedigte Grundstücke gilt, ist unberührt geblieben. Die Einfriedigung muß die in Art. 6 Abschn. 4 Titel I des Ruralgesetzes vorgeschriebenen Eigenschaften haben. (Vgl. unten Bem. 2 zu § 49 und Dalloz Rép. v^{0} commune 779.)

c) Während nach dem Ruralgesetz der Weinlesebann nur in solchen Gemeinden zulässig war, wo er von alters her im Gebrauch stand, ist der Weinlesebann nach dem FPStG. in allen Gemeinden, auch da, wo er bisher nicht üblich war, zulässig. Dagegen sind Verordnungen über die Erntezeit hinsichtlich anderer Bodenerzeugnisse auch nach dem FPStG. nur in solchen Gemeinden gestattet, wo sie bisher üblich waren. Von praktischer Wirksamkeit ist diese Bestimmung nur für den Heuerntebann (ban de fauchaison), der sich in vielen Gemeinden Elsaß-Lothringens erhalten hat. Das Bedürfnis, ihn auch weiter zu sanktionieren, ergab sich daraus, daß Wiesen häufig von fremden Wiesen rings eingeschlossen sind, so daß ihre Bewirtschaftung wesentlich erschwert ist. Der Eigentümer ist in solchem Falle darauf angewiesen, die fremden Wiesen zur Zu- und Abfahrt in Anspruch zu nehmen. Um den damit verbundenen Schaden zu vermeiden, ist es zweckmäßig, daß die Wiesen tunlichst zur gleichen Zeit abgemäht werden. Dies wird in vielen Gemeinden dadurch erreicht, daß die Ortspolizeibehörde den Beginn der Heuernte und damit den Zeitpunkt festsetzt, von welchem ab die Durchfahrt gestattet ist. Anderen Zwecken als der Regelung der Zu- und Abfahrt darf nach der ausdrücklichen Bestimmung des Gesetzes der Heuerntebann nicht dienen. (Vgl. auch KB. S. 410, 412.)

3. Die Strafbestimmungen zum Schutze der in § 48 geregelten Polizeiverordnungen sind in Ansehung des Weinlesebannes enthalten in § 368 Ziff. 1 StGB., in Ansehung des Heuerntebannes in § 38 dieses Gesetzes.

Einschränkungen des Weidgangs und der Koppelhütung.

§ 49.

Durch Verordnung des Bezirkspräsidenten kann nach Einholung des Gutachtens des Bezirkstags für den ganzen Bezirk oder einzelne Gemeinden bezüglich der Ausübung des Weidgangs und der Koppelhütung bestimmt werden:

1. daß auch andere als die in Artikel 6 Abschnitt 4 Titel I des Gesetzes, betreffend die ländlichen Güter und Gebräuche und die Feldpolizei, vom 28. September — 6. Oktober 1791 bezeichneten Arten der Einfriedigung geeignet sein sollen, die Ausübung auszuschließen:
2. daß außer den durch Artikel 9 Abschnitt 4 Titel I des Gesetzes vom 28. September — 6. Oktober 1791 und § 6 des Gesetzes vom 11. Mai 1877 (GBl. S. 12) bereits ausgenommenen Grundstücken auch solche Grundstücke von der Ausübung ausgenommen sein sollen, welche durch besondere Bearbeitung des Bodens in Wiesen umgewandelt sind, oder auch solche Wiesen, auf welchen künstliche Ent- oder Bewässerungsanlagen in der Ausführung begriffen sind;
3. daß die Ausübung während bestimmter Zeiten verboten sein soll.

1. Vor Einführung des FPStG. waren nur die Gemeinderäte auf Grund des Art. 19 Nr. 8 des Gesetzes vom 18. Juli 1837, der heute durch § 56 Ziff. 12 der Gemeindeordnung ersetzt ist, befugt, Anordnungen über die Ausübung des Weidgangs und der Koppelhütung zu treffen. Ihre Anordnungen fanden ihre gesetzliche Schranke in den in Abschnitt 4 Tit. I des Ruralgesetzes von 1791 enthaltenen Bestimmungen. Die seit 1791 wesentlich veränderten wirtschaftlichen Verhältnisse haben das Bedürfnis gezeitigt, die der Landwirtschaft häufig schädliche Ausübung des Weidgangs und der Koppelhütung noch in weiterem Maße einzuschränken, als es nach dem Ruralgesetze möglich war. Dies geschah schon durch § 6 des Gesetzes, betr. die Abänderung der Gesetzgebung hinsichtlich des Wasserrechts (s. unten zu Ziff. 2). Dem gleichen Zwecke dienen die in § 49 FPStG. zugelassenen Bezirkspolizeiverordnungen, die zum Teil mit den in § 56 Ziff. 12 Gem.-O. vorgesehenen Reglements der Gemeinderäte konkurrieren, zum Teil aber auch über diese hinaus ergänzend und abändernd auf die Vorschriften des Ruralgesetzes selbst einwirken können. Das Verordnungsrecht des Bezirkspräsidenten hinsichtlich aller in § 49 erwähnten Gegenstände ist durch das Gesetz, welches bei diesen Angelegenheiten der Stimme der beteiligten Kreise Gehör geben will, an die vorherige Einholung eines Gutachtens des Bezirkstages geknüpft. (Vgl. Bem. 1 zu § 47.) Die einschlägigen Verordnungen des Bezirkspräsidenten können nach der ausdrücklichen Bestimmung des Gesetzes nicht nur für den ganzen Bezirk oder eine bestimmte Gruppe von Gemeinden erlassen werden, sondern auch für einzelne Gemeinden. Hierdurch ist

eine beachtenswerte Ausnahme von dem Grundsatz des französischen Verwaltungsrechts, wonach der Präfekt zum Erlaß von Verordnungen für eine einzelne Gemeinde nicht zuständig ist, geschaffen worden.

2. Im einzelnen ist zu bemerken:

Zu Ziff. 1:

Der Weidgang und die Koppelhut dürfen nach Art. 7 und 6 Abschn. 4 Titel I des Ruralgesetzes im „enclos rural" nicht ausgeübt werden. Als eingefriedigt ist nach dem letztgenannten Artikel ein Grundstück anzusehen, wenn es umgeben ist von einer 4 Fuß hohen Mauer mit Schlagbäumen oder Türen, oder mit Pallisaden oder mit Gitterwerk oder mit einer lebendigen oder toten Hecke oder mit einem 4 Fuß tiefen und 2 Fuß breiten Graben. Der Zweck der Bestimmung über die Beschaffenheit der Einfriedigung ist der, daß die letztere geeignet sei, fremdes Vieh vom Eindringen in das Grundstück abzuhalten. Durch die ins einzelne gehende Beschreibung der Einfriedigungen im Artikel 6 sind nun manche Arten derselben, die denselben Zweck bei billigerer Herstellung erfüllen, z. B. Drahteinfriedigungen, ausgeschlossen. Um diesen Uebelstand zu beseitigen, ist die Bestimmung unter Nr. 1 des § 49 getroffen.

Zu Ziff. 2:

Durch die Rechtsprechung des Appellhofs zu Köln und des preußischen Obertribunals wurde auf Grund des Wörterbuchs der französischen Akademie bereits in den 60er Jahren festgestellt, daß „prairies artificielles", welche Artikel 9 Abschnitt 4 Titel I des Ruralgesetzes von Weidgang und Koppelhut ausnimmt, nicht eigentlich Kunstwiesen seien, sondern Ackerfelder, auf welchen im gewöhnlichen Wege der Beackerung und Besamung Futterkräuter, wie Klee u. s. w., gezogen werden. Demzufolge wurde bereits durch § 6 des Gesetzes, betreffend Abänderung der Gesetzgebung hinsichtlich des Wasserrechts, vom 11. Mai 1877 (GBl. S. 12) bestimmt, daß unter „künstlichen Wiesen" auch alle mit Berieselungs- und Drainageanlagen versehenen Grundstücke zu verstehen sind. Das Interesse der Wiesenkultur erfordert aber einen noch weiteren Schutz, welcher unter Nr. 2 des § 49 angeordnet ist.

Zu Ziff. 3:

Wie bereits erwähnt, steht die Befugnis, den Weidgang und die Koppelhütung zu reglementieren, den Gemeinderäten zu. Da jedoch nicht mit Sicherheit vorausgesetzt werden kann, daß in dem Kreise des Gemeinderats die Interessen der Gutsbesitzer gegenüber den Weideberechtigten durchweg eine völlig unbefangene Würdigung erfahren, so erschien es zweckmäßig, über zwei besonders wichtige Punkte den Bezirkspräsidenten, denen unter Nr. 1 und 2 die erst durch das Gesetz der Reglementierung überwiesenen Punkte ausschließlich vorbehalten sind, wenigstens konkurrierend mit den Gemeinderäten ein Verordnungsrecht einzuräumen. Dies geschieht unter Nr. 3 bezüglich des Verbots der Weide zu bestimmten Zeiten des Jahres und des Tages. (Begr. S. 29.)

3. Die Strafandrohung für Zuwiderhandlungen gegen die in § 49 vorgesehenen Polizeiverordnungen ist in § 25 Ziff. 1 enthalten.

Ländliche Polizei im allgemeinen. Besichtigung der Feuerstätten.

§ 50.

Den Bürgermeistern liegt die Aufgabe ob, über die Ruhe, die Gesundheit und die Sicherheit auf dem Lande zu wachen.

Die Bürgermeister haben jährlich mindestens einmal die Feuerstätten und Schornsteine aller weniger als 200 Meter von Wohnhäusern entfernten Gebäude zu besichtigen oder besichtigen zu lassen, nachdem der Tag der Besichtigung mindestens eine Woche zuvor öffentlich bekannt gemacht ist. Nach der Besichtigung ist Reinigung, Ausbesserung oder Niederlegung derjenigen Feuerstätten oder Schornsteine anzuordnen, welche den Vorschriften des § 368 Nr. 4 oder des § 369 Nr. 3 des Strafgesetzbuchs nicht entsprechen.

1. § 50 ersetzt lediglich den durch § 51 aufgehobenen Artikel 9 Tit. II C. r., welcher lautete:

Les officiers municipaux veilleront généralement à la tranquillité, à la salubrité et à la sûreté des campagnes; ils seront tenus particulièrement de faire, au moins une fois par an, la visite des fours et cheminées, de toutes maisons et de tous bâtiments éloignés de moins de cent toises d'autres habitations; ces visites seront préalablement annoncées huit jours d'avance.

Après la visite ils ordonneront la réparation ou démolition des fours et cheminées qui se trouveront dans un état de délabrement qui pourrait occasionner un incendie ou d'autres accidents.

2. § 368 Ziff. 4 und § 369 Ziff. 3 sind im Anhang abgedruckt. Der Bürgermeister hat bei der ihm in § 50 Abs. 2 zur Pflicht gemachten Besichtigung der Feuerstätten und Schornsteine sein Augenmerk darauf zu richten, ob die Feuerstätten in brandsicherem Zustande unterhalten und ordnungsmäßig — in den durch etwa vorhandene Polizeiverordnungen bestimmten Zeitabschnitten oder sonst zur rechten Zeit — gereinigt sind. Bei Gewerbetreibenden, die in Feuer arbeiten, hat er insbesondere darauf zu achten, ob sie die Vorschriften befolgen, welche von der Polizeibehörde wegen Anlegung und Verwahrung ihrer Feuerstätten, sowie wegen der Art und der Zeit, sich des Feuers zu bedienen, erlassen sind. Neben der Anordnung, die nicht entsprechenden Feuerstätten und Schornsteine zu reinigen, auszubessern oder niederzulegen, hat der Bürgermeister die Säumigen gegebenenfalls zur Anzeige behufs Bestrafung nach § 368 Ziff. 4 und § 369 Ziff. 3 StGB. zu bringen.

Vierter Titel.

Schlußbestimmungen.

Aufgehobene Gesetze.

§ 51.

Es werden aufgehoben:

1. das arrêt du Conseil d'État, betreffend das Abweiden der Ränder großer Straßen, vom 16. Dezember 1759;
2. Artikel 8 Abschnitt 7 Titel I und Titel II des Gesetzes, betreffend die ländlichen Güter und Gebräuche, vom 28 September — 6. Oktober 1791;
3. das Gesetz über das Abraupen der Bäume vom 26. Ventose Jahr IV (Bull. des lois sér. II n° 242);
4. Artikel 457, 471 Nr. 7, 9, 10, Artikel 472, 473, 475 Nr. 1 und 15, Artikel 479 Nr. 10, 11, 12 des Code pénal.

Die Vorschriften des Gesetzes, betreffend das Forststrafrecht und das Forststrafverfahren, vom 28. April 1880 (GBl. S. 75) werden durch das gegenwärtige Gesetz nicht berührt.

1. Hier sind nur diejenigen Gesetzesvorschriften erwähnt, welche durch den Entwurf vollständig aufgehoben werden; berührt werden von demselben auch andere, z. B. die Ordonnanz vom 4. August 1731 (durch §§ 26 und 27), Art. 2 des Dekrets vom 4. August 1789 (durch § 22), das Gesetz vom 30. März 1887 (s. oben S. X). Vollständig ersetzt ist durch das Gesetz (vergl. Bem. 5 zu § 20 und Bem. 3 zu § 26) auch Art. 372 der Bezirks-Vizinalwegeordnungen.

Zu Ziff. 1. Vergl. Bem. 5 zu § 20.

Zu Ziff. 2. Die hier nicht aufgeführten Bestimmungen des Ruralgesetzes sind durch das FPStG. nicht aufgehoben worden und zum Teil noch fortdauernd in Geltung, so insbesondere der vierte Abschnitt des ersten Titels über Herden, Einfriedigung, Koppelhütung und Weidgang, ferner der fünfte Abschnitt des ersten Titels über die Ernten.

Der Art. 8 Abschn. 7 Tit. I Code rural, die Verjährung betreffend, ist durch § 1 ersetzt.

Was den Tit. II C. r. anbetrifft, so waren schon vor Einführung des Gesetzes aufgehoben bzw. ersetzt die Art. 1, 2, 5, 6, 10, 11, 16, 19, 20,

21 Satz 1 und 2, 23, 24, 27, 29, 31, 32, 33 Satz 2, 34, 35, 38, 39, 40, 42, 44; es handelt sich hier nur noch um die Artikel 3, 4, 7, 8, 9, 12, 13, 14, 15, 17, 18, 21 Satz 3, 22, 25, 26, 28, 30, 33 Satz 1, 36, 37, 41, 43.

Art. 3 und 4 sind, soweit neben dem Gesetze vom 30. März 1887 noch bestehend, durch § 1 beseitigt. Bezüglich der im Art. 4 verordneten Zuweisung der Geldstrafe an die Gemeindekasse ist nach wie vor Art. 466 C. p. und Art. 4 der Ordonnanz vom 30. Dezember 1823 maßgebend. (Vgl. oben Bem. 4 zu § 1.)

Art. 7 und 8 sind durch §§ 3 und 4 ersetzt, Art. 9 durch § 50, Art. 12 durch §§ 42 ff., Art. 13 durch § 34, Art. 14 und 17 durch §§ 9 ff., 26 ff., Art. 15 durch § 30, Art. 18 durch § 24, Art. 21 Satz 3 durch § 19, Art. (24) 25, 26 durch §§ 20 ff., Art. 28 durch §§ 9 ff., 26 ff., Art. 33 Satz 1 durch § 17, Art. 36, 37, 43 durch §§ 9 ff., 26 ff., Art. 41 durch § 29 Abs. 2.

Nicht ersetzt sind Art. 30, welcher die Verletzung von Tieren bestraft, da die Vorschrift des gemeinen Rechts über Sachbeschädigung genügt, und Art. 22 als eine nicht mehr zu rechtfertigende Eigentumsbeschränkung (Begr. S. 30).

Zu Ziff. 3. Zum Ersatz dient der § 47, welcher den Erlaß besonderer Polizeiverordnungen über das Abraupen der Bäume zur Voraussetzung hat.

Zu Ziff. 4. Von den feldpolizeilichen Artikeln des Code pénal durften schon vor Erlaß des Gesetzes als aufgehoben, bzw. ersetzt gelten: die Art. 444—456, 458—461, 471 Nr. 1, 4, 8, 13, 14, 475 Nr. 9, 10, (479 Nr. 10, 11); hier kommen demnach nur in Betracht: Art. 457, 471 Nr. 7, 9, 10 mit 472 und 473, 475 Nr. 1 und 15, 479 Nr. 10, 11, 12.

Ersetzt sind: Art. 457 durch §§ 30, 31, Art. 471 Nr. 9 durch §§ 9 ff., Art. 471 Nr. 10 durch § 19, ebenso Art. 473, Art. 475 Nr. 1 durch § 38, Art. 471 Nr. 15 durch §§ 9 ff., Art. 479 Nr. 10 durch §§ 20 ff., Art. 479 Nr. 11 durch § 27.

Nicht ersetzt ist als unpraktisch Art. 471 Nr. 7 mit Art. 472 C. p., welche es verbieten, Werkzeuge, die den Dieben nützlich sein können, im Freien liegen zu lassen. Auch ist eine ausdrückliche Anerkennung des in Art. 479 Nr. 12 C. p. ausgesprochenen Satzes, daß die Entnahme von Erde und Materialien aus Gemeindeländereien straflos ist, wenn dieselbe einem Gebrauche entspricht, in das Gesetz nicht aufgenommen, da anzunehmen ist, daß auch ohne solchen Ausspruch der Richter bei Anwendung des § 370 Nr. 2 StGB. eine „unbefugte" Entnahme nicht als vorliegend erachten wird, wenn dieselbe auf Grund eines Gebrauchs geschehen ist. (Begr. S. 30.)

2. Durch Abs. 2 wird außer Zweifel gestellt, daß auch diejenigen Paragraphen des Forststg., deren Tatbestand als in den gewisser Paragraphen des Gesetzes aufgegangen erscheinen könnte, neben letzteren in voller Geltung verbleiben (z. B. §§ 21, 22 Forststg. neben §§ 20, 21; § 25 Nr. 3 Forststg. neben § 23; § 28 Forststg. neben § 29. (Begr. S. 31.)

Inkrafttreten des Gesetzes.

§ 52.

Das gegenwärtige Gesetz tritt am 1. Januar 1889 in Kraft.

Urkundlich unter Unserer Höchsteigenhändigen Unterschrift und beigedrucktem Kaiserlichen Insiegel.

Gegeben Marmor-Palais, den 9. Juli 1888.

Wilhelm

[L. S.] Fürst von Hohenlohe.

Anhang.

A. Im Strafgesetzbuch enthaltene Bestimmungen feldpolizeilichen oder verwandten Charakters.

§ 366.

Mit Geldstrafe bis zu sechzig Mark oder mit Haft bis zu vierzehn Tagen wird bestraft:

2. wer in Städten oder Dörfern übermäßig schnell fährt oder reitet, oder auf öffentlichen Straßen oder Plätzen der Städte oder Dörfer mit gemeiner Gefahr Pferde einfährt oder zureitet;

3. wer auf öffentlichen Wegen, Straßen, Plätzen oder Wasserstraßen das Vorbeifahren anderer mutwillig verhindert;

4. wer in Städten mit Schlitten ohne feste Deichsel oder ohne Geläute oder Schelle fährt;

5. wer Tiere in Städten oder Dörfern, auf öffentlichen Wegen, Straßen oder Plätzen, oder an anderen Orten, wo sie durch Ausreißen, Schlagen oder auf andere Weise Schaden anrichten können, mit Vernachlässigung der erforderlichen Sicherheitsmaßregeln stehen läßt oder führt;

7. wer Steine oder andere harte Körper oder Unrat auf Menschen, auf Pferde oder andere Zug- oder Lasttiere, gegen fremde Häuser, Gebäude oder Einschließungen, oder in Gärten oder eingeschlossene Räume wirft;

8. wer nach einer öffentlichen Straße oder Wasserstraße, oder nach Orten hinaus, wo Menschen zu verkehren pflegen, Sachen, durch deren Umstürzen oder Herabfallen jemand beschädigt werden kann, ohne gehörige Befestigung aufstellt oder aufhängt, oder Sachen auf eine Weise ausgießt oder auswirft, daß dadurch jemand beschädigt werden kann;

9. wer auf öffentlichen Wegen, Straßen, Plätzen oder Wasserstraßen Gegenstände, durch welche der freie Verkehr gehindert wird, aufstellt, hinlegt oder liegen läßt;

10. wer die zur Erhaltung der Sicherheit, Bequemlichkeit, Reinlichkeit und Ruhe auf den öffentlichen Wegen, Straßen, Plätzen oder Wasserstraßen erlassenen Polizeiverordnungen übertritt.

§ 367.

Mit Geldstrafe bis zu einhundert fünfzig Mark oder mit Haft wird bestraft:

8. wer ohne polizeiliche Erlaubnis an bewohnten oder von Menschen besuchten Orten Selbstgeschosse, Schlageisen oder Fußangeln legt, oder an solchen Orten mit Feuergewehr oder anderem Schießwerkzeuge schießt, oder Feuerwerkskörper abbrennt;

12. wer auf öffentlichen Straßen, Wegen oder Plätzen, auf Höfen und überhaupt an Orten, an welchen Menschen verkehren, Brunnen, Keller, Gruben, Oeffnungen oder Abhänge dergestalt unverdeckt oder unverwahrt läßt, daß daraus Gefahr für andere entstehen kann.

§ 368.

Mit Geldstrafe bis zu sechzig Mark oder mit Haft bis zu vierzehn Tagen wird bestraft:

1. wer den polizeilichen Anordnungen über die Schließung der Weinberge zuwiderhandelt;

2. wer das durch gesetzliche oder polizeiliche Anordnungen gebotene Raupen unterläßt;

3. wer ohne polizeiliche Erlaubnis eine neue Feuerstätte errichtet oder eine bereits vorhandene an einen anderen Ort verlegt;

4. wer es unterläßt, dafür zu sorgen, daß die Feuerstätten in seinem Hause in baulichem und brandsicherem Zustande unterhalten, oder daß die Schornsteine zur rechten Zeit gereinigt werden;

5. wer Scheunen, Ställe, Böden oder andere Räume, welche zur Aufbewahrung feuerfangender Sachen dienen, mit unverwahrtem Feuer oder Licht betritt, oder sich denselben mit unverwahrtem Feuer oder Licht nähert;

6. wer an gefährlichen Stellen in Wäldern oder Heiden, oder in gefährlicher Nähe von Gebäuden oder feuerfangenden Sachen Feuer anzündet;

7. wer in gefährlicher Nähe von Gebäuden oder feuerfangenden Sachen mit Feuergewehr schießt oder Feuerwerke abbrennt;

8. wer die polizeilich vorgeschriebenen Feuerlöschgerätschaften überhaupt nicht oder nicht in brauchbarem Zustande hält oder andere feuerpolizeiliche Anordnungen nicht befolgt;

9. wer unbefugt über Gärten oder Weinberge, oder vor beendeter Ernte über Wiesen oder bestellte Aecker, oder über solche Aecker, Wiesen, Weiden oder Schonungen, welche mit einer Einfriedigung versehen sind, oder deren Betreten durch Warnungszeichen untersagt ist, oder auf einem durch Warnungszeichen geschlossenen Privatwege geht, fährt, reitet oder Vieh treibt.

§ 369.

Mit Geldstrafe bis zu einhundert Mark oder mit Haft bis zu vier Wochen werden bestraft:

3. Gewerbetreibende, welche in Feuer arbeiten, wenn sie die Vorschriften nicht befolgen, welche von der Polizeibehörde wegen Anlegung und Verwahrung ihrer Feuerstätten, sowie wegen der Art und der Zeit, sich des Feuers zu bedienen, erlassen sind.

§ 370.

Mit Geldstrafe bis zu einhundertfünfzig Mark oder mit Haft wird bestraft:

1. wer unbefugt ein fremdes Grundstück, einen öffentlichen oder Privatweg oder einen Grenzrain durch Abgraben oder Abpflügen verringert;

2. wer unbefugt von öffentlichen oder Privatwegen Erde, Steine oder Rasen, oder aus Grundstücken, welche einem anderen gehören, Erde, Lehm, Sand, Grand oder Mergel gräbt, Plaggen oder Bülten haut, Rasen, Steine, Mineralien, zu deren Gewinnung es einer Verleihung, einer Konzession oder einer Erlaubnis der Behörde nicht bedarf, oder ähnliche Gegenstände wegnimmt;

B. Verordnung des Ministeriums betr. die Jagdpolizei.

Vom 31. Dezember 1888.

(GBl. 1889, S. 1.)

Auf Grund des § 2 des Jagdpolizeigesetzes vom 7. Mai 1883 in Verbindung mit § 24 Nr. 2 des Feldpolizeistrafgesetzes vom 9. Juli 1888 wird hierdurch bestimmt, was folgt:

Tauben, welche zur Saat- oder Erntezeit dem von der Ortspolizei bekannt gemachten Verbote zuwider nicht eingeschlossen gehalten werden, dürfen von dem Eigentümer, Besitzer oder Pächter der Grundstücke, auf welchen sie getroffen werden, mit den in § 3 der Verordnung des Ministeriums vom 20. Juni 1883 (GBl. S. 65)*) bezeichneten Mitteln vertilgt werden. Die Anwendung von Schußwaffen zu dem angegebenen Zweck ist von dem Kreisdirektor oder dem Polizeidirektor gemäß § 4 der erwähnten

*) jetzt § 3 der Verordnung des Ministeriums, betr. das schädliche Wild vom 16. Juli 1890 (GBl. S. 49), wo es heißt: „Die Vertilgung der . . . . bezeichneten Vögel kann erfolgen unter Anwendung von Fangkörben, Fangnetzen und Pfahleisen . . . .“.

Verordnung*) nur unter der Voraussetzung zu gestatten, daß die Tauben scharenweise in ein Grundstück einfallen.

Tauben, von welchen den Umständen nach anzunehmen ist, daß es Brieftauben sind, sollen geschont werden.

C. Reichsgesetz vom 28. Mai 1894, betr. den Schutz der Brieftauben und den Brieftaubenverkehr im Kriege.

(RGBl. 1894 S. 463 f.)

§ 1.

Die Vorschriften der Landesgesetze, nach welchen das Recht, Tauben zu halten, beschränkt ist, und nach welchen im Freien betroffene Tauben der freien Zueignung oder der Tötung unterliegen, finden auf Militärbrieftauben keine Anwendung.

Dasselbe gilt von landesgesetzlichen Vorschriften, nach welchen Tauben, die in ein fremdes Taubenhaus übergehen, dem Eigentümer des letzteren gehören.

§ 2.

Insoweit auf Grund landesgesetzlicher Bestimmungen Sperrzeiten für den Taubenflug bestehen, finden dieselben auf die Reiseflüge der Militärbrieftauben keine Anwendung. Die Sperrzeiten dürfen für Militärbrieftauben nur einen zusammenhängenden Zeitraum von höchstens je zehn Tagen im Frühjahr und Herbst umfassen. Sind längere als zehntägige Sperrzeiten eingeführt, so gelten für Militärbrieftauben immer nur die ersten zehn Tage.

*) jetzt §§ 4 und 6 der in der vorigen Anmerkung bezeichneten Verordnung. § 4 lautet: „Sind die . . . . zugelassenen Vertilgungsmittel nach Lage des Falles für nicht ausreichend zu erachten, so kann der Kreisdirektor bezw. Polizeidirektor einzelnen Eigentümern, Besitzern oder Pächtern für ihre Person oder für ihre Beauftragten die Anwendung von Schußwaffen auf ihren Ländereien zum Zwecke der Vertilgung erlauben.“ § 6 lautet: „Personen, welchen nach den Bestimmungen der §§ 10 und 11 Nr. 3 des Jagdpolizeigesetzes die Ausstellung eines Jagdscheines versagt werden muß oder versagt werden kann, darf die Anwendung von Schußwaffen zum Zwecke der Vertilgung des schädlichen Wilds nicht gestattet werden. — Die Erlaubnis ist schriftlich und nur auf bestimmte Zeit zu erteilen. Der Erlaubnisschein muß den Namen desjenigen, für den er erteilt ist, und die Bezeichnung des Wilds enthalten, dessen Vertilgung bezweckt wird. — Die Anwendung von Schußwaffen während der Nachtzeit (§ 7 des Jagdpolizeigesetzes) ist nicht gestattet, wenn die Erlaubnis sich nicht ausdrücklich auch auf die Nachtzeit erstreckt. — Wer auf Grund der erteilten Erlaubnis schädlichem Wild mit der Schußwaffe nachstellt, muß den Erlaubnisschein bei sich führen.“

§ 3.

Als Militärbrieftauben im Sinne dieses Gesetzes gelten Brieftauben, welche der Militär-(Marine-)Verwaltung gehören oder derselben gemäß den von ihr erlassenen Vorschriften zur Verfügung gestellt und welche mit dem vorgeschriebenen Stempel versehen sind.

Privatpersonen gehörige Militärbrieftauben genießen den Schutz dieses Gesetzes erst dann, wenn in ortsüblicher Weise bekannt gemacht worden ist, daß der Züchter seine Tauben der Militärverwaltung zur Verfügung gestellt hat.

§ 4.

Für den Fall eines Krieges kann durch Kaiserliche Verordnung bestimmt werden, daß alle gesetzlichen Vorschriften, welche das Töten und Einfangen fremder Tauben gestatten, für das Reichsgebiet oder einzelne Teile desselben außer Kraft treten, sowie daß die Verwendung von Tauben zur Beförderung von Nachrichten ohne Genehmigung der Militärbehörde mit Gefängnis bis zu drei Monaten zu bestrafen ist.

D. Amtliches Formular für Anzeigen wegen Feldfrevel.

Nr. der Strafprozeßliste.

Anzeige wegen Feldfrevels.

Gemeinde:

Amtsgerichtsbezirk:

Der Unterzeichnete bringt hiermit folgendes zur Anzeige:

Am [1] 190

um Uhr mittags

hat [2]

an dem Orte genannt

auf dem Gemeindebanne von

[3]

1) Zeit der Tat.

2) Vor- und Zunamen, Alter (bei Minderjährigen Jahr und Tag der Geburt), Religion, Beruf, Geburts- und Wohnort des Täters oder der Täter.

Bei den unter der Gewalt, der Aufsicht oder im Dienste eines andern stehenden Tätern außerdem: Vor- und Zunamen, Beruf und Wohnort der Eltern bzw. des Ehemanns, Vormunds, Dienstherrn, Arbeitgebers, und, falls die Täter nicht zur Hausgenossenschaft dieser Personen gehören, auch Angabe dieses Umstandes.

3) Genaue Bezeichnung der strafbaren Handlung unter Angabe der Erschwerungsgründe, sowie des Geschädigten nach Vor- und Zunamen, Stand und Wohnort.

Besteht die strafbare Handlung in der Übertretung einer Ortspolizeiverordnung, so ist letztere anzuführen.

4) Die nicht zutreffenden Worte sind zu streichen.

4) der Wert des Entwendeten beträgt:

4) der angerichtete Schaden beträgt:

Der Tatbestand wurde festgestellt:

4) durch eigene Wahrnehmung;

4) auf Grund der Aussage folgender Zeugen 5)

5) Vor- und Zunamen, Beruf und Wohnort der Zeugen.

4) durch eine in Begleitung von 6)

4) vorgenommene Haussuchung.

6) Vor- und Zunamen, Beruf und Wohnort der bei der Haussuchung zugezogenen Personen.

In Verwahr genommen sind: 7)

7. Bezeichnung der in Beschlag genommenen Frevelwerkzeuge bzw. der gepfändeten Tiere unter Angabe des Verwahrungsorts.

8) , den 190

Der

8) Ort und Datum, sowie Amtscharakter und Unterschrift des Anzeigenden.

9)

9) Wenn die Strafverfolgung nur auf Antrag zulässig ist, oder wenn der Geschädigte eine an ihn zu zahlende Buße verlangt, so ist der gestellte Antrag hier zu vermerken und von dem Geschädigten zu unterzeichnen.

Urschriftlich

der Staatsanwaltschaft beim Kaiserl. Amtsgericht

in

zur weiteren Veranlassung zu übersenden.

10)

10) Bemerkungen über die Vermögensverhältnisse und die Zahlungsfähigkeit der Täter und der haftbaren Personen.

, den 190

Der Bürgermeister

Sachregister.

(Die Zahlen bezeichnen die Seiten.)

Systematische Sammlung

der in

Elsaß-Lothringen geltenden Gesetze.

Herausgegeben im Auftrag des Kaiserlichen Ministeriums.

Erster und zweiter Band

Privatrecht

bearbeitet von

Karl Paffrath und **Franz Großmann,**

Landgerichtsräte in Straßburg.

I. Band: Lex. 8° XXII, 660 S. 1900. Geheftet ℳ 15.—,
in solidem Halbfranzband ℳ 17.—.

II. Band: Lex. 8° XXXIV, 1200 S. 1901. Geheftet ℳ 24.50,
in solidem Halbfranzband ℳ 26.50.

In Vorbereitung:

Dritter Band

Öffentliches Recht, 1. Teil:

Allgemeine Bestimmungen über das Finanzwesen, die direkten Steuern und Verkehrssteuern, sowie verwandte Materien

bearbeitet von **Karl Jacob**, Ministerialrat.

Lex. 8° ca. 35 Bogen.

Die Einführung des Bürgerlichen Gesetzbuchs hat auch eine neue Bearbeitung der alten Möller'schen Gesetzsammlung nötig gemacht, da sie durch die neue Gesetzgebung und die damit verbundene Aufhebung einer großen Anzahl alter Gesetze nur noch in beschränktem Umfange gebrauchsfähig war. Auf Anweisung des Kaiserlichen Ministeriums erscheint die neue Ausgabe nicht mehr chronologisch, sondern systematisch nach den einzelnen Rechtsmaterien geordnet.

Die Sammlung wird mit dem Privatrecht eröffnet, das in zwei Bänden vollständig vorliegt. Der erste Band enthält das Bürgerliche Gesetzbuch, das Gesetz über die Angelegenheiten der freiwilligen Gerichtsbarkeit, die Grundbuchordnung, das Handels-

Systematische Sammlung der in Elsaß-Lothringen geltenden Gesetze (Fortsetzung).

gesetzbuch, die Wechselordnung, das Gerichtsverfassungsgesetz, die Civilprozeßordnung, die Konkursordnung, das Gesetz über die Zwangsversteigerung und die Zwangsverwaltung, ferner die sämtlichen zu diesen Gesetzen gehörigen Ein- und Ausführungsgesetze, die Gerichtskostengesetze und die Gebührenordnungen: alles mit auf Elsaß-Lothringen bezüglichen kurzen Erläuterungen versehen; der zweite Band enthält in der gleichen Bearbeitung die sämtlichen übrigen Gesetze und Verordnungen privatrechtlichen Charakters bis auf die neueste Zeit. — Ein ausführliches Register am Schlusse des zweiten Bandes gewährt einen vollständigen und sicheren Überblick über den heutigen Rechtszustand.

Alte Möller'sche Gesetzsammlung.

Sammlung der in Elsaß-Lothringen geltenden Gesetze. Auf Anregung des Wirkl. Geheimen Raths Dr. von Möller bearbeitet und herausgegeben in Verbindung mit anderen reichsländischen Juristen von F. Althoff, ordentlichem Professor, R. Förtsch, Landgerichtsdirektor, A. Harseim, Justizrat und Gouvernementsauditeur, A. Keller, Oberlandesgerichtsrat und A. Leoni, Landgerichtsrat; 6 Bände und Register.

I. Band: Verfassungsrecht und Gesetzbücher. Mit Anhang: Kostengesetze und Gebührenordnungen. Lex. 8°. XXXI, 761 Seiten. 1880. (ℳ 18.—) ℳ 12.—

II. Band: Französische Einzelgesetze. Lex. 8°. V, 1007 Seiten. 1881. (ℳ 18.—) ℳ 12.—

III. Band: Deutsche Einzelgesetze. Lex. 8°. 1183 Seiten. 1881. (ℳ 21.—) ℳ 14.—

Registerband zum I.—III. Band: 1. Sachliche Übersicht der Gesetze. — 2. Alphabetisches Register. Lex. 8°. 181 Seiten. 1881. (ℳ 3.50) ℳ 2.50

IV. Band, die Gesetze von 1881—1885 enthaltend, mit alphabetischem Register für alle 4 Bände. Bearbeitet von R. Förtsch. Lex. 8°. XXXII, 815 Seiten. 1886. (ℳ 18.—) Vergriffen.

V. Band, die Gesetze von 1886—1890 enthaltend, mit Register. Bearbeitet von R. Förtsch, Reichsgerichtsrat. Lex. 8°. XXXVIII, 981 Seiten. 1892. ℳ 17.—

VI. Band, die Gesetze von 1891—1895 enthaltend, mit Gesamtregister für alle Bände. Bearbeitet von C. Jacob, Ministerialrat. Lex. 8°. XXXV, 1073 Seiten. 1896. Geheftet ℳ 20.—, geb. ℳ 22.—

Praktische Handausgabe des Bürgerlichen Gesetzbuches für Elsaß-Lothringen.

Das

Bürgerliche Gesetzbuch

mit den

dazu gehörigen Nebengesetzen

für

Elsaß-Lothringen.

Bürgerliches Gesetzbuch
Gesetz über die Angelegenheiten der freiwilligen Gerichtsbarkeit
Grundbuchordnung
Gesetz über die Zwangsversteigerung und Zwangsverwaltung
nebst den Einführungsgesetzen und den für Elsaß-Lothringen
erlassenen Ausführungsgesetzen.

Bearbeitet von

Karl Paffrath und **Franz Großmann,**
Landgerichtsräte in Straßburg.

XVI, 848 S. 1901. Geheftet *M* 3.—, in elegantem Halbfranzband *M* 4.—

Vorwort: Die vorliegende Ausgabe ist ein Abdruck aus der Systematischen Sammlung der in Elsaß-Lothringen geltenden Gesetze. Das Bedürfniß einer Handausgabe des Bürgerlichen Gesetzbuchs bedarf keiner weiteren Begründung. Daß dieses Bedürfnis zur Zeit schon mehr oder weniger Befriedigung erfahren hat, dürfte der vorliegenden Ausgabe ihre Berechtigung nicht nehmen. Sie enthält neben dem Bürgerlichen Gesetzbuch auch noch die wichtigeren Nebengesetze und die für Elsaß-Lothringen erlassenen Ausführungsgesetze und beschränkt sich nicht auf die bloße Wiedergabe des Gesetzestextes, sondern verweist in den Bemerkungen auf verwandte und ergänzende Bestimmungen und erleichtert dadurch den praktischen Gebrauch der Gesetze.

Ausführungsgesetze.

Das Gesetz, betreffend die Ausführung des Bürgerlichen Gesetzbuchs in Elsaß-Lothringen vom 17. April 1899 in der Fassung der Bekanntmachung des Kaiserlichen Statthalters vom 22. Dezember 1899. Erläutert von Hugo Molitor, Kaiserl. Landgerichtsrat in Straßburg. 8°. XXIV, 595 S. 1900. (M 10.50) Vergriffen.
Neue Auflage in Vorbereitung.

Das elsaß-lothringische Gesetz, betreffend die Ausführung des Reichsgesetzes über die Angelegenheiten der freiwilligen Gerichtsbarkeit vom 6. November 1899 nebst den zugehörigen Ausführungsverfügungen. Auf amtliche Veranlassung erläutert von Hugo Molitor, Oberlandesgerichtsrat. 8°. X, 242 S. 1901.
Geheftet M 4.50, in Leinwand gebunden M 5.—.

Die Elsaß-Lothringischen Ausführungsgesetze und Verordnungen zu den Reichs-Justizgesetzen nebst einigen weiteren Gesetzen über Gewerbegerichte, Forststrafrecht, Erbbescheinigungen, Feld- und Fährpolizei-Übertretungen, Vormundschaften und Zwangsenteignung. Textausgabe mit französischer Übersetzung und Anmerkungen. Zweite verbesserte und vermehrte Auflage. — Les lois et ordonnances d'exécution des lois de justice de l'Empire d'Allemagne et les lois concernant les Prud'hommes, de Droit pénal forestier, les Certificats de Qualité d'héritier, les Contraventions en matière de police rurale et des bacs et bâteaux, les Tutelles et l'Expropriation pour cause d'utilité publique. Texte allemand et français avec annotations. 2e édit. revue et augmentée. 8°. 445 S. 1887.
Geheftet M 5.—, gebunden M 6.30.

Siehe auch Justizgesetze.

Siehe auch Bürgerliches Gesetzbuch S. 3.

Baurecht.

Das Elsaß-Lothringische Baurecht, enthaltend eine systematische Darstellung der auf Bauten bezügl. Vorschriften des öffentlichen und Privatrechts, sowie eine Zusammenstellung der zugehörigen Gesetze und Verordnungen in deutscher Übersetzung. Herausgegeben von R. Förtsch, Kammerpräsident in Metz, und M. Caspar, Abteilungsbaumeister in Straßburg. 8°. XII, 394 S. 1879. (M 5.—) Vergriffen.
Neue Bearbeitung in Vorbereitung.

Bergrecht.

Entstehung und Beendigung des Bergwerkseigentums nach dem in Elsaß-Lothringen geltenden Rechte, von Dr. Otto Kusenberg. 8°. 115 S. 1905. Geheftet ℳ 2.—.

Bürgerliches Gesetzbuch.

Siehe S. 3 und Ausführungsgesetze S. 4.

Erbbescheinigungen.

Gesetz, betreffend die Ausstellung gerichtlicher Erbbescheinigungen und die Zuständigkeit der Amtsgerichte vom 10. Mai 1886. — Loi sur la délivrance de certificats judiciaires constatant la qualité d'héritier et la compétence des tribunaux cantonaux du 10 mai 1886. Texte allemand et français. 8°. 11 S. 1886. Kartoniert ℳ —.50.

Erbschaftssteuergesetz.

Das Erbschaftssteuergesetz für Elsaß-Lothringen. Im amtlichem Auftrage erläutert und herausgegeben von Karl Jacob, Ministerialrat. Zweite, gänzlich umgearbeitete Auflage unter Zugrundlegung der Fassung vom 17. Juni 1900. 8°. VIII, 236 S. 1900. Geheftet ℳ 4.50, gebunden in Ganz-Leinwand ℳ 5.—.

Feld- und Fährpolizei.

Feldpolizeistrafgesetz vom 9. Juli 1888. Erläutert von Dr. A. Stieve, Gerichtsassessor. 8°. XIV, 100 S. 1907. Kartoniert ℳ 1.60.

Gesetz, betr. die Bestrafung von Zuwiderhandlungen gegen die Vorschriften der Feld- und Fährpolizei, vom 30. März 1887.

Siehe Ausführungsgesetze zu den Reichsjustizgesetzen.

Fischereigesetz.

Das Gesetz für Elsaß-Lothringen, betr. die Fischerei, vom 2. Juli 1891. Im Auftrag des Ministeriums, Abtheilung für Finanzen, Landwirthschaft und Domänen, erläutert und herausgegeben von Freiherrn von Bibra, Ministerialrat, und Lichtenberg, Regierungsassessor. 8°. VI, 111 S. 1893. Amtliche Ausgabe. Geheftet ℳ 3.—, in Ganz-Leinwand gebunden ℳ 3.50.

Forststrafrecht. S.: Ausführungsgesetze zu den Reichsjustizgesetzen.

Freiwillige Gerichtsbarkeit.

Siehe: Ausführungsgesetze S. 4 und Bürgerliches Gesetzbuch Seite 3.

Fuhrwesen.

Die in Elsaß-Lothringen geltenden Gesetze über das Fuhrwesen. Übersetzt und herausgeg. im Auftrage des kaiserl. Oberpräsidiums. 8°. 52 S. 1877. Kart. M —.60.

Gemeindegesetzgebung.

Die Gemeindeordnung für Elsaß-Lothringen vom 6. Juni 1895. Zweite auf Grund des Halleyschen Kommentars neu bearbeitete Auflage von Dr. Ernst Bruck. 8°. X, 444 S. 1905. Geheftet M 5.—.
Gebunden in Ganz-Leinwand M 5.50.

Gewerbegerichte. S.: Ausführungsgesetze zu den Reichsjustizgesetzen.

Grundbuchordnung. Siehe: Bürgerliches Gesetzbuch S. 3.

Grundeigentum und Hypothekenwesen.

Die elsaß-lothringischen Gesetze betreffend Grundeigenthum und Hypothekenwesen sowie die Einrichtung von Grundbüchern nebst den Kostengesetzen und Ausführungsverordnungen. Mit Erläuterungen herausgegeben von A. Leoni, k. Ministerialrat. 8°. VIII, 248 S. 1892. In Ganz-Leinwand gebunden M 4.—.

Jagdgesetze.

Die Jagd- und Vogelschutzgesetzgebung in Elsaß-Lothringen. Von Dr. Ernst Bruck, Gerichtsassessor. 8°. 205 Seiten Text und 1 Bogen Register. 1907.
In Halbleinwand geb. M 3.50.

Das Elsaß-Lothringische Jagdgesetz. Textausgabe mit französischer Uebersetzung und Anmerkungen von einem Mitgliede des Landesausschusses. — Loi sur l'exercice du droit de chasse en Alsace-Lorraine. Texte allemand et français avec annotations par un membre de la Délégation. 8°. 31 S. 1881. Kart. M —.80.

Das Elsaß-Lothringische Jagdpolizeigesetz nebst der darauf bezüglichen Verordnung. Textausgabe mit französischer Uebersetzung und Anmerkungen von einem Mitgliede des Landesausschusses 2. Auflage. — Loi et ordonnance ministérielle sur la police de la chasse. Texte allemand et français avec annotations par un membre de la Délégation. 2e édit. 8°. 41 S. 1883.
Kartoniert M —.80.

Justizgesetze.

Das Gesetz für Elsaß-Lothringen, betreffend die Aenderung verschiedener Justizgesetze, vom 13. Februar 1905 (Justiznovelle) nebst Ausführungs-Verordnungen und Materialien. Auf amtliche Veranlassung mit Erläuterungen herausgegeben von Hugo Molitor, Geh. Regierungsrat im Ministerium für Elsaß-Lothringen. 8°. VIII, 166 S. 1905. In Halbleinwand geb. ℳ 3.—.

Inhalt: I. Gesetz betr. die Änderung verschiedener Justizgesetze (Justiznovelle): Art. 1 (Notariats-Gesetz v. 23. XII. 1873 §§ 9, 10). — Art. 2 (Ausf.-Ges. zum Gerichtsverfassungsgesetz v. 4. XI. 1878). — Art. 3 (Ausf.-Ges. zum B.G.B. § 40, § 40a, Exkurs). — Art. 4 (Ausf.-Ges. zum Reichsges. über die Angel. der freiw. Gerichtsbarkeit, § 28, § 28a, § 56). — Art. 5 (Ausf.-Ges. zur Z.P.O. und K.O. § 7). — Art. 6 (Gerichtskostengesetz für Elf.-Lothr.). — Art. 7 (Gebührenordnung für die Notare). — Art. 8 (Schlußbestimmung).

II. Ausführungsverordnung. Verfügung des Ministeriums, betr. das Notariat und das Verfahren in Angel. der freiw. Gerichtsbarkeit, v. 28. II. 1905.

III. Materialien.

Katastergesetz.

Das Elsaß-Lothringische Katastergesetz vom 31. März 1884 und das Gesetz vom 6. April 1892 betreffend Abänderung des Katastergesetzes sowie Ausgleichung der Grundsteuer. Text mit französischer Uebersetzung und Anmerkungen von Dr. Wilhelm Gunzert, Mitglied des Landesausschusses. — Loi sur le renouvellement du cadastre en Alsace-Lorraine du 31 mars 1884 et sur la Péréquation des impôts fonciers du 6 avril 1892. Texte allemand et français avec annotations par Guill. Gunzert, député à la Délégation. 8°. 102 S. 1892. Kartoniert ℳ 1.—.

Kirchenrecht.

Das Staatskirchenrecht in Elsaß-Lothringen. Zusammenstellung der betreffenden Gesetze, Verordnungen, Staatsrathsgutachten und Ministerialverfügungen. Im Auftrage des Oberpräsidenten übersetzt und herausgegeben von E. Dursy, Kaiserl. Ministerialrat.

I. Katholischer Cultus. 8°. 1876. (ℳ 5.—) Vergriffen.

II. Protestantischer Cultus. 8°. XXII, 624 S. 1879. ℳ 8.—.

Strafgesetze.

Schöffenbüchlein für Elsaß-Lothringen. Ein Hilfsbuch für Schöffen, Amtsrichter, Amtsanwälte, Rechtsanwälte, Bürgermeister und Polizeikommissare, von R. Förtsch, Landgerichtsdirektor in Metz. — Manuel des échevins, d'Alsace-Lorraine. Guide pour Messieurs les échevins, juges et procureurs cantonaux, avocats, maires et commissaires de police. Traduction française par G. Gunzert, président de Chambre à Strasbourg. 8°. 432 S. 1881. Kartoniert ℳ 2.40.

Unschädlichkeitserklärung.

Gesetz über das Unschädlichkeitszeugnis vom 19. Juni 1906. Erläutert von H. Molitor, Ministerialrat. Erscheint im Laufe des Jahres 1907.

Verwaltungsrecht.

Theorie des französischen Verwaltungsrechts. Von O. Mayer, Professor des französischen Rechts an der Universität Strassburg. Gr. 8°. X, 533 S. 1886. (ℳ 10.—) Vergriffen.

Vormundschaften.

Siehe: Ausführungsgesetze zu den Reichsjustizgesetzen.

Wassergesetz.

Das Gesetz für Elsaß-Lothringen, betr. Wasserbenutzung und Wasserschutz, vom 2. Juli 1891. Im Auftrag des Ministeriums, Abtheilung für Finanzen, Landwirthschaft und Domänen, erläutert und herausgegeben von den Ministerialräten Jacob und Fecht. Mit Anhang, enthaltend die sonstigen wassergesetzlichen Vorschriften, die Ausführungsbestimmungen und eine Uebersicht über die Eintheilung der Wasser- und der Meliorationsbauverwaltung. 8°. IV, 180 S. 1892. Amtliche Ausgabe. Brosch. ℳ 3.—, in Ganz-Leinwand gebunden ℳ 3.50.

Zwangsenteignung und Zwangsvollstreckung.

Siehe: Ausführungsgesetze zu den Reichsjustizgesetzen.

Zwangsversteigerung und Zwangsverwaltung.

Siehe: Bürgerliches Gesetzbuch S. 3.

Zu § 15 Bem. 2.

Die Bestimmung des Abs. 2 ist, wie gegenüber mehrfachen in letzter Zeit in landwirtschaftlichen Kreisen und auch in der Presse hervorgetretenen Bedenken betont zu werden verdient, durchaus zweckmäßig, da anderenfalls jeder noch so geringfügige Felddiebstahl, z. B. wenn ein Spaziergänger einige Kirschen pflückt, ohne Antrag des Eigentümers, ja sogar gegen dessen ausdrücklichen Willen zur Bestrafung gebracht werden muß (vgl. Unterstaatssekretär v. Puttkamer in LA. S. 382 und Abg. Dr. Gunzert in LA. S. 547). Den zunehmenden Felddiebstählen kann durch energisches Einschreiten seitens der Feldhüter und durch rücksichtslose strafrechtliche Verfolgung in denjenigen Fällen, in denen kein Antrag erforderlich ist, wirksam entgegengetreten werden. In dieser Beziehung sei ausdrücklich darauf hingewiesen, daß es eines Antrages nur bei Felddiebstählen an Nahrungs- und Genußmitteln von unbedeutendem Wert oder geringer Menge bedarf. Da es sich im FPStG. überhaupt nur um verhältnismäßig geringfügige Straftaten handelt, wird es z. B. bei einem Felddiebstahl an Nahrungs- oder Genußmitteln im Werte von etwa 40 oder 50 Pfennig keines Antrages mehr bedürfen. In zweifelhaften Fällen ist es zweckmäßig, daß die Bürgermeister die Anzeige zunächst auch ohne Strafantrag an den Amtsanwalt abgeben und diesem die vorläufige Entscheidung überlassen, ob ein Antrag notwendig ist.

FSC
www.fsc.org
MIX
Papier | Fördert
gute Waldnutzung
FSC® C083411

Kolibri 360 GmbH
Ferdinand-Jühlke-Straße 7
99095 Erfurt, Deutschland
produktsicherheit@kolibri360.de